WORKSHEETS
WITH THE MATH COACH

DEVELOPMENTAL MATHEMATICS: PREALGEBRA, BEGINNING ALGEBRA, AND INTERMEDIATE ALGEBRA

John Tobey
North Shore Community College

Jeffrey Slater
North Shore Community College

Jamie Blair
Orange Coast College

Jennifer Crawford
Normandale Community College

PEARSON

Boston Columbus Indianapolis New York San Francisco Upper Saddle River
Amsterdam Cape Town Dubai London Madrid Milan Munich Paris Montreal Toronto
Delhi Mexico City São Paulo Sydney Hong Kong Seoul Singapore Taipei Tokyo

Reproduced by Pearson from electronic files supplied by the author.

Publishing as Pearson, 75 Arlington Street, Boston, MA 02116.

ISBN-13: 978-0-321-88027-7
ISBN-10: 0-321-88027-7

RRDPrintCode

www.pearsonhighered.com

Worksheets with the Math Coach

Developmental Mathematics

Table of Contents

Name: ______________________________ Date: ________________

Instructor: ___________________________ Topic: ________________

Module 1 Whole Numbers and Introduction to Algebra
Topic 1 Understanding Whole Numbers

Vocabulary

whole numbers • digits • place-value system • period • round-off place • place-holder
expanded notation • word names • number line • inequality symbol • rounding

1. The whole numbers 0, 1, 2, 3, 4, 5, 6, 7, 8, 9 are called ____________.

2. When we write very large numbers, we place a comma after every group of three digits, called a(n) ____________, moving from right to left.

3. A number system in which the position, or placement, of the digits in a number tells the value of the digits is called a(n) ____________.

4. ____________ is a process that approximates a number to a specific round-off place.

Example	Student Practice
1. In the number 573,025: **(a)** In what place is the digit 7? 5$\underline{7}$3,025 The digit 7 is in the ten thousands place. **(b)** In what place is the digit 0? The digit 0 is in the hundreds place.	**2.** In the number 4,905,817: **(a)** In what place is the digit 9? **(b)** In what place is the digit 5?
3. Write 1,340,765 in expanded notation. Write 1 followed by a zero for each of the remaining digits. We continue in this manner for each digit. Since there is a zero in the thousands place, it is not included in the sum. $\underline{1},340,765$ as $\underline{1},000,000+\underline{3}00,000+$ $\underline{4}0,000+\underline{7}00+\underline{6}0+\underline{5}$	**4.** Write 5,302,769 in expanded notation.

Vocabulary Answers: 1. digits 2. period 3. place-value system 4. rounding

Example	Student Practice
5. Jon withdraws \$493 from his account. He requests the minimum number of bills in one-, ten-, and hundred-dollar bills. Describe the quantity of each denomination of bills the teller must give Jon. If we write \$493 in expanded notation, we can easily describe the denominations needed. 400 + 90 + 3 4 9 3 Thus, the teller must give Jon 4 hundred-dollar bills, 9 ten-dollar bills, and 3 one-dollar bills.	**6.** Michael withdraws \$389 from his account. He requests the minimum number of bills in one-, ten-, and hundred-dollar bills. Describe the quantity of each denomination of bills the teller must give Michael.
7. Write a word name for each number. **(a)** 2135 Look at a place value chart if you need help identifying the period. The number begins with 2 in the thousands place. The word name is two thousand, one hundred thirty-five. Notice that we use a hyphen in "thirty-five." **(b)** 300,460 The number begins with 3 in the hundred thousands place. The word name is three hundred thousand, four hundred sixty. We place a comma here to match the comma in the number.	**8.** Write a word name for each number. **(a)** 8592 **(b)** 5,230,089

Example

Student Practice

9. Replace each question mark with the inequality symbol $<$ or $>$.

(a) 1 ? 6

1 is less than 6.

$1<6$

(b) 8 ? 7

$8>7$

10. Replace each question mark with the inequality symbol $<$ or $>$.

(a) 3 ? 7

(b) 9 ? 5

11. Rewrite using numbers and an inequality symbol.

(a) Five is less than eight.

Five	is less than	eight.
↓	↓	↓
5	$<$	8

(b) Nine is greater than four.

$9>4$

12. Rewrite using numbers and an inequality symbol.

(a) Four is less than nine.

(b) Eight is greater than two.

13. Round 57,441 to the nearest thousand.

The round-off place digit is in the thousands place.

Identify the round-off place digit, $5\boxed{7},441$.

The digit to the right of 7 is less than 5, $5\boxed{7},\underline{4}41$.

Do not change the round-off place digit. Replace all digits to the right with zero, $57,000$.

14. Round 65,745 to the nearest hundred.

Example	Student Practice
15. Round 4,254,423 to the nearest hundred thousand. The round-off place digit is in the hundred thousands place. Identify the round-off place digit, $4,\boxed{2}54,423$. The digit to the right of 2 is 5 or more, $4,\boxed{2}\underline{5}4,423$. Increase the round-off place digit by 1. Replace all digits to the right with zeros. $4,300,000$ Thus, 4,254,423 rounded to the nearest thousand is 4,300,000.	**16.** Round 5,678,231 to the nearest ten thousand.

Extra Practice

1. Write the word name for $80,059$. Then write the number in expanded notation.

2. Replace the question mark with $<$ or $>$.

$11,032 \ ? \ 10,032$

3. Rewrite using numbers and inequality symbols.

Ninety-eight is less than one hundred fourteen

4. Round 2940 to the nearest hundred.

Concept Check

Explain how to round 8937 to the nearest hundred.

Name: ______________________________ Date: ______________

Instructor: ___________________________ Topic: ______________

Module 1 Whole Numbers and Introduction to Algebra
Topic 2 Adding Whole Number Expressions

Vocabulary

addition • addends • sum • variable • algebraic expression • variable expression
identity property of zero • addition facts • commutative property of addition • square
associative property of addition • evaluate • rectangle • perpendicular • triangle
right angle • simplify • perimeter • inductive reasoning

1. A letter that represents a number is called a(n) ___________.

2. The ___________ states that two numbers can be added in either order to produce the same result.

3. To ___________ an algebraic expression, we replace the variables in the expression with their corresponding values and simplify.

4. The distance around an object is called the ___________.

Example	Student Practice
1. Translate the English phrase using numbers and symbols.	**2.** Translate the English phrase using numbers and symbols.
(a) The sum of six and eight The words "the sum of" indicate addition, "+." $6+8$	**(a)** Six increased by two
(b) A number increased by four A number → x; increased by → $+$; four → 4 Although we used the variable x to represent the unknown quantity, any letter could have been used.	**(b)** The sum of a number and ten

Vocabulary Answers: 1. variable 2. commutative property of addition 3. evaluate 4. perimeter

Example	Student Practice
3. Express 4 as a sum of two whole numbers. Write all possibilities. How many addition facts must we memorize? Why? Write all the sums equal to 4 and observe any patterns. $4+0=4$ $3+1=4$ $2+2=4$ $1+3=4$ $0+4=4$ Notice that the last two rows of the pattern are combinations of the same numbers listed in the first two rows. We need to learn only 2 addition facts for the number 4, $3+1$ and $2+2$. The remaining facts are either a repeat of these or use the addition property of zero.	**4.** Express 6 as a sum of two whole numbers. Write all possibilities. How many addition facts must we memorize? Why?
5. Use the associative property and/or commutative property as necessary to simplify the expression $5+(n+7)$. Apply the commutative property. $5+(n+7)=5+(7+n)$ Regroup using the associative property and simplify. Then rewrite with the variable first. $5+(7+n)=(5+7)+n$ $=12+n$ $=n+12$	**6.** Use the associative property and/or commutative property as necessary to simplify the expression $(8+n)+4$.

Example	Student Practice

7. Evaluate $x+y+3$ for the given values of x and y.

x is equal to 6 and y is equal to 1

$x+y+3$

$6+1+3$

10

8. Evaluate $x+y+5$ for the given values of x and y.

x is equal to 9 and y is equal to 14

9. A market research company surveyed 1870 people to determine the type of beverage they order most often at a restaurant. The results of the survey are shown in the table. Find the total number of people whose responses were iced tea, soda, or coffee.

Type of Beverage	Number of Responses
Soda	577
Orange juice	475
Coffee	84
Iced tea	357
Milk	286
Other	91

We add whenever we must find the "total" amount. First add in the ones column, $7+7+4=18$. Since 18 is 1 ten and 8 ones, carry the tens by placing a 1 at the top of the tens column. Repeat the process for the tens and hundreds columns to complete the addition.

$$\begin{array}{r} {}^{2\,1}\\ 357 \\ 577 \\ +\ 84 \\ \hline 1018 \end{array}$$

A total of 1018 people responded iced tea, soda, or coffee.

10. A market research company surveyed 1738 people to determine the type of meal they order most often at a restaurant. The results of the survey are shown in the table. Find the total number of people whose responses were fish, beef, or chicken.

Type of Meal	Number of Responses
Beef	678
Chicken	598
Fish	271
Vegetarian	191

Example

11. Find the perimeter of the triangle. (The abbreviation "ft" means feet.)

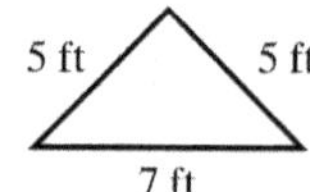

We add the lengths of the sides to find the perimeter.

$5 \text{ ft} + 5 \text{ ft} + 7 \text{ ft} = 17 \text{ ft}$

The perimeter is 17 feet.

Student Practice

12. Find the perimeter of the triangle. (The abbreviation "m" means meters.)

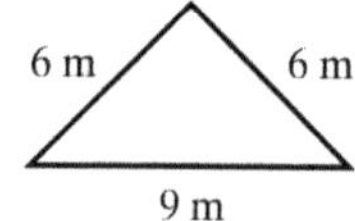

Extra Practice

1. Use the indicated property to rewrite the sum, then simplify if possible.

$(a+2)+7$; associative property of addition

2. Evaluate the expression using the given values of x and y.

$x+200+y$ when $x=645$ and $y=1000$

3. Add.

$$\begin{array}{r} 3754 \\ 358 \\ +6839 \\ \hline \end{array}$$

4. Find the perimeter of the figure.

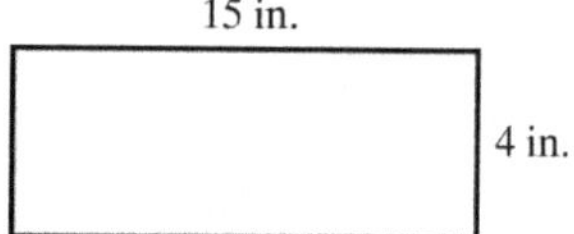

Concept Check

(a) When we carry, what is the value of the 1 that is placed above the 9?
(b) When we carry, what is the value of the 1 that is placed above the 3?

$$\begin{array}{r} {}^{1\,1} \\ 395 \\ +\ \ 28 \\ \hline 423 \end{array}$$

Name: ______________________________ Date: ______________

Instructor: ___________________________ Topic: ______________

Module 1 Whole Numbers and Introduction to Algebra
Topic 3 Subtracting Whole Number Expressions

Vocabulary

subtraction • minus sign • minuend • subtrahend • difference • borrowing

1. The number being subtracted in a subtraction problem is called the __________.

2. The result of a subtraction problem is called a __________.

3. The symbol used to indicate subtraction is called a __________ "–."

4. We do __________ when we take objects away from a group.

Example	Student Practice
1. Subtract. **(a)** $9-5$ $9-5=4$ **(b)** $15-0$ $15-0=15$	**2.** Subtract. **(a)** $8-5$ **(b)** $19-19$
3. $800-50=750$. Use this fact to find $800-53$. Since we know $800-50=750$, we can use subtraction patterns to find $800-53$. $800-50=750$ $800-51=749$ $800-52=748$ $800-53=747$ Observe the patterns in the second and the last columns. In the second column, the numbers increase by 1 and in the third column, they decrease by 1.	**4.** $700-50=650$. Use this fact to find $700-53$.

Vocabulary Answers: 1. subtrahend 2. difference 3. minus sign 4. subtraction

Example	Student Practice
5. Translate using numbers and symbols. Four less than seven The words "less than" indicate subtraction, "$-$". Note that the numbers are reversed because of the way the phrase is worded. $7-4$	**6.** Translate using numbers and symbols. The difference of eight and five
7. Evaluate $7-x$ for $x=2$. Replace x with 2 and simplify. $7-x=7-2=5$ When x is equal to 2, $7-x$ is equal to 5.	**8.** Evaluate $9-a$ for $a=5$.
9. Subtract $304-146$. We must borrow since we cannot subtract 6 ones from 4 ones. Also we cannot borrow a ten since there are 0 tens, so we must borrow from 3 hundreds. Rewrite 3 hundreds as hundreds and tens, $3\text{ hundreds}=2\text{ hundreds}+10\text{ tens}$. Again, rewrite 10 tens as tens and ones, $10\text{ tens}=9\text{ tens}+10\text{ ones}$. Thus, $4\text{ ones}+10\text{ ones}=14\text{ ones}$. $\begin{array}{rrrr} & & 9 & \\ & 2 & \not{1}\not{0} & 14 \\ & \not{3} & \not{0} & \not{4} \\ - & 1 & 4 & 6 \\ \hline & 1 & 5 & 8 \end{array}$	**10.** Subtract $709-248$.

Example	Student Practice

11. Find the perimeter of the shape consisting of rectangles.

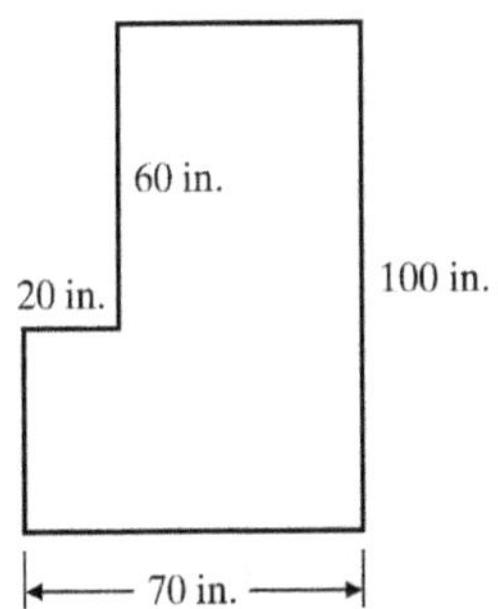

To find the perimeter we must find the distance around the figure. Therefore, we must find the measures of the unlabeled sides.

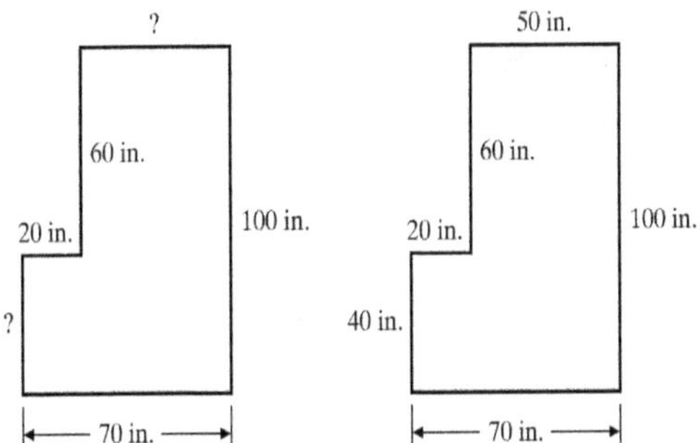

To find the unlabeled side on the top subtract $70-20=50$. Thus, the unlabeled top side is 50 inches.

To find the unlabeled left side, subtract $100-60=40$. Thus, the unlabeled left side is 40 inches.

To find the perimeter, add the lengths of the six sides.

$$50+100+70+40+20+60=340$$

Thus, the perimeter is 340 in.

12. Find the perimeter of the shape consisting of rectangles.

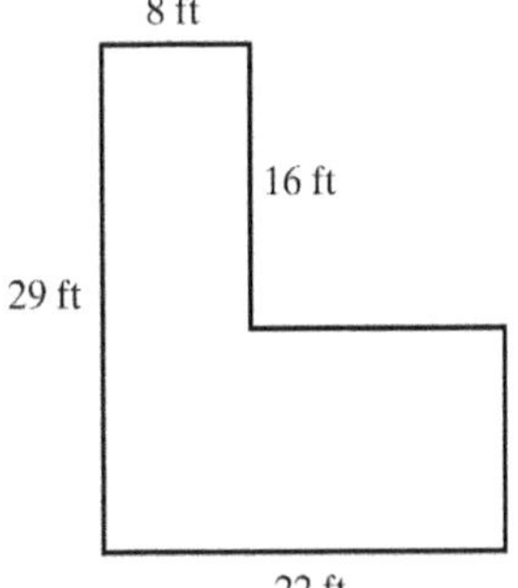

Extra Practice

1. Translate the following into numbers and symbols.

three hundred one subtracted from four hundred

2. Evaluate the expression using the given values of x and y.

$x - y - 99$ when $x = 700$ and $y = 101$

3. Subtract and check.

$$\begin{array}{r} 109{,}000 \\ -\ \ 67{,}898 \\ \hline \end{array}$$

4. Barbara Ann received \$175 in birthday money from her relatives. She spent \$35 on a new hairstyle, \$26 to fix the front tire on her bicycle, and \$59 on new clothes. How much money was left after all of these purchases?

Concept Check

Explain why when we subtract $800 - 35$, we change 8 to 7 in the borrowing process.

Name: ______________________ Date: ______________

Instructor: ______________________ Topic: ______________

Module 1 Whole Numbers and Introduction to Algebra
Topic 4 Multiplying Whole Number Expressions

Vocabulary

multiplication • array • factors • product • trailing zeros
commutative property of multiplication • multiplication property of 0
identity property of 1 • associative property if multiplication

1. The result of a multiplication problem is called a(n) __________.

2. The __________ states that when any number is multiplied by 0, the product is 0.

3. The __________ states that changing the order of factors does not change the product.

4. The numbers or variables we multiply are called __________.

Example	Student Practice
1. Draw two arrays that represent the multiplication 3 times 4. There are two arrays consisting of twelve items that represent the multiplication 3 times 4. One array has 4 rows and 3 columns, and the other one has 3 rows and 4 columns. ★ ★ ★ ★ 3 ★ ★ ★ ★ ★ ★ ★ ★ 4 12 items ★ ★ ★ ★ ★ ★ 4 ★ ★ ★ ★ ★ ★ 3 12 items	**2.** Draw two arrays that represent the multiplication 2 times 5.
3. Identify the product and the factors in the given equation. $5(4) = 20$ 5 and 4 are the factors and 20 is the product.	**4.** Identify the product and the factors in the given equation. $3b = 24$

Vocabulary Answers: 1. product 2. multiplication property of 0 3. commutative property of multiplication 4. factors

Example	Student Practice
5. Translate using numbers and symbols. The product of four and a number $4 \cdot n = 4n$	**6.** Translate using numbers and symbols. Five times a number
7. Multiply $4 \cdot 2 \cdot 4 \cdot 5$. Use the commutative property to change the order of factors so that one factor is 10. $4 \cdot 2 \cdot 4 \cdot 5 = (4 \cdot 4) \cdot (2 \cdot 5)$ Since $4 \cdot 4 = 16$ and $2 \cdot 5 = 10$, we have that $(4 \cdot 4) \cdot (2 \cdot 5) = 16 \cdot 10$. To multiply 16(10), write 16 and attach a zero at the end. $16 \cdot 10 = 160$	**8.** Multiply $2 \cdot 3 \cdot 5 \cdot 6$.
9. Simplify $2(3)(n \cdot 7)$. Rewrite using familiar notation, and multiply $2 \cdot 3 = 6$. $2(3)(n \cdot 7) = 2 \cdot 3 \cdot (n \cdot 7) = 6 \cdot (n \cdot 7)$ Now, change the order of factors and regroup. $6 \cdot (n \cdot 7) = 6 \cdot (7 \cdot n) = (6 \cdot 7) \cdot n$ Multiply and write in standard notation. $(6 \cdot 7) \cdot n = 42 \cdot n$ $= 42n$ Thus, $2(3)(n \cdot 7) = 42n$.	**10.** Simplify $6(7)(n \cdot 5)$.

Example	Student Practice
11. Multiply (547)(600). Since the number 600 has trailing zeros, multiply the nonzero digits and attach the trailing zeros to the right side of the product. Bring down the trailing zeros and multiply $6(7) = 42$. Place the 2 in the hundreds place and carry the 4. $\begin{array}{r} \overset{4}{5}47 \\ \times \quad 600 \\ \hline 200 \end{array}$ Next, multiply $6 \cdot 4$ and add the carried digit. Place the 8 in the thousands place and carry the 2. Repeat the process for $6 \cdot 5$ to complete the multiplication. $\begin{array}{r} \overset{2\,4}{5\,4}7 \\ \times \quad 600 \\ \hline 328{,}200 \end{array}$	**12.** Multiply (458)(300).
13. Multiply 857(43). To multiply 857(43), multiply $857(3+40)$ or $857(3)+857(40)$ using the condensed form. First multiply $3(857) = 2571$. Then to find the product of 40(857), multiply 4(857) and add one trailing zero. Then finally, add the partial products. $\begin{array}{r} 857 \\ \times \ 43 \\ \hline 2571 \\ 34280 \\ \hline 36{,}851 \end{array}$	**14.** Multiply 105(3245).

Example	Student Practice
15. Jessica drove an average speed of 60 miles per hour for 7 hours (per hour means each hour). How far did she drive?	**16.** Mike earns $8 per hour as a retail cashier. How much will he earn if he works 38 hours?

We draw a diagram and see that this is a situation that involves repeated addition, which indicates that we multiply.

60 miles	60 miles	60 miles	and so on.

60	Miles driven each hour
× 7	Number of hours driven
420	Total miles driven

Check the answer using the diagram. We see that in 3 hours, Jessica drove 180 miles $(60+60+60)$. Thus, in 6 hours she drove 360 miles $(180+180)$. Now, since she drove 60 miles the seventh hour, we add $360+60=420$ miles.

Extra Practice

1. State what property of multiplication is represented in the mathematical statement.

   $3\cdot(4\cdot2)=(3\cdot4)\cdot2$

2. Translate using numbers and symbols. Do not evaluate.

   Four doubled

3. Multiply.

   $$\begin{array}{r} 12{,}402 \\ \times \quad 607 \\ \hline \end{array}$$

4. Rita bought three boxes of blank CDs. If each box holds 25 CDs, how many CDs did Rita buy?

Concept Check

Explain what to do with the zeros when you multiply 546×2000.

Name: ______________________________ Date: ________________
Instructor: ___________________________ Topic: ______________

Module 1 Whole Numbers and Introduction to Algebra
Topic 5 Dividing Whole Number Expressions

Vocabulary
division • quotient • divisor • dividend • undefined • remainder

1. Division by 0 is said to be ___________.

2. The result of a division problem is called the ___________.

3. The number sometimes "left over" in a division problem is called the ___________.

4. The process of repeated subtraction is called ___________.

Example	**Student Practice**
1. Write the division statement that corresponds to the following situation. You need not carry out the division. 120 students in a band are marching in 5 rows. How many students are in each row? We draw a picture. We want to split 120 into 5 equal groups. 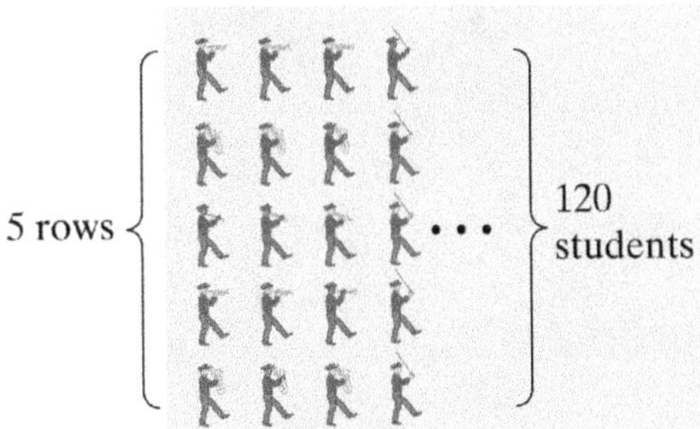 The division statement that corresponds to this situation is $120 \div 5$.	**2.** Write the division statement that corresponds to the following situation. You need not carry out the division. Ria has a total of $180 and she would like to divide it out among 9 charities. How much money will each charity receive?

Vocabulary Answers: 1. undefined 2. quotient 3. remainder 4. division

Example	Student Practice
3. Translate using numbers and symbols. The quotient of forty-six and two The phrase “the quotient” indicates division, and the order of the numbers indicates the dividend, 46, and the divisor, 2. $46 \div 2$	**4.** Translate using numbers and symbols. The quotient of four and forty-eight
5. Divide $18 \div 3$. Think, $18 = \boxed{?} \cdot 3$. $18 \div 3 = \boxed{?}$ $18 = \boxed{6} \cdot 3$ Thus, $18 \div 3 = 6$.	**6.** Divide $36 \div 9$.
7. Divide. **(a)** $0 \div 9$ 0 divided by any nonzero number is equal to 0. $0 \div 9 = 0$ **(b)** $9 \div 0$ Zero can never be the divisor in a division problem. So, $9 \div 0$ is undefined. **(c)** $\frac{16}{16}$ Any number divided by itself is 1, $16 \div 16 = 1$.	**8.** Divide. **(a)** $\frac{28}{28}$ **(b)** $19 \div 0$ **(c)** $0 \div 15$

Example	Student Practice
9. Divide and check your answer. $293 \div 41$	**10.** Divide and check your answer. $17{,}985 \div 57$

First guess (too large): First guess that 41 times 8 is close to 293. Write 8 in the quotient.

$$\begin{array}{r} 8 \\ 41\overline{)293} \\ \underline{-328} \end{array}$$

Check the first guess, $41(8) = 328$; Our guess is too large, so we must adjust.

Second guess (too small): Try 6.

$$\begin{array}{r} 6 \\ 41\overline{)293} \\ \underline{-246} \\ 47 \end{array}$$

Check the second guess, $41(6) = 246$; 246 is less than 293, but 47 is not less than 41. Our guess is too small so we must adjust.

Third guess: Try 7.

$$\begin{array}{r} 7\text{ R6} \\ 41\overline{)293} \\ \underline{-287} \\ 6 \end{array}$$

Check the third guess, $41(7) = 287$; 287 is less than 293, 6 is less than 41. We do not need to adjust our guess, and 6 is the remainder.

Verify that the answer is correct by multiplying the divisor by the quotient and then adding the remainder.

Example	Student Practice
11. Twenty-six students in Ellis High School entered their class project in a contest sponsored by the Falls City Baseball Association. The class won first place and received 250 tickets to the baseball play-offs. The teacher gave each student in the class an equal number of tickets, then donated the extra tickets to a local Boys and Girls Club. How many tickets were donated to the Boys and Girls Club?	**12.** Mike, Julie, Amber, and Ali pooled their allowances and purchased a box of 47 candy bars. They agreed to split the box evenly among the four of them and give any leftover candy to their friend Stacey. How many candy bars did Stacey receive?

Since we must split 250 equally among 26 students, we divide.

$$\begin{array}{r} 9 \text{ R } 16 \\ 26\overline{)250} \\ \underline{234} \\ 16 \end{array}$$

Since 16 tickets are left over, 16 tickets are donated to the Boys and Girls Club.

Extra Practice

1. Divide $77 \div 0$.

2. Divide $21\overline{)2063}$.

3. Translate using numbers and symbols.

   fifty-two divided by a number

4. Keith planted eight rows of tomatoes all containing the same number of plants. He planted 96 tomato plants all together. How many plants were in each row?

Concept Check

Explain the next 2 steps for this division problem.

$$\begin{array}{r} 2 \\ 13\overline{)2645} \\ \underline{26} \\ 04 \end{array}$$

Name: ______________________________ Date: ______________

Instructor: ______________________________ Topic: ______________

Module 1 Whole Numbers and Introduction to Algebra
Topic 6 Exponents and the Order of Operations

Vocabulary

factor • exponent form • exponent • base • squared • cubed • order of operations

1. The first step in the ____________ is to perform operations inside parentheses.

2. If the value of the exponent is 2, we say the base is ____________.

3. In the expression 3^5, the number 5 is called the ____________.

4. In the expression 3^5, the number 3 is called the ____________.

Example	Student Practice
1. Write in exponent form. **(a)** $2\cdot2\cdot2\cdot2\cdot2\cdot2$ $2\cdot2\cdot2\cdot2\cdot2\cdot2=2^6$ **(b)** $y\cdot y\cdot y\cdot3\cdot3\cdot3\cdot3$ $y\cdot y\cdot y\cdot3\cdot3\cdot3\cdot3=y^3\cdot3^4$, or 3^4y^3 Note that it is standard to write the number before the variable in a term. Thus $y^3 3^4$ is written 3^4y^3.	**2.** Write in exponent form. **(a)** $7\cdot7\cdot7\cdot7\cdot7\cdot7\cdot7\cdot7$ **(b)** $a\cdot a\cdot a\cdot a\cdot9\cdot9\cdot9$
3. Write as a repeated multiplication. **(a)** n^3 $n^3=n\cdot n\cdot n$ **(b)** 6^5 $6^5=6\cdot6\cdot6\cdot6\cdot6$	**4.** Write as a repeated multiplication. **(a)** y^7 **(b)** 9^6

Vocabulary Answers: 1. order of operations 2. squared 3. exponent 4. base

Example	Student Practice
5. Evaluate each expression. **(a)** 3^3 $3^3 = 3 \cdot 3 \cdot 3 = 27$ **(b)** 1^9 $1^9 = 1$ We do not need to write out this multiplication because repeated multiplication of 1 is equal to 1. **(c)** 2^4 $2^4 = 2 \cdot 2 \cdot 2 \cdot 2 = 16$	**6.** Evaluate each expression. **(a)** 4^5 **(b)** 1^{15} **(c)** 10^3
7. Evaluate 10^7. Write 1. Then, since the exponent is 7, attach 7 trailing zeros. $10^7 = 10{,}000{,}000$	**8.** Evaluate 10^9.
9. Evaluate x^3 for $x = 3$. Replace x with 3. $x^3 = (3)^3$ Write as repeated multiplication, and then multiply. $(3)^3 = 3 \cdot 3 \cdot 3 = 27$ When $x = 3$, x^3 is equal to 27.	**10.** Evaluate y^4 for $y = 5$.

Example	Student Practice
11. Translate using symbols. **(a)** Five cubed Five cubed $= 5^3$ **(b)** y to the eighth power y to the eighth power $= y^8$	**12.** Translate using symbols. **(a)** Twelve squared **(b)** Five to the seventh power
13. Evaluate $4+3(6-2^2)-7$. Always perform the calculations inside the parentheses first. Once inside the parentheses, proceed using the order of operations. Within the parentheses, exponents have the highest priority, $2^2=4$. $4+3(6-2^2)-7=4+3(6-4)-7$ Finish all operations inside the parentheses. Subtract $6-4=2$. $4+3(6-4)-7=4+3(2)-7$ Now the highest priority is multiplication, $3\cdot 2=6$. $4+3(2)-7=4+6-7$ Next, add first, $4+6=10$. $4+6-7=10-7$ Finally, subtract last, $10-7=3$. Thus, $4+3(6-2^2)-7=3$.	**14.** Evaluate $4+9(20-3\cdot 5)-6$.

Example	Student Practice
15. Evaluate $\frac{(6+6\div3)}{(5-1)}$.	**16.** Evaluate $\frac{(9+12\div4)}{(8-5)}$.

Rewrite the problem as division and then follow the order of operations.

$(6+6\div3)\div(5-1)$

First, perform operations inside parentheses. $6\div3=2$; $5-1=4$.

$$(6+6\div3)\div(5-1)=(6+2)\div4$$
$$=8\div4$$

Finally, divide.

$8\div4=2$

Extra Practice

1. Write in exponent form.

$a\times a\times a\times a\times a$

2. Translate using symbols.

y to the sixth power

3. Using the correct order of operations, evaluate the expression.

$5\times2^3-4\times(14\div7)$

4. Using the correct order of operations, evaluate the expression.

$$\frac{54(13-10)-18}{3^2-3}$$

Concept Check

Explain in what order you would do the steps to evaluate $50+3\times5^2\div25$.

Name: ______________________________ Date: ______________

Instructor: ______________________________ Topic: ______________

Module 1 Whole Numbers and Introduction to Algebra
Topic 7 More on Algebraic Expressions

Vocabulary

distributive property • parentheses • order of operations • distribute

1. It is necessary to include __________ when the phrases "sum of" or "difference of" are used or the answer obtained may be wrong.

2. The __________ states that if a, b, and c are numbers or variables, then $a(b+c)=ab+ac$ and $a(b-c)=ab-ac$.

3. We __________ a over addition and subtraction by multiplying every number or variable inside the parentheses by a.

4. When we translate phrases into numbers and symbols we must take care to preserve the __________ indicated by the phrase.

Example	Student Practice
1. Translate using numbers and symbols.	**2.** Translate using numbers and symbols.
(a) Two times x plus seven Two times x plus seven ↓ ↓ ↓ ↓ ↓ $2 \cdot x + 7$	**(a)** Eight times a plus eleven
(b) Two times the sum of x and seven The key phrase "sum of" indicates that $x+7$ is placed in parentheses. Two times the sum of x and seven ↓ ↓ $2\cdot$ $(x+7)$ Thus, "two times the sum of x and seven" translates to $2(x+7)$.	**(b)** Four times the sum of n and six

Vocabulary Answers: 1. parentheses 2. distributive property 3. distribute 4. order of operations

Example	Student Practice
3. Evaluate $\frac{(2a+3)}{7}$ for $a=9$. Replace a with 9. $\frac{(2a+3)}{7}=\frac{(2\cdot 9+3)}{7}$ Now, multiply. $\frac{(2\cdot 9+3)}{7}=\frac{(18+3)}{7}$ Next, complete the operations within the parentheses. Then, divide. $\frac{(18+3)}{7}=\frac{21}{7}$ $=3$	**4.** Evaluate $\frac{(7y-5)}{4}$ for $y=3$.
5. Evaluate $\frac{(x^2-2)}{y}$ for $x=4$ and $y=2$. Replace x with 4 and y with 2. $\frac{(x^2-2)}{y}=\frac{(4^2-2)}{2}$ Now square 4 and then subtract: $4^2=16,\ 16-2=14$. $\frac{(4^2-2)}{2}=\frac{14}{2}$ Finally, divide. $\frac{14}{2}=7$	**6.** Evaluate $\frac{(a^4-9)}{b}$ for $a=3$ and $b=12$.

Example	Student Practice
7. Use the distributive property to simplify. $3(x-2)$ First, multiply 3 times x, then multiply 3 times 2. $3(x-2)=3\cdot x-3\cdot 2$ Finally, simplify your answer. $3\cdot x-3\cdot 2=3x-6$ Thus, $3(x-2)=3x-6$.	**8.** Use the distributive property to simplify. $4(8+y)$
9. Simplify. $2(y+1)+4$ First use the distributive property to multiply $2(y+1)$ and then simplify. $2(y+1)+4=2\cdot y+2\cdot 1+4$ $=2y+2+4$ Now, simplify the resulting expression. $2y+2+4=2y+6$ Thus, $2(y+1)+4=2y+6$.	**10.** Simplify. $9(y+4)+5$

Extra Practice

1. Translate using numbers and symbols.

Eight times the difference of x and one

2. Evaluate for the given values.

$\frac{(m^2-5)}{4}$ when $m=5$

3. Evaluate for the given values.

$3mn+2m+5n$ when $m=4$ and $n=8$

4. Use the distributive property to simplify.

$3(x+y+4)-2(y+1)$

Concept Check

Simplify $5(x+1)$, then evaluate $5(x+1)$ for $x=2$. Compare results and state the difference in the process to simplify and to evaluate.

Name: ______________________________ Date: ________________

Instructor: ____________________________ Topic: ________________

Module 1 Whole Numbers and Introduction to Algebra
Topic 8 Introduction to Solving Linear Equations

Vocabulary

term • coefficient • constant term • variable term • like terms
expression • equation • solution • solve • combine like terms

1. Terms that have identical variable parts are called ____________.

2. In a(n) ____________, we use an equals sign $(=)$ to indicate that two expressions are equal in value.

3. The numerical part of a term is called the ____________ of the term.

4. A(n) ____________ is a number, a variable, or a product of a number and one or more variables.

Example	Student Practice
1. Write a term that represents the following. Two y's Two $y\text{'s} = 2y$	**2.** Write a term that represents the following. $b+b+b+b+b+b$
3. Identify like terms, then combine like terms. $4xy+8y+2xy$ Identify and group like terms. $4xy+8y+2xy=(4xy+2xy)+8y$ Add the numerical coefficients of like terms. $(4xy+2xy)+8y=(4+2)xy+8y$ Thus, $4xy+8y+2xy=6xy+8y$. We cannot combine $8y$ with $6xy$ since the variable parts are not the same.	**4.** Identify like terms, then combine like terms. $8y+2x+7y+9x$

Vocabulary Answers: 1. like terms 2. equation 3. coefficient 4. term

Example	Student Practice

Example

5. Write the perimeter of the rectangular field as an algebraic expression and simplify.

$4a + 7b$

$2a + 3b$

Since the figure is a rectangle, opposite sides are equal.

$4a + 7b$

$2a + 3b$ $2a + 3b$

$4a + 7b$

We add all sides to find the perimeter.

$$(2a+3b)+(4a+7b)+(2a+3b)+$$
$$(4a+7b)$$

Now use the associative and commutative properties to change the order of addition and regroup. Then finally, combined like terms.

$$(2a+2a+4a+4a)+(3b+3b+7b+7b)$$
$$=12a+20b$$

Thus, the algebraic expression for the perimeter is $12a+20b$.

Student Practice

6. Write the perimeter of the triangular figure as an algebraic expression and simplify.

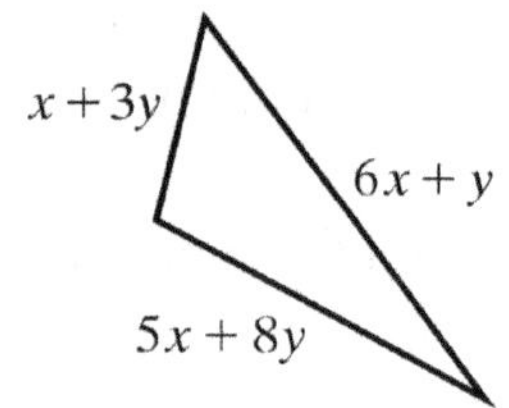

Example

7. Is 2 a solution to $6-x=9$?

If 2 is a solutions to $6-x=9$, when we replace x with the value $x=2$ we will get a true statement.

Replace the variable with 2 and simplify.

$$6-2\overset{?}{=}9$$
$$4\overset{?}{=}9$$

This is a false statement. Since $4=9$ is not a true statement, 2 is not a solution to $6-x=9$.

Student Practice

8. Is 7 a solution to $x+5=14$?

Example	Student Practice
9. Solve the equation $3+n=10$ and check your solution. To solve the equation $3+n=10$, answer the question: "Three plus what number is equal to ten?" Using addition facts we see that the answer, or solution, is 7. To check the solution, we replace n with the value $n=7$ and verify that we get a true statement. $3+n = 10$ $3+7 \stackrel{?}{=} 10$ $10 = 10$ True Since we get a true statement, the solution to $3+n=10$ is 7 and is written $n=7$.	**10.** Solve the equation $n+6=11$ and check your solution.
11. Simplify using the associative and commutative properties and then find the solution to the equation $(5+n)+1=9$. First, apply the commutative property. Next, apply the associative property and simplify. $(n+5)+1=9$ $n+(5+1)=9$ $n+6=9$ Now, solve $n+6=9$. Answer the question "What number plus 6 is equal to 9?" $n=3$ Thus, the solution to the equation is 3.	**12.** Simplify using the associative and commutative properties and then find the solution to the equation $(6+a)+3=13$.

Example	Student Practice
13. Translate, then solve. Double what number is equal to eighteen? "Double" translates to "2 times" or "$2\cdot$." "What number" translates to "n." "is equal to" translates to "$=$." Thus, the equation is $2\cdot n=18$. Use multiplication facts to find n. Since $2\cdot 9=18$, $n=9$.	**14.** Translate, then solve. Triple what number is equal to twenty-one?

Extra Practice

1. Combine like terms.

$14ab+8+4ab+3$

2. Translate the mathematical symbols using words.

$x+8=12$

3. Solve and check your answer.

$(a+5)+2=16$

4. Translate to an equation. Do not solve the equation.

Three plus what number equals twelve?

Concept Check

Explain the difference in the process you must use to complete **(a)** and **(b)**.

(a) Combine like terms. $3x+x+2x$

(b) Solve. $3x+x+2x=12$

Name: ______________________________ Date: ________________
Instructor: ___________________________ Topic: ________________

Module 1 Whole Numbers and Introduction to Algebra
Topic 9 Solving Applied Problems Using Several Operations

Vocabulary
estimate • understand the problem • calculate and state the answer • check the answer

1. The first step in solving applied problems is to ____________, which involves reading the problem and organizing the information.

2. To ____________ a sum or difference, round each number to the same round-off place and then find the sum or difference.

Example	Student Practice
1. Some sample sale prices for 2010 Ford motor vehicles are listed below.	**2.** Use the figure in example **1** to answer the following. Estimate the difference in the price between an F-150 SVT Raptor and a Focus SE Coupe by rounding each price to the nearest thousand.

Manufacturers Suggested Retail Prices on 2010 Ford Vehicles

Vehicle	Price
F-150 SVT Raptor	$38,020
Taurus SEL	$28,195
Focus SE Coupe	$15,895
Mustang V6 Premium Coupe	$26,695

Source: www.fordvehicles.com

Estimate the difference in the price between an F-150 SVT Raptor and a Mustang V6 Premium Coupe by rounding each price to the nearest thousand.

The exact value for the price of the F-150 SVT Raptor is 38,020 and the rounded value is 38,000. The exact value for the price of the Mustang V6 Premium Coupe is 26, 695 and the rounded value is 27,000. To estimate the difference in the cost of the two vehicles, subtract the two rounded figures, $38,000-27,000=11,000$.
Thus, the estimated difference in price is $11,000.

Vocabulary Answers: 1. understand the problem 2. estimate

Example	Student Practice
3. A frequent-flyer program offered by many major airlines to first-class passengers awards 3 frequent-flyers mileage points for every 2 miles flown. When customers accumulate a certain number of frequent-flyer points, they can cash them in for free air travel, ticket upgrades, or other awards. How many frequent-flyer points would a customer accumulate after flying 3500 miles in first class? Sometimes, drawing charts or pictures can help us understand the problem as well as plan our approach to solving the problem. $2 \text{ miles} + 2 \text{ miles} \cdots = 3500 \text{ miles}$ $3 \text{ points} + 3 \text{ points} \cdots = ? \text{ points}$ Organize the information and make a plan using the Mathematics Blueprint. We divide to find how many groups of 2 are in 3500, $3500 \div 2 = 1750$. Then, we multiply 1750 times 3 to find the total points earned, $1750 \cdot 3 = 5250$ points. Check: If the customer earned 4 points (instead of 3) for every 2 miles traveled, we could just double the mileage to find the points earned, 2 miles earns 4 points and 3500 miles earns 7000 points. Since the customer earned a little less than 4 points, the total should be less than 7000. It is: $5250 < 7000$. The customer also earned more points than miles traveled, so the total points should be more than the total miles traveled. It is: $5250 > 3500$. Our answer is reasonable.	**4.** A frequent-flyer program offered by many major airlines to first-class passengers awards 3 frequent-flyers mileage points for every 2 miles flown. When customers accumulate a certain number of frequent-flyer points, they can cash them in for free air travel, ticket upgrades, or other awards. How many frequent-flyer points would a customer accumulate after flying 5500 miles in first class?

Example	Student Practice
5. Koursh was offered two different jobs: a 40-hour-a-week store assistant management position that pays $14 per hour and an executive secretary position paying a monthly salary of $2600. Which job pays more per year?	**6.** Anny works 40 hours a week and earns $17 an hour as a payroll clerk. She is considering accepting a job offer to work as an assistant office manager earning a salary of $3500 per month. Which job pays more per year?

Read the problem carefully and create a Mathematics Blueprint. We write out a process to find the answer.

$\$14 \times 40 = \$560$ Pay for 1 week (management)

$\$560 \times 52 = \$29{,}120$ Pay for 1 year (management)

$\$2600 \times 12 = \$31{,}200$ Pay for 1 year (secretary)

The yearly pay is $29,120 for the management position and $31,200 for the secretary's position.

The secretary's position pays more per year.

Check: We estimate the assistant manager's pay per year by rounding $14 per hour to $10 and 52 weeks to 50 weeks.

$\$10 \times 40 \text{ hr} = \$400$ per week;
$\$400 \times 50 \text{ weeks} = \$20{,}000$ per year

We estimate the secretary's pay per year by rounding 12 months per year to 10.

$\$10 \times \$2600 = \$26{,}000$ per year

Since $\$26{,}000 > \$20{,}000$, the secretary position pays more.

Extra Practice

1. Karen wants to buy the following four items shown in the chart. Round the price of each item to the nearest hundred dollars.

Item	Price
Bedspread	$189
Digital Camera	$331
Fountain	$94
Stained Glass	$453

2. Using the chart and rounded values from extra practice **1**, estimate the total cost of all four items.

3. Last week, Maureen worked 53 hours. She earns $10 per hour for the first 40 hours worked, and then earns $15 per hour every hour after 40 hours worked. How much money should Maureen expect to be paid for last week's work?

4. A textbook author estimates that she can write 3 pages per day. How many pages can she expect to write in 30 days? 60 days?

Concept Check

At the end of January, Sahara had $200 left in her vacation savings account and $1000 in her household savings account. Each month for the next six months, Sahara plans to put $100 in her vacation account and $200 in her household account. In addition, she plans to split her $900 tax return equally between both accounts. Explain how to determine if Sahara will have enough money in her vacation account at the end of six months to take a $1500 vacation.

MATH COACH

Mastering the skills you need to do well on the test.

Watch the MATH COACH videos in MyMathLab® or on YouTube™ while you work the problems below. These helpful hints will help you avoid making common errors on test problems.

Subtract Whole Numbers with Borrowing—Problem 7(b)

Subtract. $\begin{array}{r} 20{,}105 \\ -7{,}826 \\ \hline \end{array}$

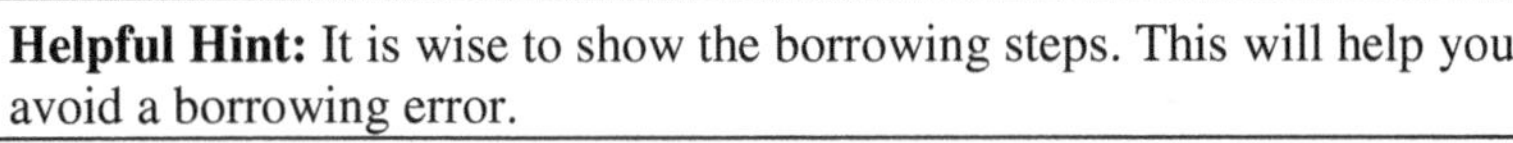

Helpful Hint: It is wise to show the borrowing steps. This will help you avoid a borrowing error.

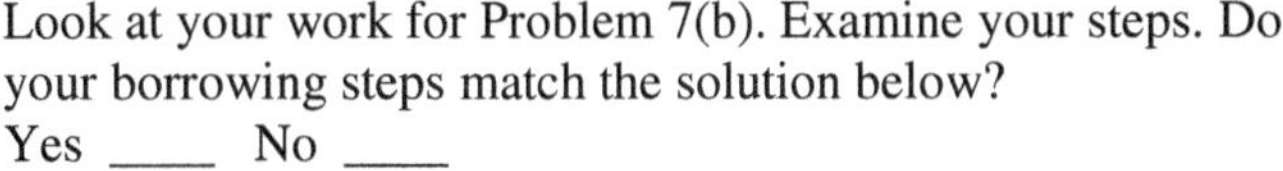

Look at your work for Problem 7(b). Examine your steps. Do your borrowing steps match the solution below?
Yes ____ No ____

If you answered No, be sure to write your borrowing steps carefully and then check each subtraction step for errors. Stop now and rework the problem.

If you answered Yes, and still got an incorrect answer, check each subtraction step for errors.

Write out the borrowing steps: $\begin{array}{rrrrr} \overset{1}{\cancel{2}} & \overset{9}{\cancel{0},} & \overset{10}{\cancel{1}} & \overset{9}{\cancel{0}} & \overset{15}{\cancel{5}} \\ - & 7 & 8 & 2 & 6 \\ \hline \end{array}$

If you answered Problem 7(b) incorrectly, go back and rework the problem using these suggestions.

Performing Long Division with Whole numbers—Problem 11(b) Divide $5523 \div 46$.

Helpful Hint: When performing long division, be sure to line up each column exactly as shown below. Accuracy in alignment will increase your chance of completing the problem correctly. Remember that a remainder results when there are no more numbers to bring down from the dividend.

Look at your work from Problem 11(b). Compare each line of your work with the problem below.

Did you align numbers correctly?
Yes ____ No ____

Did you obtain the correct value each time you subtracted?
Yes ____ No ____

$$\begin{array}{r} 46\overline{)5523} \\ \underline{46} \\ 92 \\ \underline{92} \\ 03 \end{array}$$

If you answered No to either question, stop and rework the problem now.

Did you get 12 R3 for your answer?

Yes ____ No ____

If you answered Yes, then you stopped dividing too soon. Because 46 cannot be divided into 3, we must write 0 in the quotient and continue to divide.

Now go back and rework the problem using these suggestions.

Evaluate Algebraic Expression—Problem 16(a)

Evaluate $2x-3y$ if x is equal to 16 and y is equal to 4.

Helpful Hint: Take time to show the step of replacing the variable with the appropriate number before you do any calculations. To avoid errors, do not skip steps or complete calculations mentally.

As your first step, did you replace $2\boxed{x}-3\boxed{y}$ with $2\cdot\boxed{16}-3\cdot\boxed{4}$? Yes ____ No ____

If you answered No, stop and complete this step.

Did you multiply $2\cdot\boxed{16}$ for your next step?
Yes ____ No ____

If you answered No, then you *did not* follow the proper order of operations. Remember to complete all multiplications before you subtract.

If you answered Problem 11(b) incorrectly, go back and rework the problem using these suggestions.

Solving Applied Problems Using Several Operations—Problem 26(a) and (b)

Tickets to a play were \$25 for adults and \$18 for children. 412 adult tickets were sold and 280 child tickets were sold.

(a) Find the total income from the sale of tickets.
(b) If the expenses for the play were \$7350, how much profit was made?

Helpful Hint: Remember to use pictures, charts, or the Mathematics Blueprint for Problem Solving. First, try to **understand the problem** by rereading the problem carefully. What are the facts? What am I asked to do? What type of calculation(s) must I perform? Next, **calculate and state the answer**. If possible, take the time to **check** your answer.

Reread Problem 26 carefully, then fill in the information needed to solve the problem.

What are the facts? There are ____ adult tickets. They cost \$____ each. There are ____ child tickets. They cost \$____ each.

What am I asked to do for Part (a)? You must find the total cost of tickets (income).

What type of calculation must I perform? Did you multiply $412\times\$25$ to find the income from adult tickets and $280\times\$18$ to find the income from child tickets? Then did you add these results? Yes ____ No ____

If you answered No, stop now and use these hints to complete Part (a) correctly.

What am I asked to do for Part (b)?
You must find the profit if the expenses were \$7350.

What type of calculation must I perform?
Did you calculate
Income − Expenses = Profit ?
Yes ____ No ____

If you answered No, stop now and use these hints to complete Part (b) correctly.

Try to write out all the facts of the situation to help you keep track of the details and determine what type of calculations are needed to solve the problem.

Now go back and rework the problem using these suggestions.

Name: ______________________________ Date: ________________

Instructor: ____________________________ Topic: _______________

Module 2 Integers
Topic 1 Understanding Integers

Vocabulary

negative numbers • positive numbers • number line • signed numbers

integers • opposites • absolute value

1. Positive numbers are to the right of 0 and negative numbers are to the left of 0 on the ____________.

2. Signed numbers, ..., $-3,\ -2,\ -1,\ 0,\ 1,\ 2,\ \ldots$, are also called ____________.

3. Numbers that are the same distance from zero but lie on the opposite sides of zero on the number line are called ____________.

4. Numbers that are less than 0 are called ____________.

Example	Student Practice
1. Graph -5, -3, 1, and 5 on a number line. We draw a dot in each of the correct locations on the number line. We start at zero for each and count five places in the negative direction for -5, three places in the negative direction for -3, one place in the positive direction for 1, and five places in the positive direction for 5. 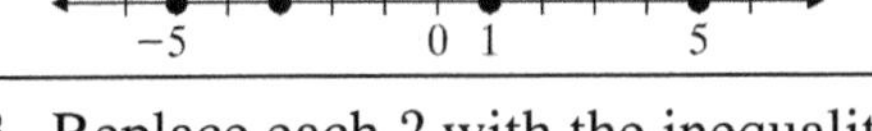 	**2.** Graph $-3, -1,\ 0,$ and 2 on a number line. 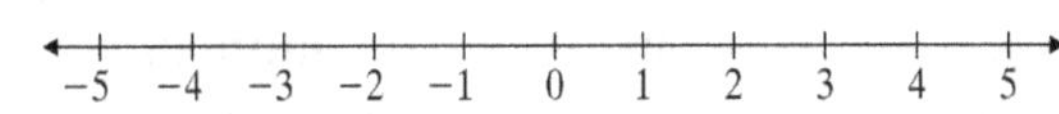
3. Replace each ? with the inequality symbol < or >. **(a)** $-3\ ?\ -1$ $-3 < -1$ **(b)** $4\ ?\ -5$ $4 > -5$	**4.** Replace each ? with the inequality symbol < or >. **(a)** $-6\ ?\ -3$ **(b)** $-5\ ?\ 2$

Vocabulary Answers: 1. number line 2. integers 3. opposites 4. negative numbers

Example	Student Practice
5. Fill in each blank with the appropriate symbol, + or −, to describe either an increase or a decrease.	**6.** Fill in each blank with the appropriate symbol, + or −, to describe either an increase or a decrease.
(a) A discount of \$5: __ \$5 A discount of \$5 results in the price decreasing: <u>−\$5</u>	**(a)** A rent increase of \$50: __ \$50
(b) The temperature rises 10°F: __ 10°F The temperature rises, or increases, by 10°F: <u>+10°F</u>	**(b)** A rock falling 30 ft: __ 30 ft
7. Label −5 and the opposite of −5 on a number line. 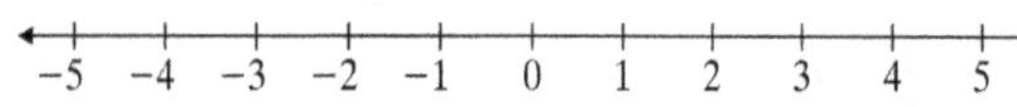 We start at −5 and locate the number that is the same distance from zero but lies on the opposite side of zero. 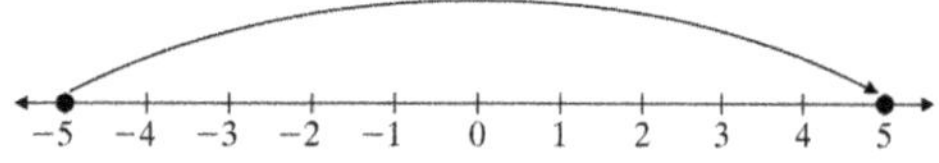 Thus, the opposite of −5 is 5.	**8.** Label −2 and the opposite of −2 on a number line.
9. Evaluate $-(-x)$ for $x=-9$. To avoid errors involving negative signs, we can place parentheses around the variables and their replacements. $-(-x)=-(-(x))$ $=-(-(-9))$ $=-(9)$ $=-9$	**10.** Evaluate $-(-(-b))$ for $b=-2$.

Example	Student Practice
11. Replace the ? with the symbol $<$, $>$, or $=$. $\lvert -15\rvert \ ? \ \lvert 6\rvert$ $\lvert -15\rvert \ ? \ \lvert 6\rvert$ $15 \ ? \ 6$ $15 > 6$ $\lvert -15\rvert > \lvert 6\rvert$ Note that when we say -15 has a larger absolute value, we mean that -15 is a greater distance from 0 than 6 is.	**12.** Replace the ? with the symbol $<$, $>$, or $=$. $\lvert -105\rvert \ ? \ \lvert 5\rvert$
13. Simplify. $-\lvert -7\rvert$ We must find the opposite of the absolute value of -7. $-\lvert -7\rvert = -(7)$ $= -7$	**14.** Simplify. $-\lvert -(-4)\rvert$
15. The line graph below indicates the low temperatures for selected cities on a typical winter day. 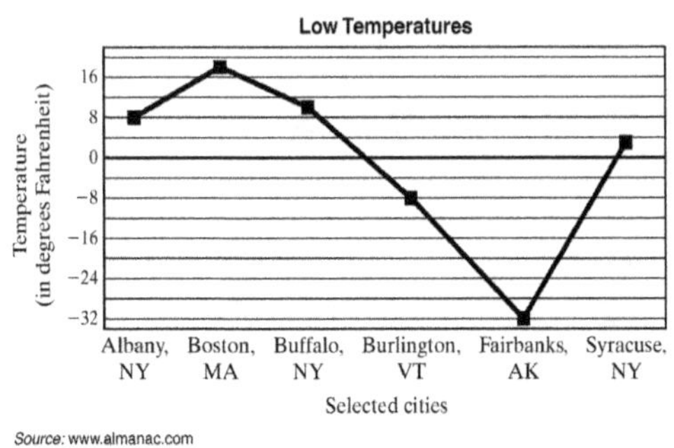 In which city was the temperature colder, Syracuse or Burlington? The low temperature was 3°F in Syracuse and −8°F in Burlington. It was colder in Burlington.	**16.** Use the line graph in example **15** to answer the following. **(a)** In which city was the temperature colder, Boston or Albany? **(b)** Which three cities recorded the highest temperatures for the day?

Extra Practice

1. Graph the following values on a number line.

$-5, -2, 0, 3, 4$

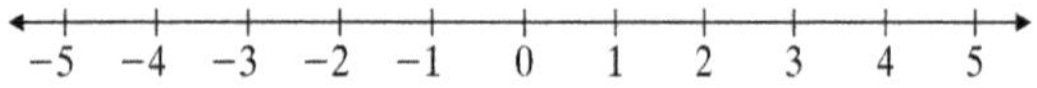

2. State both the opposite and the absolute value.

-21

3. Evaluate $-(-(-x)) - (-|x|)$ for $x = -21$.

4. Which number has the greater opposite, -4 or 5? Explain your thinking.

Concept Check

Explain how to rearrange the following numbers in order from smallest to largest:
$-1, -6, -4, -10, 0$

Name: ______________________________ Date: ______________
Instructor: ___________________________ Topic: ______________

Module 2 Integers
Topic 2 Adding Integers

Vocabulary
additive inverse property • additive inverses • the same sign • different signs

1. 3 and -3 are called ___________.

2. The ___________ states that for any number a, $a+(-a)=0$ and $(-a)+a=0$. The sum of any number and its opposite is 0.

3. Adding numbers with ___________ involves finding the difference between the larger absolute value and the smaller absolute value.

4. Adding numbers with ___________ involves adding the absolute values.

Example	Student Practice
1. Answer parts **(a)** through **(c)**.	**2.** Answer parts **(a)** and **(c)**.

Example

1. Answer parts **(a)** through **(c)**.

(a) Begin at 0 on the number line and move 3 units to the left followed by another 2 units to the left.

−5 −4 −3 −2 −1 0 1 2 3 4 5

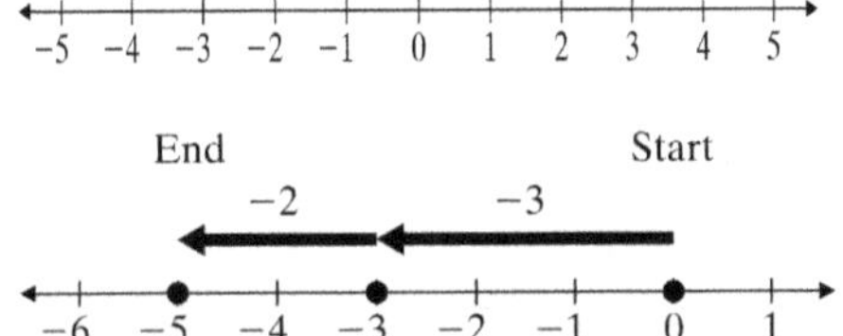

(b) Write the math symbols that represent the situation.

$-3+(-2)$

(c) Use the number line to find the sum.

We end at -5, which is the sum.

$-3+(-2)=-5$

Student Practice

2. Answer parts **(a)** and **(c)**.

(a) Begin at 0 on the number line and move 2 units to the left followed by another 1 unit to the left.

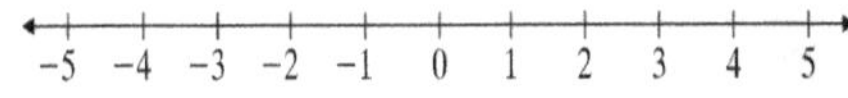

(b) Write the math symbols that represent the situation.

(c) Use the number line to find the sum.

Vocabulary Answers: 1. additive inverses 2. additive inverse property 3. different signs 4. the same sign

Example	Student Practice
3. Add. $-1+(-3)$ We are adding two numbers with the same sign, so keep the common sign and then add the absolute values. $-1+(-3)=-$ $-1+(-3)=-4$	**4.** Add. $-5+(-6)$
5. One night the temperature on Long Island, New York, dropped to $-25°F$. At dawn the temperature had risen $10°F$. Find the sum. Use the thermometer to help determine the sum. 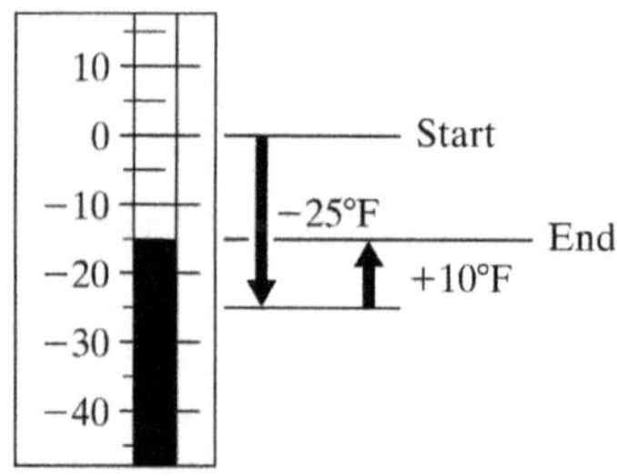 The final temperature is $-15°F$, which is the sum: $-25°F+10°F=-15°F$.	**6.** The temperature was $25°F$ below zero. At dawn it had risen $18°F$. Find the sum. 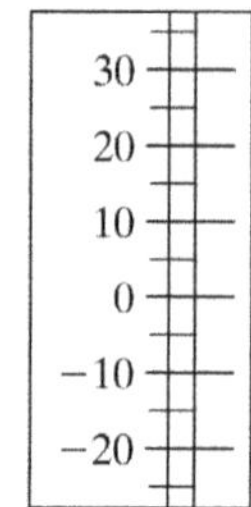
7. Add. $2+(-3)$ We are adding numbers with different signs, so we keep the sign of the larger absolute value and subtract. $2+(-3)=-$ $2+(-3)=-1$	**8.** Add. $-6+8$
9. Find x. $x+21=0$ The sum of additive inverses is 0. Thus if $x+21=0$, then since $-21+21=0$.	**10.** Find x. $-18+x=0$

Example	Student Practice
11. Add.	**12.** Add.
$-3+9+(-4)+12$	$8+(-4)+3+(-10)$
Change the order of addition and regroup.	
$[(-3)+(-4)]+(9+12)$	
Then, add the negative numbers.	
$[(-3)+(-4)]+(9+12)=-7+(9+12)$	
Now, add the positive numbers.	
$-7+(9+12)=-7+21$	
Finally, add the result.	
$-7+21=14$	
13. Evaluate.	**14.** Evaluate.
$-7+a+b$ for $a=-3$ and $b=9$	$x+(-4)+y$ for $x=6$ and $y=-1$
We place the parentheses around the variables, and then replace each variable with the appropriate values.	
$-7+(a)+(b)=-7+(-3)+(9)$	
Then, add the negative numbers.	
$-7+(-3)+(9)=-10+9$	
Finally, add the results.	
$-10+9=-1$	

Example	Student Practice
15. Information on Micro Firm Computer Sales' profit and loss situation is given on the graph. What was the company's overall profit or loss at the end of the third quarter?	**16.** Use the bar graph in Example 15. What was Micro Firm Computer Sales' overall profit or loss at the end of the fourth quarter?

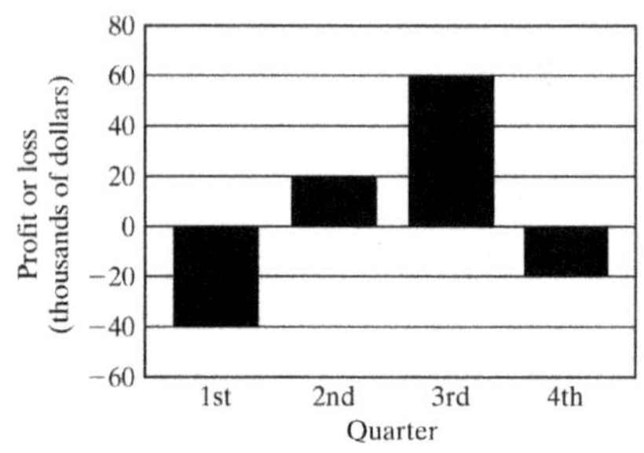

$\boxed{\text{1st quarter loss}} + \boxed{\text{2nd quarter profit}}$

$+ \boxed{\text{3rd quarter profit}} = \boxed{\text{net profit}}$

$-\$40,000 + \$20,000 + \$60,000$
$= \$40,000$

At the end of the third quarter the company had a net profit of $40,000.

Extra Practice

1. Add using the rules for addition of integers. $63+(-21)$
2. Add using the rules for addition of integers. $(-15)+(-20)+(-6)+40$
3. Evaluate $-20+(-x)+(-1)$ for $x=-15$.
4. On Friday morning, Kevin began his camping trip at an altitude of 400 feet. He descended 350 feet on Friday, descended another 75 feet on Saturday, and then climbed up 200 feet on Sunday. Represent his altitude on Sunday night as an integer.

Concept Check

Without completing the calculations, explain how you can determine whether the answer is a positive or negative number. Evaluate $132+x+y+z$ for $x=-1$, $y=-3$, and $z=-2$.

Name: ______________________________ Date: ______________

Instructor: ______________________________ Topic: ______________

Module 2 Integers
Topic 3 Subtracting Integers

Vocabulary

subtraction • addition • difference • opposite

1. When solving subtraction problems, the operation is usually changed to __________.
2. When subtracting a negative number from another number, the __________ is larger.
3. __________ of integers is not commutative or associative.
4. To subtract, add the __________ of the second number to the first.

Example	Student Practice
1. Subtract. **(a)** \$15 − \$20 If we have \$15 and want to spend \$20, we are short \$5, or −\$5. \$15 − \$20 = −\$5 **(b)** $3-4$ If we have 3 items and try to take 4 items, we are short 1 item, or −1. $3-4=-1$	**2.** Subtract. **(a)** \$25 − \$50 **(b)** $4-8$
3. Rewrite each subtraction as addition of the opposite. **(a)** $40-10=30$ $40-10=30 \rightarrow 40+(-10)=30$ **(b)** $6-2=4$ $6-2=4 \rightarrow 6+(-2)=4$	**4.** Rewrite each subtraction as addition of the opposite. **(a)** $50-25=25$ **(b)** $8-5=3$

Vocabulary Answers: 1. addition 2. difference 3. subtraction 4. opposite

Example	Student Practice
5. Subtract. **(a)** $-8-3$ We replace the second number by its opposite and then add using the rules for adding numbers with the same sign. $-8-3=-8+(-3)$ $=-11$ **(b)** $-6-(-4)$ We replace the second number by its opposite and then add using the rules for adding numbers with different signs. $-6-(-4)=-6+4$ $=-2$	**6.** Subtract. **(a)** $-2-3$ **(b)** $-7-(-1)$
7. Subtract. **(a)** $8-9$ $8-9=8+(-9)$ $=-1$ **(b)** $-3-16$ $-3-16=-3+(-16)$ $=-19$ **(c)** $5-(-4)$ $5-(-4)=5+4$ $=9$	**8.** Subtract. **(a)** $4-6$ **(b)** $-6-12$ **(c)** $7-(-3)$

Example	Student Practice
9. Perform the necessary operations. $4-7-5-3$ First, write all subtraction as addition of the opposite, and then group like signs. $4-7-5-3=4+(-7)+(-5)+(-3)$ $=4+\left[(-7)+(-5)+(-3)\right]$ Then add all like signs before adding unlike signs. $4+\left[(-7)+(-5)+(-3)\right]=4+(-15)$ $=-11$	**10.** Perform the necessary operations. $5-7-2-9$
11. Evaluate $-x-y-4$ for $x=-3$ and $y=-1$. Place parentheses around the variables and replace x with -3 and y with -1. $-(x)-(y)-4=-(-3)-(-1)-4$ Then simplify $-(-3)$ and change each subtraction to addition of the opposite. $-(-3)-(-1)-4=3-(-1)-4$ $=3+1+(-4)$ Finally, add. $3+1+(-4)=4+(-4)$ $=0$	**12.** Evaluate $-p-q-6$ for $p=-5$ and $q=-8$.

Example	Student Practice
13. A portion of the Dead Sea is 1286 feet below sea level. What is the difference in altitude between Mount Carmel in Israel, which has an altitude of 1791 feet, and the Dead Sea?	**14.** Find the difference in altitude between a mountain 2986 feet high and a desert valley 734 feet below see level.

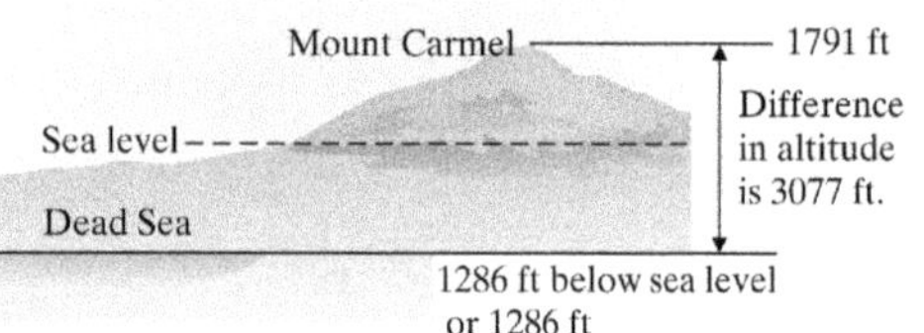

We want to find the difference, so we must subtract.

$$1791 \text{ ft} - (-1286 \text{ ft}) = 1791 \text{ ft} + 1286 \text{ ft}$$
$$= 3077 \text{ ft}$$

The difference in altitude is 3077 ft.

Extra Practice

1. Subtract. $-12-(-20)$

2. Subtract. $-7+(-11)-(-15)$

3. Evaluate $-5-(-x)+(-4)-y-7$ for $x=1$ and $y=-1$.

4. Martha left New York by plane at 2:00 P.M. Her flight to San Francisco took six hours, but the local time in San Francisco is three hours earlier than in New York. What was the local time in San Francisco when Martha's plane landed?

Concept Check

Is the following problem completed correctly? Why or why not?

$$-6-(-3)+(-7) = -6-(-10)$$
$$= -6+10$$
$$= 4$$

Name: ______________________________ Date: ________________
Instructor: ____________________________ Topic: ________________

Module 2 Integers
Topic 4 Multiplying and Dividing Integers

Vocabulary
odd numbers • even number • positive • negative • exponent

1. To find each consecutive ____________, add 2 to the previous number. The number 0 is the first of these.

2. When multiplying or dividing, the answer will be ____________ if the problem has an even number of negative signs.

3. Whole numbers that are not even numbers are ____________.

4. When multiplying or dividing, the answer will be ____________ if the problem has an odd number of negative signs.

Example	**Student Practice**
1. Find the product by writing as repeated addition. $2(-3)$ $2(-3)=-3+(-3)=-6$	**2.** Find the product by writing as repeated addition. $4(-8)$
3. Multiply. **(a)** $6(7)$ The number of negative signs, 0, is even. Thus, the answer is positive. $6(7)=+42$ **(b)** $6(-7)$ The number of negative signs, 1, is odd. Thus, the answer is negative. $6(-7)=-$ $=-42$	**4.** Multiply. **(a)** $-3(-5)$ **(b)** $-3(5)$

Vocabulary Answers: 1. even numbers 2. positive 3. odd numbers 4. negative.

Example	Student Practice
5. Multiply $(-3)(2)(-1)(4)(-3)$. The answer is negative since there are 3 negative signs and 3 is an odd number. Multiply the absolute values. $(-3)(2)(-1)(4)(-3)$ $=-[3(2)(1)(4)(3)]$ $=-72$	**6.** Multiply $4(-3)(-2)(-5)(-2)$.
7. Evaluate $(-4)^3$. The answer is negative since the exponent, 3, is odd. $(-4)^3=(-4)(-4)(-4)$ $=-64$	**8.** Evaluate $(-2)^2$.
9. Evaluate. **(a)** -3^2 -3^2, "the opposite of three squared" The base is 3; we use 3 as the factor for repeated multiplication and take the opposite of the product. $-3^2=-(3\cdot 3)=-9$ **(b)** $(-3)^2$ $(-3)^2$, "negative three squared" The base is -3; we use -3 as the factor for repeated multiplication. $(-3)^2=(-3)(-3)=9$	**10.** Evaluate. **(a)** -4^2 **(b)** $(-4)^2$

Example	Student Practice
11. Divide. **(a)** $36 \div 6$ $36 \div 6 = 6$ **(b)** $36 \div (-6)$ $36 \div (-6) = -6$ **(c)** $-36 \div (-6)$ $-36 \div (-6) = 6$	**12.** Divide. **(a)** $16 \div (-4)$ **(b)** $-16 \div (-4)$ **(c)** $-16 \div 4$
13. Perform the operation indicated. $\frac{-72}{-8}$ $\frac{-72}{-8} = -72 \div (-8)$ $= 9$	**14.** Perform the operation indicated. $-15(-4)$
15. Evaluate. **(a)** $\frac{m}{-n}$ for $m = -16$ and $n = -2$ We place parentheses around the variables and then we replace m with -16 and n with -2. $\frac{(m)}{-(n)} = \frac{(-16)}{-(-2)} = \frac{-16}{2} = -8$ **(b)** x^4 for $x = -2$ We place parentheses around the variable and then we replace x with -2. $(x)^4 = (-2)^4 = 16$	**16.** Evaluate. **(a)** $\frac{-c}{d}$ for $c = -64$ and $d = -8$ **(b)** z^7 for $z = -1$

Extra Practice

1. Find the product by writing as repeated addition.

   $4(-7)$

2. Perform the operation indicated.

   $(-3)^2$

3. Perform the operation indicated.

   $-100 \div 100$

4. Evaluate $(-x)^2 \div y$ for $x = 4$ and $y = -2$.

Concept Check

When we evaluate $(-x)(y)$ for $x = -6$ and $y = 2$, we obtain a positive number. Is this true? Why or why not?

Name: ______________________________ Date: ________________

Instructor: ______________________________ Topic: ________________

Module 2 Integers

Topic 5 The Order of Operations and Applications Involving Integers

Vocabulary

replace • calculate • order of operations • identify

1. When solving problems with more than one operation, you must first ___________ the operation with the highest priority.

2. Once the operation with the highest priority is known, ___________ the result of this operation.

3. Use the result of the operation to ___________ the original expression of this operation.

4. When there is more than one operation in a problem, we must follow the ___________.

Example	Student Practice
1. Simplify $12-30\div 5(-3)^2-2$. Identify that the highest priority is exponents, $(-3)^2$. Calculate, $(-3)^2=9$. Replace $(-3)^2$ with 9. $12-30\div 5(-3)^2-2=12-30\div 5(9)-2$ Identify that the next highest priority is division. Calculate, $30\div 5=6$. Replace $30\div 5$ with 6. $12-30\div 5(9)-2=12-6(9)-2$ Continue to repeat this process, following the order of operations, to complete the simplification. $12-6(9)-2=12-54-2$ $=12+(-54)+(-2)$ $=-44$	**2.** Simplify $-7+(-6)^2\div 9+3$.

Vocabulary Answers: 1. identify 2. calculate 3. replace 4. order of operations.

Example	**Student Practice**
3. Simplify $\frac{\left[-15+5(-3)\right]}{(13-18)}$.	**4.** Simplify $\frac{\left[3^4+5(-4)+3\right]}{\left(2^2+4\right)}$.

We perform the operations inside the parentheses and brackets first. We multiply; $5(-3)=-15$.

$$\frac{\left[-15+5(-3)\right]}{(13-18)}=\frac{\left[-15+(-15)\right]}{(13-18)}$$

Then, we add, $-15+(-15)=-30$.

$$\frac{\left[-15+(-15)\right]}{(13-18)}=\frac{-30}{(13-18)}$$

We subtract; $13-18=-5$.

$$\frac{-30}{(13-18)}=\frac{-30}{-5}$$

We divide last; $-30\div(-5)=6$.

$$\frac{-30}{-5}=6$$

5. Simplify $-24\div\left\{-3\cdot\left[4\div(-2)\right]\right\}$.	**6.** Simplify $20\div\left\{5\left[24\div(-6)\right]\right\}$.

We perform operations within the innermost grouping symbols first and work our way outward.

$$\begin{aligned}-24\div\left\{-3\left[4\div(-2)\right]\right\}&=-24\div\left\{-3(-2)\right\}\\&=-24\div 6\\&=-4\end{aligned}$$

Example	Student Practice
7. Ions are atoms or groups of atoms with positive or negative electrical charges. An oxide ion has an electrical charge of -2, while a magnesium ion has a charge of $+2$. Find the total charge of 8 oxide and 3 magnesium ions. We summarize the information. The total charge is equal to the number of oxide ions times the charge of one oxide ion plus the number of magnesium ions times the charge of one magnesium ion. $\text{Total charge} = 8\times(-2)+3\times(+2)$ $= -16+6$ $= -10$	**8.** Use the information from example **7** to answer the following. Find the total charge of 4 oxide ions and 7 magnesium ions.

Extra Practice

1. Simplify.

$$2(-5)(7-5)-7$$

2. Simplify.

$$\frac{16(-1)-(-4)(-5)}{2\left[-12\div(-3-3)\right]}$$

3. Evaluate $(x+2)^2-(-y-3)^3$ for $x=-5$ and $y=-2$.

4. When Isaac decorated his dining room, he bought a table for \$570, 2 armchairs for \$125 each, and 4 chairs for \$75 each. Write an expression that represents this situation and evaluate it to find out how much Isaac spent.

Concept Check

Explain in what order to do the operations to obtain the answer to the problem $3^2+5(2-4)$.

Name: ______________________________ Date: ________________
Instructor: ____________________________ Topic: ______________

Module 2 Integers
Topic 6 Simplifying and Evaluating Algebraic Expressions

Vocabulary
integers • commutative • distributive property • order of operations

1. To multiply $-3(x+1)$, we use the ___________.

2. Simplifying algebraic expressions with ___________ differs from doing so with whole numbers only in that we must consider the sign of the number when simplifying.

3. To evaluate expressions, we replace the variables with the given numbers and then perform the indicated operations following the ___________.

4. Since subtraction is not ___________, we must first change all subtractions to additions of the opposite and then rearrange the terms.

Example	Student Practice
1. Simplify by combining like terms. $-4x+7y+2x$ Rearrange terms, then add numerical coefficients of like terms. $-4x+7y+2x=-4x+2x+7y$ $=(-4+2)x+7y$ $=-2x+7y$ Note: $-2x$ and $7y$ are not like terms.	**2.** Simplify by combining like terms. $-8y+5x+3y$
3. Simplify $3x+5y-x$. First we change subtraction to addition of the opposite. Then we rearrange terms and add like terms. Note that $-x=-1x$. $3x+5y-x=3x+5y+(-x)$ $=3x+(-1x)+5y$ $=2x+5y$	**4.** Simplify $5m+2n-m$.

Vocabulary Answers: 1. distributive property 2. integers 3. order of operations 4. commutative.

Example	Student Practice
5. Perform each operation indicated. **(a)** $2-3+6$ $2-3+6=2+(-3)+6$ $=-1+6$ $=5$ **(b)** $2x-3x+6x$ $2x-3x+6x=2x+(-3x)+6x$ $=-1x+6x$ $=5x$ Notice the similarities with part **(a)**.	**6.** Perform each operation indicated. **(a)** $-7+3+9$ **(b)** $-7x+3x+9x$
7. Simplify $3a+8b-9a+3ab-10b$. First we change subtraction to addition of the opposite. $3a+8b+(-9a)+3ab+(-10b)$ Next we rearrange the terms to group like terms. $3a+(-9a)+8b+(-10b)+3ab$ Then we add coefficients of like terms. $[3+(-9)]a+[8+(-10)]b+3ab$ $=-6a+(-2b)+3ab$ Finally, simplify by rewriting addition of the opposite as subtraction. $-6a-2b+3ab$	**8.** Simplify $3m+7n-5mn-6m-9n$.

Example	Student Practice
9. Evaluate $\frac{(x^2-y)}{4}$ for $x=-1$ and $y=-3$. Place parentheses around each variable, then replace x with -1 and y with -3. $\frac{[(x)^2-(y)]}{4}=\frac{[(-1)^2-(-3)]}{4}$ Begin by calculating $(-1)^2$. Then, write subtraction as addition of the opposite and simplify. $\frac{[(-1)^2-(-3)]}{4}=\frac{[1-(-3)]}{4}$ $=\frac{(1+3)}{4}$ $=\frac{4}{4}$ $=1$	**10.** Evaluate $\frac{(m^2-n)}{-5}$ for $m=-3$ and $n=-1$.
11. Simplify $-2(y-4)$. We distribute the -2 over subtraction. $-2(y-4)=-2y-(-2)(4)$ Next, multiply: $(-2)(4)$. $-2y-(-2)(4)=-2y-(-8)$ We simplify by writing subtraction as addition of the opposite. $-2(y-4)=-2y+8$	**12.** Simplify $-7(x-2)$.

Example	Student Practice
13. To find the speed of a free-falling skydiver, we use the formula given below. The speed of skydiver, s, is equal to the initial velocity, v, minus the time since start of free fall, $32t$; $s = v - 32t$. Find the speed of a skydiver at time $t = 5$ seconds if her initial downward velocity (v) is -7 feet per second. We evaluate the formula for the values given: $v = -7$ and $t = 5$. $s = v - 32t$ $= -7 - 32(5)$ $= -7 - 160$ $= -167$ A negative speed means that the object is moving in a downward direction. Therefore, the skydiver is falling 167 feet per second.	**14.** Use the formula in example **13** to answer the following. Find the speed of the skydiver at $t = 6$ seconds if her initial downward velocity (v) is -2 feet per second.

Extra Practice

1. Simplify by combining like terms.
   $100 - 3a - 7ab + 5b + 9 + ab$

2. Evaluate $12p + 3p^9 - q^2$ for $p = -1$ and $q = 9$.

3. Simplify by using the distributive property and combining like terms.
   $7(a + 5) + 7$

4. Simplify by using the distributive property and combining like terms.
   $-6(-2a) + 6(a - 3) + 2(-3a) + 19 + a$

Concept Check

State two other ways we can write $-3b + 7$.

MATH COACH

Mastering the skills you need to do well on the test.

Watch the MATH COACH videos in MyMathLab® or on YouTube™ while you work the problems below. These helpful hints will help you avoid making common errors on test problems.

Adding and Subtracting Integers—Problem 14

$-14-3+(-6)-1$

> **Helpful Hint:** You may find it easier to change all subtraction to addition of the opposite as your first step. This allows you to add numbers in any order. Write out your steps carefully to avoid careless errors.

Look at your work for Problem 14. Examine your steps. As your first step, did you change $-14-3$ to $14+(-3)$?
Yes ____ No ____

As your second step, did you rewrite $(-6)-1$ as a related addition problem? Yes ____ No ____

If you answered No to either question, stop now and complete these steps.

Did you get $-14+(-3)+(-6)+(-1)$ as your next step?
Yes ____ No ____

If you answered No, think about how to rewrite subtraction as addition of the opposite. Recall that the subtraction sign changes to addition, and the number following the subtraction sign is changed to its opposite.

If you answered Problem 14 incorrectly, go back and rework the problem using these suggestions

---

Multiplying Tow of More Integers—Problem 17 $(-5)(-2)(-1)(3)$

> **Helpful Hint:** As a first step, write down the sign of the product or quotient. Do this *before* you complete any calculations to be sure that you do not forget the sign of the answer.

As a first step, did you identify that there are three negative signs in the problem? Yes ____ No ____

If you answered No, go back and count the number of negative signs again.

Next, did you determine that the product will be negative?
Yes ____ No ____

If you answered No, recall that an odd number of negative signs indicates a negative answer, while an even number of negative signs indicates a positive answer.

As your final step, did you multiply the absolute values of the numbers and add the negative sign to your answer?
Yes ____ No ____

If you answered No, stop and complete this step again.

Now go back and rework the problem using these suggestions.

Combining Like Terms with Integer Coefficients—Problem 28

Simplify $-3x+7xy+8y-12x-11y$.

Helpful Hint: Make sure you *do not* combine two terms with different variables or two terms with a different number of variables. You may only combine like terms. Be careful to avoid sign errors.

Look at your work for Problem 28. Examine your steps.

Did you combine $7xy$ with any other terms?
Yes ____ No ____

If you answered Yes, stop and look at each term carefully. Notice that the only term with xy as the variable part is $7xy$, so we cannot combine it with any other terms.

Did you combine $-3x$ and $-12x$ and combine $8y$ and $-11y$? Yes ____ No ____

If you answered No, go back and check your steps. Make sure you did not combine either of the two x-terms with either of the y-terms.

If you answered Yes, make sure you did not make a *sign error.*

If you answered Problem 28 incorrectly, go back and rework the problem using these suggestions.

Using the Distributive Property with Integers—Problem 29

Simplify $-6(a+7)$.

Helpful Hint: Be sure to multiply every number or term inside the parentheses by the number outside the parentheses. Then simplify the result. Be careful to avoid sign errors.

Did you multiply both a and 7 by -6?

Yes ____ No ____

If you answered No, then it may help with accuracy to write arrows as follows: $-6(a+7)$.

Are both of your terms in the final answer negative?

Yes ____ No ____

If you answered No, then you made at least one sign error. Try this step again.

Now go back and rework the problem using these suggestions.

Name: ______________________________ Date: ______________

Instructor: ______________________________ Topic: ______________

Module 3 Introduction to Equations and Algebraic Expressions
Topic 1 Solving Equations of the Form $x+a=c$ and $x-a=c$

Vocabulary

opposites • additive inverse property • addition principle of equality • line
ray • angle • line segment • vertex • supplementary angles • adjacent angles

1. A(n) ____________ is formed whenever two rays meet at the same endpoint.

2. In geometry a ____________ extends indefinitely.

3. A ____________ starts at a point and extends indefinitely in one direction.

4. A portion of a line, called a ____________, has a beginning and an end.

Example	Student Practice
1. Fill in the box with the number that gives the desired result. $x+8+\square=x+0=x$ We want the sum of $8+\square$ to equal 0. $x+8+(-8)=x+0=x$ $\uparrow \quad \uparrow$ $\boxed{8+(-8) \;=\; 0}$ Thus, $x+8+(-8)=x+0=x$.	**2.** Fill in the box with the number that gives the desired result. $y-5+\square=y+0=y$
3. Solve. $x-22=-14$ We want an equation of the form $x=\text{some number}$. Therefore, to get x alone on one side of the equation, add the opposite of -22 to both sides of the equation. $x-22+22=-14+22$ $x+0=8$ $x=8$	**4.** Solve. $x-17=-43$

Vocabulary Answers: 1. angle 2. line 3. ray 4. line segment

Example	Student Practice
5. Solve and check your solution. $2-6=y-7+12$ First, simplify each side of the equation separately. $2-6=y-7+12$ $-4=y+5$ Then, add -5 to both sides of the equation to get y alone on the right side: some number = y. $-4=y+5$ $+\ -5 \quad -5$ $-9=y$ Replace y with -9 in the original equation and verify that a true statement results.	**6.** Solve and check your solution. $7-3=y-4+17$
7. Answer parts **(a)** and **(b)**. **(a)** Translate into symbols. Angle y measures $40°$ more than angle x. The phrase "more than" indicates addition, "+". Thus, $\angle y=40°+\angle x$. **(b)** Find the measure of $\angle x$ if the measure of $\angle y$ is $95°$. First, write an equation and replace $\angle y$ with $95°$. Then, solve the equation. $\angle y = 40°+\angle x$ $95° = 40°+\angle x$ $+\ -40° \quad -40°$ $55° = \angle x$	**8.** Answer parts **(a)** and **(b)**. **(a)** Translate into symbols. Angle a measures $40°$ less than angle b. **(b)** Find the measure of $\angle b$ if the measure of $\angle a$ is $60°$.

Example	**Student Practice**

9. Find x and the measure of $\angle b$ for the pair of supplementary angles in the figure.

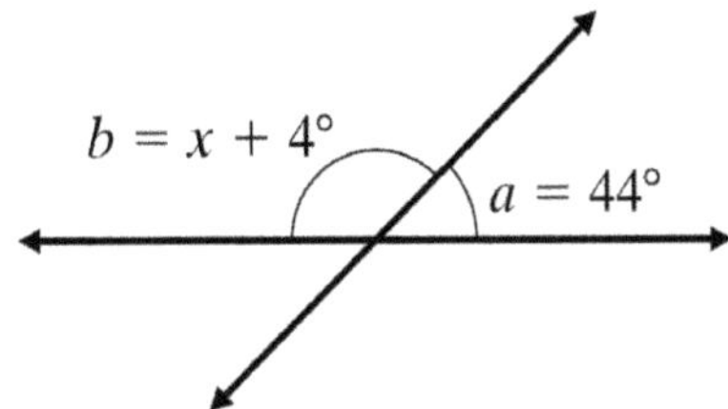

Since $\angle a$ and $\angle b$ are supplementary angles, their sum is 180°. Write an equation as follows.

$\angle a + \angle b = 180°$

Replace $\angle a$ with 44° and $\angle b$ with $x + 4°$.

$44° + (x + 4°) = 180°$

Now, simplify $44° + 4° + x = 48° + x$ and solve the equation.

$$\begin{array}{rcrcr} 48° & + & x & = & 180° \\ + \; -48° & & & & -48° \\ \hline & & x & = & 132° \end{array}$$

Since $\angle b = x + 4°$, substitute 132° for x to find the measure of $\angle b$.

$\angle b = x + 4°$

$\angle b = 132° + 4° = 136°$

Therefore, $x = 132°$ and $\angle b = 136°$.

10. Find x and the measure of $\angle b$ if the measure of $\angle a$ is 56° for the pair of supplementary angles in the figure.

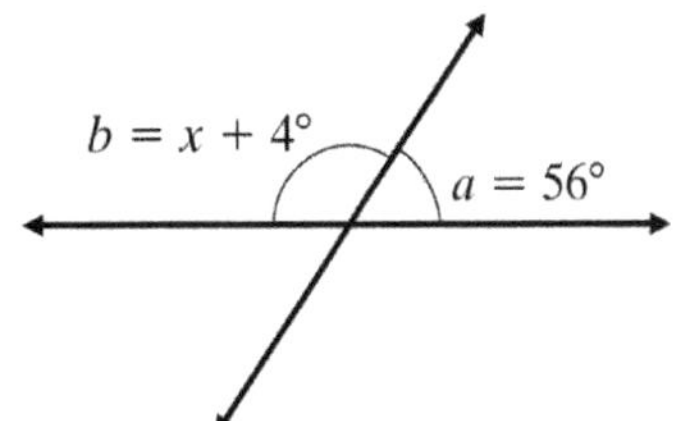

Extra Practice

1. Fill in the box with the number that gives the desired result.

   $n+124+\square=n+0=n$

2. Solve and check your solution.

   $8n-32-7n=6$

3. Solve and check your solution.

   $-2+4+x=-3+17$

4. Find the measure of the unknown angle for the given pair of supplementary angles $(\angle a+\angle b=180°)$.

   $\angle a=?,\ \angle b=103°$

Concept Check

To solve the equation $-9+x=-15$, Damien subtracted 9 from both sides of the equation. Is this correct? Why or why not?

Name: ______________________________ Date: ________________

Instructor: ______________________________ Topic: ________________

Module 3 Introduction to Equations and Algebraic Expressions
Topic 2 Solving Equations of the Form $ax = c$

Vocabulary

division principle of equality • variables • equation • multiplication

1. When we translate a statement into an equation, we must first define the ____________ that we are going to use.

2. The ____________ states that if both sides of an equation are divided by the same nonzero number, the results on both sides are equal in value.

3. Division undoes ____________.

4. Whatever is done to one side of a(n) ____________ must be done to the other side.

Example	**Student Practice**
1. Fill in the box with the number that gives the desired result.	**2.** Fill in the box with the number that gives the desired result.
$\frac{-5x}{\square} = 1 \cdot x = x$	$\frac{-17x}{\square} = 1 \cdot x = x$
We want the quotient of $\frac{-5}{\square}$ to equal 1.	
$\frac{-5x}{-5} = 1 \cdot x = x$ $\uparrow \quad \uparrow$ $\boxed{\frac{-5}{-5} = 1}$	

Vocabulary Answers: 1. variables 2. division principle of equality 3. multiplication 4.

Example	Student Practice
3. Solve and check your solution. $7x=-147$ We want to make $7x=-147$ into a simpler equation, $x=\text{some number}$. The variable, x, is multiplied by 7. Dividing both sides of the equation by 7 undoes the multiplication by 7. $7x=-147$ $\frac{7x}{7}=\frac{-147}{7}$ $x=-21$ Check by replacing x with -21 and verifying that a true statement results.	**4.** Solve and check your solution. $-8x=152$
5. Solve. $3(x\cdot 5)=\frac{450}{5}$ First, simplify the equation. Divide: $450\div 5=90$. Then change the order of the factors, regroup the factors, and simplify. $3(x\cdot 5)=\frac{450}{5}$ $3(x\cdot 5)=90$ $3(5x)=90$ $(3\cdot 5)x=90$ $15x=90$ Finally, dividing both sides by 15 undoes the multiplication by 15. $\frac{15x}{15}=\frac{90}{15}$ $x=6$	**6.** Solve. $6(y\cdot 2)=\frac{480}{4}$

Example	Student Practice
7. Translate the statement into an equation. There are five times as many dimes (D) as pennies (P) in a coin collection. It is helpful to write a sentence that compares the two quantities; "there are more dimes (D) than pennies (P)." Think of a simple comparison of pennies and dimes, such as the case when there is only 1 penny. If there is 1 penny in the collection, then there are 5 dimes. Now rephrase the statement and translate into an equation; "the number of dimes is five times the number of pennies." $D = 5 \cdot P$	**8.** Translate the statement into an equation. There are seven times as many nickels (N) as quarters (Q) in a tip jar.
9. The number of peanuts (P) is triple the number of cashews (C). **(a)** Translate the statement into an equation. The word triple means three times. $P = 3 \cdot C$ **(b)** Find the number of cashews if there are 27 peanuts. Use the equation from part **(a)**. Replace P with 27 and divide both sides by 3. $P = 3C$ $27 = 3C$ $\frac{27}{3} = \frac{3C}{3}$ $9 = C$	**10.** The number of cars (C) is eight times the numbers of bikes (B). **(a)** Translate the statement into an equation. **(b)** Find the number of bikes if there are 128 cars in the road.

Example	Student Practice
11. Lena purchased x shares of stock at \$35 per share. She sold all the stock for \$56 per share and made a profit of \$546. How many shares of stock did Lena purchase?	**12.** Ann purchased x shares of stock at \$28 per share. She sold all the stock for \$50 per share and made a profit of \$374. How many shares of stock did Ann purchase?
Read the problem carefully and create a Mathematics Blueprint.	
$\boxed{\text{profit}} = \boxed{\text{sale price}} - \boxed{\text{purchase price}}$ $546 = 56x - 35x$	
Now combine like terms. Then, divide both sides by 21.	
$546 = 21x$ $\frac{546}{21} = \frac{21x}{21}$ $26 = x$	
Lena purchased 26 shares of stock. Estimate to see if the answer is reasonable.	

Extra Practice

1. Translate the given statement into an equation.
   Miriam is seven times older than her daughter Edie.

2. Solve and check your solution.

   $4(5x) = 60$

3. Solve and check your solution.

   $-14 - 28 = 10x - 3x$

4. Matthew moved to an apartment building that's 11 times taller than the house he used to live in. If the apartment building is 253 feet tall, what's the height of the house?

Concept Check

Explain in words the steps that are needed to solve $4x + 3(2x) = -20$.

Name: ______________________________ Date: ______________

Instructor: ____________________________ Topic: ______________

Module 3 Introduction to Equations and Algebraic Expressions
Topic 3 Equations and Geometric Formulas

Vocabulary

perimeter • area • parallelogram • parallel lines • height • base • volume

1. The ____________ of a rectangular solid is the product of the length times the width times the height.

2. ____________ are straight lines that are always the same distance apart.

3. The ____________ of a square is the length of one side squared.

4. A ____________ is a four-sided figure in which both pairs of opposite sides are parallel.

Example	Student Practice
1. Find the perimeter of a rectangle with $L=8$ feet and $W=6$ feet. We complete the four-step process. Begin by drawing a picture. $L = 8$ ft $W = 6$ ft Now, write the formula. $P=2L+2W$ Then, replace L with 8 ft and W with 6 ft. $P=2(8\text{ ft})+2(6\text{ ft})$ Finally, simplify to find the perimeter. $P=16\text{ ft}+12\text{ ft}$ $=28\text{ ft}$ The perimeter is 28 feet.	**2.** Find the perimeter of a square with sides 14 yards in length.

Vocabulary Answers: 1. volume 2. parallel lines 3. area 4. parallelogram

Example	Student Practice

3. The length of a rectangle is three times the width. If the perimeter of the rectangle is 24 feet, find the width.

$L = 3W$

W

Since we are given a picture, we start by writing the formula, $P = 2L + 2W$. Now, we replace P with 24 and L with the given value $3W$ and solve.

$$24 = 2(3W) + 2W$$
$$24 = 8W$$
$$\frac{24}{8} = \frac{8W}{8}$$
$$3 = W$$

The width of the rectangle is 3 feet.

4. The length of a rectangle is five times the width. If the perimeter of the rectangle is 45 feet, find the width.

$L = 5W$

W

5. What is the area of the rug pictured below?

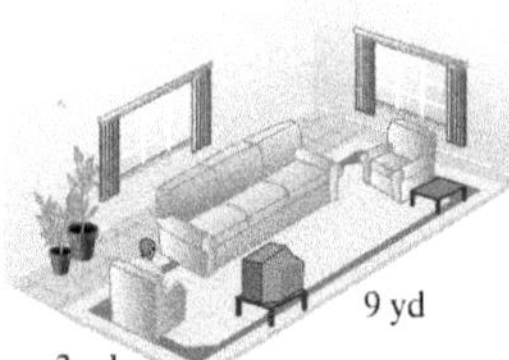

Think of an array with 3 rows and 9 columns.

9

3

1 yd → This is one square yard.

1 yd

Just as we multiplied the number of rows times the number of columns to find the number of items in an array, we multiply the length times the width to find the area of the rug.

3 yards × 9 yards = 27 square yards

6. What is the area of the garden pictured below?

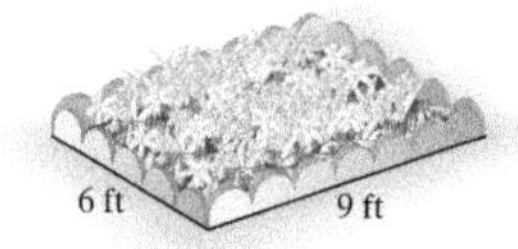

Example	Student Practice
7. Find the area of a rectangle with a length of 3 feet and a width of 2 feet. We complete the four-step process. Begin by drawing a picture. 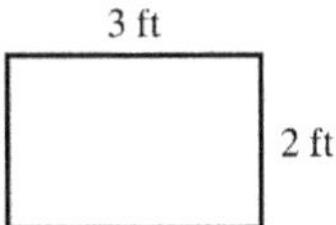 Now, write the formula, $A = L \cdot W$. Then, replace L and W with the given values, $A = 3\text{ ft} \cdot 2\text{ ft}$. Finally, simplify. Remember to multiply the units. $A = (3 \cdot 2)(\text{ft} \cdot \text{ft})$ $= 6\text{ ft}^2$ This is read as "six square feet."	**8.** Find the area of a square with a side of 9 feet.
9. Find the height of a parallelogram with $\text{base} = 8$ meters and area $\text{area} = 24\text{ m}^2$. Begin by drawing a picture. 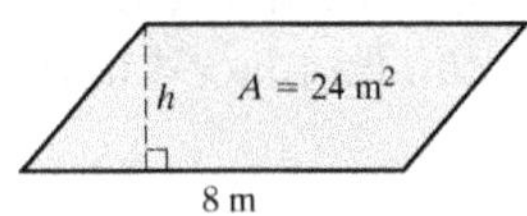 Next, write the formula, $A = bh$. Then, replace A and b with the values given and solve the equation for h . $24 = 8h$ $\frac{24}{8} = \frac{8h}{8}$ $3 = h$	**10.** Find the base of a parallelogram with area $= 115\text{ ft}^2$ and height $= 5$ ft.

Example	Student Practice
11. Find the unknown side of the rectangular solid.	**12.** Find the unknown side of the rectangular solid.

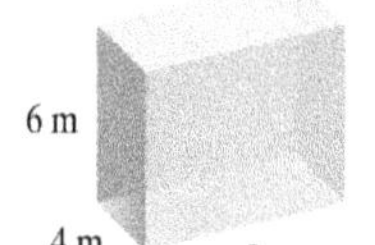

Since the picture is drawn for us, we first write the formula. Then, replace the variables with the values given, simplify, and solve for L.

$$V = L \cdot W \cdot H$$
$$216 = L \cdot 4 \cdot 6$$
$$216 = 24L$$
$$\frac{216}{24} = \frac{24L}{24}$$
$$9 = L$$

The length of the box is 9 m.

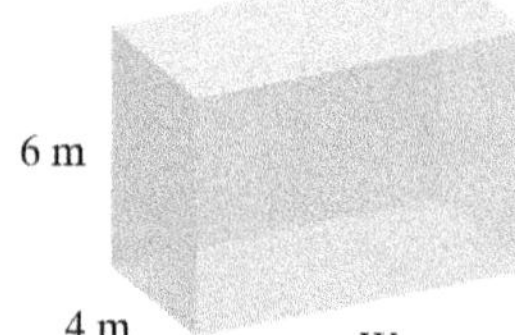

Extra Practice

1. What are the perimeter and area of a square with sides of length 15 yards?

2. Find the area of a parallelogram with a height of 15 inches and base of 19 inches.

3. Find the volume of a rectangular solid with $L = 8$ feet, $W = 5$ feet, and $H = 2$ feet.

4. A plastic feed bin is 4 feet long, 2 feet wide and 2 feet high. How many cubic feet of animal feed can be placed in this bin?

Concept Check

Hanna purchased a redwood box at the garden store and filled it with sand.

(a) To determine how much sand is needed to fill the box, do you find the area, perimeter, or volume?

(b) State the formula you must use.

(c) The volume of the box is 200 ft^3, the height is 5 in., and the length of the box is double the height. Explain in words how you would find the width of the box.

Name: ______________________________ Date: ________________

Instructor: ______________________________ Topic: ________________

Module 3 Introduction to Equations and Algebraic Expressions
Topic 4 Performing Operations with Exponents

Vocabulary

product rule for exponents • numerical coefficient • coefficient
polynomial • monomial • binomial • trinomial

1. A ____________ is an expression that contains terms with variable parts that have only whole number exponents.

2. A ____________ has two terms.

3. The ____________ states that to multiply constants or variables in exponent form that have the same base, add the exponents but keep the base unchanged.

4. A ____________ has three terms.

Example	Student Practice
1. Write $5^4 \cdot 5^3$ as repeated multiplication and then rewrite it with one exponent. $\underbrace{5\cdot5\cdot5\cdot5}_{5^4} \cdot \underbrace{5\cdot5\cdot5}_{5^3} = 5^7$ Since five appears as a factor 7 times, the exponent is 7, $5^4 \cdot 5^3 = 5^7$.	**2.** Write $7^2 \cdot 7^6$ as repeated multiplication and then rewrite it with one exponent.
3. Multiply and write the product in exponent form. **(a)** $x^3 \cdot x^6$ $x^3 \cdot x^6 = x^{3+6} = x^9$ **(b)** $x^7 \cdot x$ $x^7 \cdot x = x^7 \cdot x^1 = x^{7+1} = x^8$	**4.** Multiply and write the product in exponent form. **(a)** $3^2 \cdot 3^5$ **(b)** $3^2 \cdot 2^5$

Vocabulary Answers: 1. polynomial 2. binomial 3. product rule for exponents 4. trinomial

Example	Student Practice
5. Multiply. $(5x^6)(7y^4)(2x^4)$ Change the order of factors and regroup them. $(5x^6)(7y^4)(2x^4)=(5\cdot7\cdot2)(x^6\cdot y^4\cdot x^4)$ Multiply the numerical coefficients and variables separately. $(5\cdot7\cdot2)(x^6\cdot y^4\cdot x^4)=70x^{10}y^4$ We cannot use the product rule for exponents to simplify $x^{10}y^4$ because the bases, x and y, are not the same.	**6.** Multiply. $(7a^2)(8b^5)(3a^6)$
7. Multiply. $(-7x^2)(-4y^4)$ $(-7x^2)(-4y^4)=(-7)(-4)(x^2\cdot y^4)$ $=28x^2y^4$	**8.** Multiply. $(7a^3)(-8b^6)$
9. Identify each polynomial as a monomial, binomial, or trinomial. **(a)** $2x^2+5$ $2x^2+5$ is a binomial because there are two terms. **(b)** $8x$ $8x$ is a monomial because there is one term. **(c)** $5x^3+8x-1$ $5x^3+8x-1$ is a trinomial.	**10.** Identify each polynomial as a monomial, binomial, or trinomial. **(a)** $9x^2-7x-1$ **(b)** $2x^5$ **(c)** $5x^9-10$

Example	Student Practice
11. Use the distributive property and then simplify. $3x^2(x^5+8x)+2x^7$	**12.** Use the distributive property to simplify. $6x^3(x^4-4x)+5x^8$

Use the distributive property to multiply $3x^2(x^5+8x)$.

$3x^2 \cdot x^5 + 3x^2 \cdot 8x + 2x^7$

Write x^5 as $1x^5$ and x as x^1.

$3x^2 \cdot 1x^5 + 3x^2 \cdot 8x^1 + 2x^7$

Multiply numerical coefficients.

$(3 \cdot 1) \cdot (x^2 \cdot x^5) + (3 \cdot 8) \cdot (x^2 \cdot x^1) + 2x^7$

Multiply variables by adding their exponents.

$3x^{2+5} + 24x^{2+1} + 2x^7$

Simplify and combine the like terms.

$3x^7 + 24x^3 + 2x^7 = 5x^7 + 24x^3$

Thus, $3x^2(x^5+8x)+2x^7 = 5x^7 + 24x^3$.

Example	Student Practice
13. Write the area of the rectangle below as an algebraic expression and then simplify.	**14.** Write the area of the parallelogram below as an algebraic expression and then simplify.

x^2

$3x^3 - 2$

$h = x^4$

$5x^5 - 4x$

First write the formula for area and then replace L and W with the given expressions.

$$A = LW$$
$$= \left(3x^3 - 2\right)\left(x^2\right)$$

Then distribute x^2 and simplify.

$$\left(3x^3 - 2\right)\left(x^2\right) = \left(3x^3\right)\left(x^2\right) - (2)\left(x^2\right)$$
$$A = 3x^5 - 2x^2$$

Extra Practice

1. Multiply and write the product in exponent form. $10^3 \cdot 10^5 \cdot 10^9 \cdot 10^0 \cdot 10^1$

2. Simplify. $\left(3a^2 - 7a\right)\left(-6a^4\right)$

3. Simplify.

   $-4x^3\,(2x + y - 5) + 3x^3y + 4x^4 + y$

4. What is the volume of a box with a length of 20, a width of $3n$, and a height of $n - 1$?

Concept Check

Explain in words the steps to simplify each of the following.

(a) $2x^4\left(x^2 + y\right)$

(b) $\left(-3x^3\right)\left(x^3\right)$

(c) Using the results for **(a)** and **(b)** explain the extra steps needed to complete the problem $2x^4\left(x^2 + y\right) + \left(-3x^3\right)\left(x^3\right)$, and then state the answer.

MATH COACH

Mastering the skills you need to do well on the test.

Watch the MATH COACH videos in MyMathLab® or on YouTube™ while you work the problems below. These helpful hints will help you avoid making common errors on test problems.

Solving Equations Using the Addition Principle of Equality—Problem 4

Solve $3x-2x-7=-1+6$.

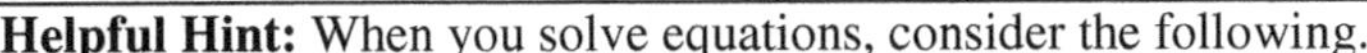

Helpful Hint: When you solve equations, consider the following.

- Did you simplify each side before you began the process to solve?
- Did you use the correct principle?
- Did you remember that whatever you do on one side of the equation, you must do on the other side?
- Did you take the time to double-check your calculations and + and − signs?

Look at your work for Problem 4. Examine your steps. Did you simplify each side of the equation as your **first step**?
Yes ____ No ____

If you answered No, stop and complete this step.

After simplifying, did you obtain the equation $x-7=5$?
Yes ____ No ____

If you answered No, consider how to identify and combine like terms. Notice that there are like terms to combine on both sides of the equation. Be careful to avoid sign errors.

Did you use the correct principle and add 7 to both sides of the equation?
Yes ____ No ____

If you answered No, go back and complete this step.

If you answered Problem 4 incorrectly, go back and rework the problem using these suggestions.

Solving Equations Using the Division Principle of Equality—Problem 7

Solve $2(4x)=-72$.

Helpful Hint: When you solve equations, consider the following.

- Did you simplify each side before you began the process to solve?
- Did you use the correct principle?
- Did you remember that whatever you do on one side of the equation, you must do on the other side?
- Did you take the time to double-check your calculations and + and − signs?

Look at your work for Problem 7. Examine your steps. Did you simplify the left side of the equation as your **first step**?
Yes ____ No ____

If you answered No, stop and complete this step.
After simplifying, did you obtain the equation $8x=-72$?
Yes ____ No ____

If you answered No, go back and multiply $2(4x)$ again.

Did you use the correct principle and divide both sides of the equation by 8? Yes ____ No ____

If you answered No, stop and complete this step. Be careful when working with + and − signs.

Now go back and rework the problem using these suggestions.

Multiplying Algebraic Expressions with Exponents—Problem 24

Multiply. Leave your answer in exponent form. $\left(-8x^2\right)\left(-9x^4\right)$

Helpful Hint: First multiply the numerical coefficients. Then apply the product rule for exponents by adding the exponents of x.

Did you multiply the numerical coefficients -8 and -9?

Yes ____ No ____

If you answered No, go back and complete this step. Be careful to avoid sign errors.

Did you add the exponents $2+4$?

Yes ____ No ____

If you answered No, consider how the product rule for exponents works again. Notice that the two variables being multiplied have the same base but different exponents. Using the product rule, we can add the two exponents.

If you answered Problem 24 incorrectly, go back and rework the problem using these suggestions.

Using the Distributive Property to Multiply a Monomial and a Binomial—Problem 26

Simplify $\left(3x^2-5x\right)6x^3$.

Helpful Hint: To help with accuracy, you may find it easier to use the commutative property of multiplication to rewrite the problem with the monomial on the left side of the parentheses. Remember to consider the following:

- If a variable does not have an exponent, then the exponent is understood to be 1.
- Multiply both terms inside the parentheses by the monomial.

Look at your work for Problem 26. Examine your steps.

Did you remember to multiply both terms by $6x^3$?
Yes ____ No ____

If you answered No, then stop now and complete these calculations.

Did you multiply $6x^3$ times $3x^2$ to get $18x^5$?
Yes ____ No ____

If you answered No, remember to multiply the numerical coefficients 6 and 3 first and then use the product rule to multiply x^3 by x^2.

Did you multiply $6x^3$ times $-5x$ to get $-30x^3$?
Yes ____ No ____

If you answered Yes, you forgot that the x in $-5x$ has an exponent of 1. Stop now and apply the product rule again.

Now go back and rework the problem using these suggestions.

Name: ______________________________ Date: ______________

Instructor: ______________________________ Topic: ______________

Module 4 Fractions, Ratio, and Proportions
Topic 1 Factoring Whole Numbers

Vocabulary

divisible • divisible by 2 • divisible by 3 • divisible by 5 • prime number
composite number • factors • prime factors • division ladder • factor tree

1. A ___________ is a whole number greater than 1 that is divisible only by itself and 1.

2. A number is ___________ if its last digit is 0 or 5.

3. A ___________ is a whole number greater than 1 that can be divided by whole numbers other than itself and 1.

4. A number is ___________ if the sum of its digits is divisible by 3.

Example	Student Practice
1. Determine if the number is divisible by 2, 3, and/or 5. **(a)** 234 Divisible by 2 because 234 is even and by 3 since $2+3+4=9$ and 9 is divisible by 3. **(b)** 38,910 Divisible by 2 because 38,910 is even, by 3 since the sum of the digits is divisible by 3, and by 5 since the last digit is 0.	**2.** Determine if the number is divisible by 2, 3, and/or 5. **(a)** 225 **(b)** 690
3. State whether each number is prime, composite, or neither. 1, 4, 7, 11, 14, 15, 17, 22, 27, 31, 120 1 is neither prime nor composite. 4, 14, 15, 22, 27, and 120 are composite. 7, 11, 17, and 31 are prime.	**4.** State whether each number is prime, composite, or neither. 0, 5, 9, 11, 18, 19, 29, 34, 37, 41, 60

Vocabulary Answers: 1. prime number 2. divisible by 5 3. composite number 4. divisible by 3

Example	Student Practice
5. Express 20 as a product of prime factors. $20 = 2 \cdot 2 \cdot 5$ or $2^2 \cdot 5$ $20 = 4 \cdot 5$ is not correct because 4 is not a prime number.	**6.** Express 64 as a product of prime factors.
7. Express 28 as a product of prime factors. Since 28 is even, it is divisible by the prime number 2. We start the division ladder by dividing 28 by 2. **Step1** $2\overline{)28}$ with quotient 14 The quotient 14 is not a prime number. We must continue to divide until the quotient is a prime number. **Step2** $2\overline{)14}$ with quotient 7 The quotient 7 is a prime number. Thus all the factors are prime. We are finished dividing. This process is simplified if we write the divisions as follows, placing step 1 on the bottom and moving up the ladder as we divide. $\begin{array}{r} 7 \\ \textbf{Step2}\ 2\overline{)14} \\ \uparrow \textbf{Step1}\ 2\overline{)28} \end{array}$ Now write all the divisors and the quotient as a product of prime factors. $28 = 2 \cdot 2 \cdot 7$ or $2^2 \cdot 7$	**8.** Express 80 as a product of prime factors.

Example	Student Practice
9. Express 210 as a product of prime factors.	**10.** Express 336 as a product of prime factors.

From the divisibility rules we know that 210 is divisible by 2, 3, and 5. We can start with 5.

$$\begin{array}{ll} & 7 \\ \textbf{Step3} & 2\overline{)14} \\ \textbf{Step2} & 3\overline{)42} \\ \textbf{Step1} & 5\overline{)210} \end{array}$$

7 is prime, so we are finished dividing.

$210 = 5 \cdot 3 \cdot 2 \cdot 7 = 2 \cdot 3 \cdot 5 \cdot 7$

Note that we wrote all factors in ascending order since this is standard notation.

Example	Student Practice
11. Use a factor tree to express 48 as a product of prime factors.	**12.** Use a factor tree to express 72 as a product of prime factors.

Write 48 as the product of two factors, $48 = 6 \cdot 8$. Neither 6 nor 8 are prime, so write them as products: $6 = 2 \cdot 3$ and $8 = 2 \cdot 4$. Circle 2 and 3 since they are prime. 4 is not prime, so write it as a product: $4 = 2 \cdot 2$. Circle the factors of 4 since they are prime. Write 48 as a product of the prime factors.

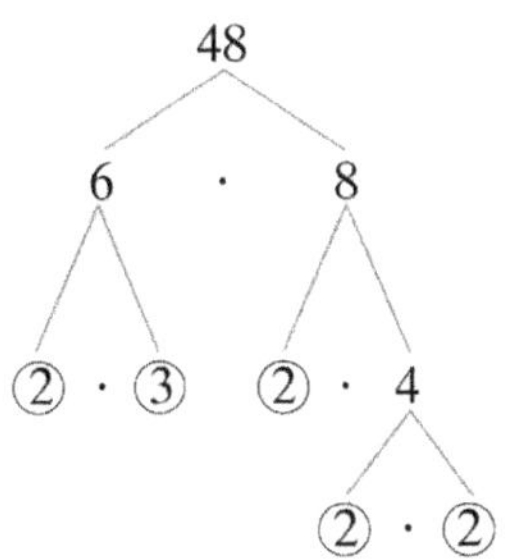

$48 = 2 \cdot 3 \cdot 2 \cdot 2 \cdot 2$ or $2^4 \cdot 3$.

Extra Practice

1. Determine if the number is divisible by 2, 3, and/or 5.

111,111

2. State whether the given number is prime, composite, or neither.

91

3. Express the given number as a product of prime factors. Write your answer as powers of prime factors.

98

4. Express the given number as a product of prime factors. Write your answer as powers of prime factors.

7200

Concept Check

Delroy is having a hard time factoring the number 318.

(a) Explain in words how Delroy can determine one factor of 318.

(b) State this factor.

Name: ______________________________ Date: ________________

Instructor: ___________________________ Topic: _______________

Module 4 Fractions, Ratio, and Proportions
Topic 2 Understanding Fractions

Vocabulary

fractions • numerator • denominator • undefined • proper fraction
improper fraction • mixed number

1. A(n) ____________ is the sum of a whole number greater than zero and a proper fraction, and is used to describe a quantity greater than 1.

2. A(n) ____________ is used to describe a quantity greater than or equal to 1.

3. A(n) ____________ is used to describe a quantity less than 1.

4. In mathematics, ____________ are a set of numbers used to describe parts of whole quantities.

Example	Student Practice
1. Use a fraction to represent the shaded part of each object.	**2.** Use a fraction to represent the shaded part of each object.
(a) One out of four parts is shaded, or $\frac{1}{4}$.	**(a)**
(b) Seven out of nine parts are shaded, or $\frac{7}{9}$.	**(b)**
(c) Three out of three parts are shaded, or $\frac{3}{3}=1$.	**(c)**

Vocabulary Answers: 1. mixed number 2. improper fraction 3. proper fraction 4. fractions

Example	Student Practice
3. Divide, if possible. **(a)** $\frac{23}{0}$ Division by 0 is undefined. **(b)** $\frac{0}{23}$ $\frac{0}{23}=0$ Any fraction with 0 in the numerator and a nonzero denominator equals 0.	**4.** Divide, if possible. **(a)** $\frac{a}{a}$, $a \neq 0$ **(b)** $\frac{0}{19}$
5. The approximate number of inches of rain that falls during selected periods of one year in Seattle, Washington, is shown by the circle graph. What fractional part of the total yearly rainfall does not occur from July to September? 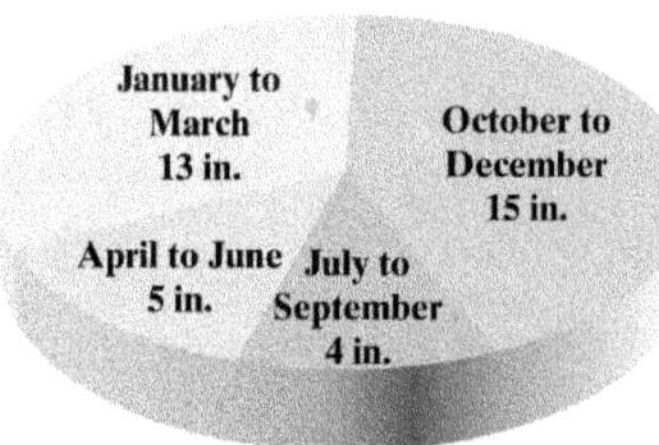 First find the total rainfall for 1 year. $13\text{ in.}+15\text{ in.}+4\text{ in.}+5\text{ in.}=37\text{ in.}$ From July to September there were 4 inches of rain out of a total of 37 inches. Rainfall that does not occur from July to September is $37\text{ in.}-4\text{ in.}=33\text{ in.}$ Thus, the fractional part of rainfall that does not occur is $\frac{33}{37}$.	**6.** Use the circle graph in example **5** to answer the following. What fractional part of the total yearly rainfall occurs from April to June?

Example	Student Practice
7. Identify each as a proper fraction, an improper fraction, or a mixed number.	**8.** Identify each as a proper fraction, an improper fraction, or a mixed number.
(a) $\frac{9}{8}$	**(a)** $\frac{17}{6}$
Since the numerator is larger than the denominator, it is an improper fraction.	
(b) $\frac{8}{9}$	**(b)** $\frac{13}{13}$
Since the numerator is less than the denominator, it is a proper fraction.	
(c) $7\frac{3}{4}$	**(c)** $5\frac{6}{7}$
Since a whole number is added to a proper fraction, it is a mixed number.	
9. Write $\frac{19}{7}$ as a mixed number.	**10.** Write $\frac{29}{4}$ as a mixed number.
The answer is in the form $\text{quotient}\frac{\text{remainder}}{\text{denominator}}$.	
$\frac{19}{7} \rightarrow 7\overline{)19}$ with quotient 2; $\underline{14}$; 5	
Here 2 is the quotient and 5 is the remainder. 7 is the denominator from the original fraction.	
Thus, $\frac{19}{7} = 2\frac{5}{7}$.	

Example	Student Practice
11. Change $6\frac{1}{2}$ to an improper fraction.	**12.** Change $9\frac{4}{5}$ to an improper fraction.

To change a mixed number to an improper fraction, multiply the whole number by the denominator of the fraction. Then add this product to the numerator. The result is the numerator of the improper fraction. The denominator does not change.

$$6\frac{1}{2}=\frac{(2\cdot 6)+1}{2}=\frac{12+1}{2}=\frac{13}{2}$$

We can also write the process as follows.

$$6\frac{1}{2}=\frac{2\text{ times }6\text{ plus }1}{2}=\frac{13}{2}$$

Extra Practice

1. A pizza is divided into eight slices and you eat three of them. Write a fraction to describe how much of the pizza you ate.

2. There are 13 girls and 12 boys at the meeting. Write a fraction that describes what part of the meeting are girls.

3. Change the given improper fractions to a mixed number.

$$\frac{83}{8}$$

4. Change the given mixed number to an improper fraction.

$$7\frac{11}{20}$$

Concept Check

Explain in words how you change the mixed number $2\frac{3}{4}$ to an improper fraction.

Name: ______________________________ Date: ________________

Instructor: ___________________________ Topic: ________________

Module 4 Fractions, Ratio, and Proportions
Topic 3 Simplifying Fractional Expressions

Vocabulary

equivalent fractions • reduced to lowest terms • reduce fractions • same value

1. Equivalent fractions look different but have the ____________ because they represent the same quantity.

2. To find a(n) ____________, we multiply both the numerator and denominator by the same nonzero number.

3. A fraction is considered to be ____________ (or written in simplest form) if the numerator and denominator have no common factors other than 1.

4. It is important that you ____________ using the method of removing factors of 1.

Example	Student Practice
1. Multiply the numerator and denominator of $\frac{3}{4}$ by $3x$ to find an equivalent fraction. $\frac{3}{4}=\frac{3\cdot 3x}{4\cdot 3x}=\frac{9x}{12x}$	**2.** Multiply the numerator and denominator of $\frac{7}{8}$ by $4x$ to find an equivalent fraction.
3. Write $\frac{3}{4}$ as an equivalent fraction with a denominator of $16x$. $\frac{3}{4}=\frac{\square}{16x}$ $\frac{3\cdot ?}{4\cdot ?}=\frac{\square}{16x}$ 4 times what number equals $16x$? $4x$ $\frac{3\cdot 4x}{4\cdot 4x}=\frac{12x}{16x}$	**4.** Write $\frac{5}{7}$ as an equivalent fraction with a denominator of $49x$.

Vocabulary Answers: 1. same value 2. equivalent fraction 3. reduced to lowest terms 4. reduce fractions

Example	Student Practice
5. Simplify $\frac{-72}{48}$. Write the negative sign in front of the fraction. $\frac{-72}{48}=-\frac{72}{48}$ Note that 8 is a common factor of 72 and 48. Write the remaining factors as products of primes and simplify. $-\frac{72}{48}=-\frac{\overset{1}{\cancel{8}}\cdot 9}{\underset{1}{\cancel{8}}\cdot 6}$ $=-\frac{3\cdot\overset{1}{\cancel{3}}}{2\cdot\underset{1}{\cancel{3}}}$ $=-\frac{3}{2}$	**6.** Simplify $\frac{81}{-36}$.
7. Simplify $\frac{150n^2}{200n}$. Note that 25 is a common factor of the numerator and denominator. Write all other factors as products of prime numbers to ensure that the fraction is reduced to lowest terms. $\frac{150n^2}{200n}=\frac{\overset{1}{\cancel{25}}\cdot 3\cdot\overset{1}{\cancel{2}}\cdot n\cdot\overset{1}{\cancel{n}}}{\underset{1}{\cancel{25}}\cdot 2\cdot\underset{1}{\cancel{2}}\cdot 2\cdot\underset{1}{\cancel{n}}}$ $=\frac{3\cdot n}{2\cdot 2}$ $=\frac{3n}{4}$	**8.** Simplify $\frac{60a^2}{180a}$.

Example	Student Practice

9. The yearly sales report below shows that the number and type of real estate sales made by Tri-Star Realty.

Type of Sale	Total Sales
Condominium	14
Town home	21
Single-family home	45

(a) What fractional part of the total sales were single-family homes?

First, find the total sales. Then write the fraction and simplify.

Condominium sales + Town home sales + Single-family home sales = Total sales

$14+21+45=80$

$$\frac{\text{single-family home sales}}{\text{total sales}}$$

$$=\frac{48}{80}=\frac{9\cdot \overset{1}{\cancel{5}}}{16\cdot \underset{1}{\cancel{5}}}=\frac{9}{16}$$

(b) What fractional part of the total sales were not condominiums?

Determine how many sales are not condominiums. Then write the fraction and simplify.

Find the sum: $21+45=66$.

$$\frac{\text{sales that were not condominiums}}{\text{total sales}}$$

$$=\frac{66}{80}=\frac{33\cdot \overset{1}{\cancel{2}}}{40\cdot \underset{1}{\cancel{2}}}=\frac{33}{40}$$

10. Use the table in example **9** to answer the following.

(a) What fractional part of the total sales were condominiums?

(b) What fractional part of the total sales were not town homes?

Extra Practice

1. Find an equivalent fraction with the given denominator.

$$\frac{5}{8}=\frac{\square}{64a}$$

2. Find an equivalent fraction with the given denominator.

$$\frac{2}{15}=\frac{\square}{45x}$$

3. Simplify $\frac{-45xy}{-72y}$.

4. Lane has 24 plants in his house, 18 of which are in his living room. Find a fraction to represent the fractional part of Lane's plants that are in the living room.

Concept Check

Can the fraction $\frac{105}{231}$ be reduced? Why or why not?

Name: ______________________________ Date: ______________

Instructor: ____________________________ Topic: ______________

Module 4 Fractions, Ratio, and Proportions
Topic 4 Simplifying Fractional Expressions with Exponents

Vocabulary

quotient rule • raising a power to a power • additional power rule • product to a power

1. The ____________ rule says to keep the same base and multiply the exponents.

2. The ____________ states that if a fraction in parentheses is raised to a power, the parentheses indicate that the numerator and denominator are each raised to that power.

3. To raise a(n) ____________, raise each factor to that power.

4. The ____________ states that if the bases in the numerator and denominator of a fractional expression are the same and a and b are positive integers, then $\frac{x^a}{x^b} = x^{a-b}$ if the larger exponent is in the numerator and $x \neq 0$.

Example	Student Practice
1. Simplify. Leave your answer in exponent form.	**2.** Simplify. Leave your answer in exponent form.
(a) $\frac{n^9}{n^6}$ More factors are in the numerator. $\frac{n^9}{n^6} = n^{9-6} = \frac{n^3}{1} = n^3$	**(a)** $\frac{5^{12}}{5^8}$
(b) $\frac{5^8}{5^9}$ More factors are in the denominator. $\frac{5^8}{5^9} = \frac{1}{5^{9-8}} = \frac{1}{5^1}$ or $\frac{1}{5}$	**(b)** $\frac{a^6}{a^{11}}$

Vocabulary Answers: 1. raising a power to a power 2. additional power rule 3. product to a power 4. quotient rule

Example	Student Practice
3. Simplify $\frac{16x^6y^0}{20x^8}$. Factor 16 and 20. $\frac{16x^6y^0}{20x^8}=\frac{\overset{1}{\cancel{4}}\cdot 2\cdot 2\cdot x^6y^0}{\underset{1}{\cancel{4}}\cdot 5\cdot x^8}$ Now, $\frac{x^6}{x^8}=\frac{1}{x^{8-6}}$; $y^0=1$. $\frac{\overset{1}{\cancel{4}}\cdot 2\cdot 2\cdot x^6y^0}{\underset{1}{\cancel{4}}\cdot 5\cdot x^8}=\frac{2\cdot 2\cdot 1}{5\cdot x^{8-6}}$ $=\frac{4}{5x^2}$ The leftover x factors are in the denominator.	**4.** Simplify $\frac{36a^6b^0}{72a^9}$.
5. Write $\left(2^4\right)^2$ as a product and then simplify. Leave your answer in exponent form. $\left(2^4\right)^2$ means $2^4\cdot 2^4=2^{4+4}=2^8$	**6.** Write $\left(5^3\right)^6$ as a product and then simplify. Leave your answer in exponent form.

Example	Student Practice
7. Use the rules for raising a power to a power or a product to a power to simplify. Leave your answer in exponent form.	**8.** Use the rules for raising a power to a power or a product to a power to simplify. Leave your answer in exponent form.
(a) $\left(3^3\right)^3$	**(a)** $\left(5^2\right)^4$
Multiply the exponents.	
$\left(3^3\right)^3 = 3^{(3)(3)} = 3^9$	
The base does not change when raising a power to a power.	
(b) $\left(x^2\right)^0$	**(b)** $\left(a^0\right)^9$
$\left(x^2\right)^0 = x^{(2)(0)} = x^0 = 1$	
(c) $\left(4x^4\right)^5$	**(c)** $\left(5y^3\right)^7$
We write 4 as 4^1.	
$\left(4x^4\right)^5 = \left(4^1 \cdot x^4\right)^5$	
Raise each factor to the power 5.	
$\left(4^1 \cdot x^4\right)^5 = 4^{(1)(5)} \cdot x^{(4)(5)}$	
Finally, multiply exponents.	
$4^{(1)(5)} \cdot x^{(4)(5)} = 4^5 \cdot x^{20}$	
Thus, $\left(4x^4\right)^5 = 4^5 x^{20}$.	

Example	Student Practice
9. Simplify $\left(\dfrac{2}{x}\right)^3$. We must remember to raise both the numerator and the denominator to the power. $\left(\dfrac{2}{x}\right)^3=\left(\dfrac{2^1}{x^1}\right)^3=\dfrac{2^{(1)(3)}}{x^{(1)(3)}}=\dfrac{2^3}{x^3}=\dfrac{8}{x^3}$	**10.** Simplify $\left(\dfrac{x}{5}\right)^4$.

Extra Practice

1. Simplify $\dfrac{10a^3b^7c^0}{15a^5b^5}$. Assume a and b are nonzero. Leave your answer in exponent form.

2. Simplify $\dfrac{45x^0y^{17}z^4}{63y^9z^8}$. Assume y and z are nonzero. Leave your answer in exponent form.

3. Write $\left(\dfrac{2a^2b^3}{c^4}\right)^2$ as a product and then simplify. Leave your answer in exponent form.

4. Write $\left(3y^3z^2\right)^2\cdot\left(2y^2\right)^3$ as a product and then simplify. Leave your answer in exponent form.

Concept Check

Explain in words the steps you would need to follow to simplify the expression $\left(\dfrac{6x^2}{3}\right)^3$.

Name: ______________________________ Date: ________________
Instructor: ___________________________ Topic: ________________

Module 4 Fractions, Ratio, and Proportions
Topic 5 Ratios and Rates

Vocabulary

ratio • rate • unit rate • fraction

1. A(n) ___________ is a comparison of two quantities with different units.

2. Although a ratio can be written in different forms, it is a(n) ___________ and therefore should always be simplified (reduced to lowest terms).

3. A(n) ___________ is a comparison of two quantities that have the same units.

4. When the denominator is 1, we have the rate for a single unit, which is the ___________.

Example	Student Practice
1. Write each ratio in simplest form. Express your answer as a fraction.	**2.** Write each ratio in simplest form. Express your answer as a fraction.
(a) The ratio of 20 dollars to 35 dollars 20 dollars to 35 dollars $=\frac{20\ \cancel{\text{dollars}}}{35\ \cancel{\text{dollars}}}=\frac{\cancel{5}\cdot 4}{\cancel{5}\cdot 7}=\frac{4}{7}$ Note that we treat units in the same way we do numbers and variables.	**(a)** The ratio of 30 feet to 42 feet
(b) $14:21$ $14:21=\frac{14}{21}=\frac{\cancel{7}\cdot 2}{\cancel{7}\cdot 3}=\frac{2}{3}$	**(b)** $32:72$

Vocabulary Answers: 1. rate 2. fraction 3. ratio 4. unit rate

Example	Student Practice

3. Bertha drove her car 416 miles in 8 hours. Find the unit rate in miles per hour.

$$\frac{416 \text{ miles}}{8 \text{ hours}} \quad \text{We divide: } 8\overline{)416}\text{ gives } 52.$$

$$\frac{416 \text{ miles}}{8 \text{ hours}} = \frac{52 \text{ miles}}{1 \text{ hour}} \text{ or 52 mph}$$

4. Lisa travels 95 miles on 5 gallons of gas. Find the unit rate in miles per gallon.

5. University of Chicago tornado researcher Tetsuya Theodore Fujita cataloged 31,054 tornados in the United States during the 70 years 1916-1985 and found that $\frac{7}{10}$ of the tornados occurred in the spring and early summer. Write as a unit rate: the average number of tornados per year that occurred in the month of May. Round your answer to the nearest whole number.

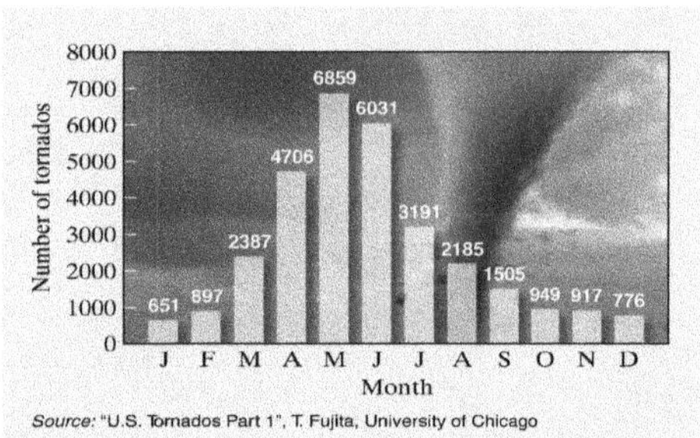

$$\frac{6859 \text{ tornados in May}}{70 \text{ years}}$$

Divide to find the unit rate.

$$70\overline{)6859} = 97\frac{69}{70} \text{ or approximately 98}$$

tornados per year in May

6. Use the bar graph in example **5** to answer the following.

Write as a unit rate: the average number of tornados per year that occurred in the month of July. Round your answer to the nearest whole number.

Example	Student Practice
7. Sunshine Preschool has a staffing policy requiring that for every 60 children, there are 3 preschool teachers, and for every 24 children, there are 2 aides.	**8.** A high school has a staffing policy requiring that for every 30 students there are 2 high school teachers, and for every 40 students, there are 2 aides.
(a) How many children per teacher does the preschool have? Children per teacher: $\frac{\text{children}}{\text{teacher}} \Rightarrow \frac{60\text{ children}}{3\text{ teachers}}$ $= \frac{20\text{ children}}{1\text{ teacher}}$ or 20 children per teacher	**(a)** How many students per teacher does the high school have?
(b) How many children per aide does the preschool have? Children per aide: $\frac{\text{children}}{\text{aide}} \Rightarrow \frac{24\text{ children}}{2\text{ aides}}$ $= \frac{12\text{ children}}{1\text{ aide}}$ or 12 children per aide	**(b)** How many students per aide does the high school have?
(c) If there are 60 students at the preschool, how many aides must there be to satisfy the staffing policy? Since every 12 children require 1 aide, we divide $60 \div 12$ to find how many aides are needed for 60 children. $60 \div 12 = 5$ aides for 60 children	**(c)** If there are 320 students at the high school, how many aides must there be to satisfy the staffing policy?

Example	Student Practice
9. The Tech Store has black cartridges on sale. A package of 6 sells for \$96, and the same brand in a package of 8 sells for \$136.	**10.** A store has its designer hand towels on sale. A package of 8 sells for \$56, and the same brand in a package of 12 sells for \$108.
(a) Find each unit price. $\frac{\$96}{6}=\$16$ per cartridge; $\frac{\$136}{8}=\$17$ per cartridge	**(a)** Find each unit price.
(b) Which is the better buy? The package of 6 cartridges is the better buy.	**(b)** Which is the better buy?

Extra Practice

1. Write the ratio as a fraction in simplest form.

   55:10

2. Write the ratio as a fraction in simplest form.

   The ratio of \$141 to \$69

3. Cecelia has five dolls and seven stuffed animals on her bed. Write a fraction that describes the ratio of dolls to stuffed animals.

4. If five pounds of chicken costs \$10.75 and three pounds of pork costs \$8.10, which type of meat costs more per pound?

Concept Check

A large furniture store determined they needed to have 8 salespeople in the store for every 160 customers. Explain how to determine how many salespeople per customer the store has.

Name: ______________________________ Date: ________________
Instructor: ____________________________ Topic: ________________

Module 4 Fractions, Ratio, and Proportions
Topic 6 Proportions and Applications

Vocabulary
proportion • equality test for fractions • cross product • fractions

1. To determine if a statement is a proportion, we must verify that the ___________ in the proportion are equal.

2. By ___________ we mean the denominator of one fraction times the numerator of the other fraction.

3. The ___________ states that if two fractions are equal, their cross products are equal.

4. A(n) ___________ states that two ratios or two rates are equal.

Example	Student Practice
1. Translate the statement into a proportion. If 6 pounds of flour cost \$2, then 18 pounds will cost \$6. We can restate as follows: 6 pounds is to \$2 as 18 pounds is to \$6. $\frac{6\text{ pounds}}{2\text{ dollars}}=\frac{18\text{ pounds}}{6\text{ dollars}}$	**2.** Translate the statement into a proportion. If it takes 5 hours to drive 150 miles, it will take 7 hours to drive 210 miles.
3. Use the equality test for fractions to see if the fractions are equal. $\frac{2}{11}\stackrel{?}{=}\frac{18}{99}$ Find the cross products. $\frac{2}{11}\nwarrow\frac{18}{99}$ $99\cdot 2=198$ $\frac{2}{11}\nearrow\frac{18}{99}$ $11\cdot 18=198$ Since $198=198$, $\frac{2}{11}=\frac{18}{99}$.	**4.** Use the equality test for fractions to see if the fractions are equal. $\frac{5}{17}\stackrel{?}{=}\frac{19}{24}$

Vocabulary Answers: 1. fractions 2. cross product 3. equality test for fractions 4. proportion

Example	Student Practice
5. Determine if the statement is a proportion. $\frac{16 \text{ points}}{35 \text{ games}} \stackrel{?}{=} \frac{48 \text{ points}}{125 \text{ games}}$ We check $\frac{16}{35} \stackrel{?}{=} \frac{48}{125}$ by forming the two cross products. $\frac{16}{35} \nwarrow \frac{48}{125}$ $125 \cdot 16 = 2000$ $\frac{16}{35} \nearrow \frac{48}{125}$ $35 \cdot 48 = 1680$ The two cross products are not equal. Thus, this is not a proportion.	**6.** Determine if the statement is a proportion. $\frac{18}{45} \stackrel{?}{=} \frac{54}{135}$
7. Find the value of n in $\frac{n}{24} = \frac{15}{60}$. First, find the cross products and form an equation. $60 \cdot n = 24 \cdot 15$ Now simplify. $60n = 360$ Finally, divide by 60 on both sides of the equation. $\frac{60 \cdot n}{60} = \frac{360}{60}$ $n = 6$ Check by replacing n with 6 in the original equation and verify if a true statement results.	**8.** Find the value of n in $\frac{n}{16} = \frac{32}{64}$.

Example	Student Practice
9. Estelle has a fence in her yard around her vegetable garden. The garden is 6 feet wide and 7 feet long. The yard's dimensions are proportional to the garden's. What is the length of the yard if the width is 25 feet?	**10.** Drew's concrete patio in his yard is 8 feet wide and 5 feet long. He wants to enlarge the patio, keeping the dimensions of the new patio proportional to those of the old patio. If he has room to increase the length to 35 feet, how wide should the patio be?

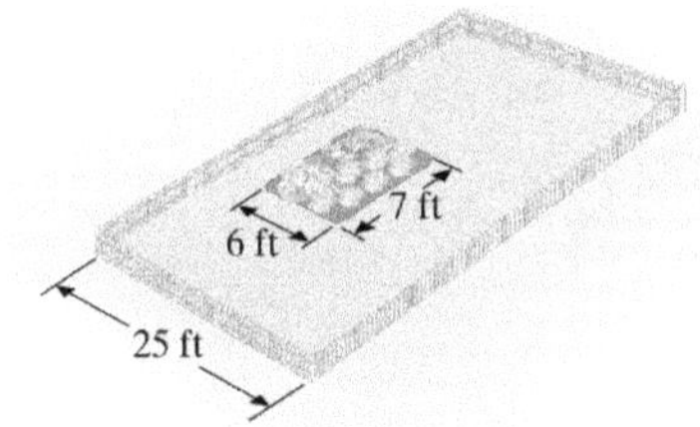

First, set up the proportion. Let the letter x represent the length of the yard.

$$\frac{\text{width of garden}}{\text{length of garden}} = \frac{\text{width of yard}}{\text{length of yard}}$$

$$\frac{6 \text{ ft}}{7 \text{ ft}} = \frac{25 \text{ ft}}{x \text{ ft}}$$

Now solve for x.

$$\frac{6}{7} = \frac{25}{x}$$

Find the cross products and then simplify.

$$6x = 7 \times 25$$
$$6x = 175$$
$$\frac{6x}{6} = \frac{175}{6}$$
$$x = 29\frac{1}{6}$$

The length of the yard is $29\frac{1}{6}$ feet.

Example	Student Practice
11. Two partners, Cleo and Julie, invest money in their small business at the ratio 3 to 5, with Cleo investing the smaller amount. If Cleo invested \$6000, how much did Julie invest?	**12.** Two partners, Mac and Alexander, invest money in their startup business at a ratio of 2 to 5, with Alexander investing the larger amount. If Alexander invested \$8000, how much did Mac invest?

The ratio 3 to 5 represents Cleo's investment to Julie's investment.

$$\frac{3}{5}=\frac{\text{Cleo's investment of \$6000}}{\text{Julie's investment of \$}x}$$

$$\frac{3}{5}=\frac{6000}{x}$$

$$3x=30,000$$

$$x=10,000$$

Extra Practice

1. Use the equality test for fractions to determine if the fractions are equal.

$$\frac{4}{36}\stackrel{?}{=}\frac{14}{117}$$

2. Find the value of x in the given proportion. Check your answer.

$$\frac{4}{13}=\frac{52}{x}$$

3. If Tamara takes 15 minutes to stock three shelves of merchandise, how long will it take her to stock 14 shelves?

4. Candace reads three books every four weeks. At that rate, how many books will she read in 20 weeks?

Concept Check

Justin and Sara share the profits from their business based on the ratio of their investment. The ratio of the investment is 5 to 7, with Sara investing the larger amount. Explain how you would determine how much profit Justin will receive if Sara gets \$840.

MATH COACH

Mastering the skills you need to do well on the test.

Watch the MATH COACH videos in MyMathLab® or on YouTube™ while you work the problems below. These helpful hints will help you avoid making common errors on test problems.

Finding the Prime Factors of Whole Numbers—Problem 3

Express as a product of prime factors. 84

> **Helpful Hint:** Choose the factoring method (division ladder or factor tree) that you prefer to solve this problem. Make sure all factors in your answer are *prime numbers*. Check your answer by multiplying all factors to be sure this product equals the number you are factoring.

Look at your work for Problem 3. Examine your steps.

Division ladder: Did you divide 84 by 4 or 6?
Yes ____ No ____

If you answered Yes, you forgot to use *only prime numbers* as divisors. Stop now and make this correction.

Did you include the final quotient as a factor?
Yes ____ No ____

If you answered No, stop and complete this step.

Factor Tree: Are all the factors in your answer prime numbers? Yes ____ No ____

Did you forget to include a prime number listed in the tree as part of your product? Yes ____ No ____

If you answered No to either of these questions, rework the problem and factor each number until all factors are prime. Circle all prime numbers as you factor to be sure you can easily see which factors to include in your answer.

Remember to check your answer by multiplying all the prime factors to be sure that the product is equal to the original number.

If you answered Problem 3 incorrectly, go back and rework the problem using these suggestions.

Finding Equivalent Fractions—Problem 11

Find an equivalent fraction with the given denominator. $\frac{4}{9}=\frac{?}{27y}$

> **Helpful Hint:** Remember to multiply *both* the numerator and denominator by the same nonzero number or expression. The equivalent fraction should have the given denominator of $27y$.

Did you choose 3 as the nonzero number or expression?
Yes ____ No ____

If you answered Yes, your resulting denominator of the equivalent fraction does not equal $27y$. Consider what expression, when multiplied by 9, equals $27y$.

Did you remember to multiply both the numerator and denominator by the same nonzero number or expression?
Yes ____ No ____

If you answered No, stop and make this correction.

Always double-check the product of the original denominator and your nonzero number or expression in the denominator. The result should be equal to the given denominator.

Now go back and rework the problem using these suggestions.

Use the Quotient Rule for Exponents—Problem 14 Simplify $\frac{y^3z^4}{y^7z}$.

Helpful Hint: For this problem, you must apply the quotient rule twice: once for the y variables and once for the z variables. Recall that when an exponent is not written, it is understood to be 1.

Did you notice that the exponent for y is larger in the denominator than in the numerator, and that the exponent for z is larger in the numerator than in the denominator?
Yes ____ No ____

If you answered No, go back and look carefully at the problem again.

Is y^4 in the denominator of your answer?
Yes ____ No ____

If you answered No, remember to subtract the exponents of y and to place the resulting y expression in the denominator since the *original exponent* for y is larger in the *denominator.*

Did you remember to subtract exponents for z and obtain the expression z^3?
Yes ____ No ____

If you answered No, make sure that you write a 1 as the exponent for z in the denominator and complete this step again.

If you answered Problem 14 incorrectly, go back and rework the problem using these suggestions.

Solving Applied Problems Involving Proportions—Problem 28 A bottle of fertilizer for your lawn states that you need 2 tablespoons to fertilize 400 square feet of lawn. How many tablespoons will you need to fertilize 1600 square feet of lawn?

Helpful Hint: When writing and setting up your proportion, remember to write the *unit names* in your proportion to avoid setting up the proportion incorrectly. Make sure that the same unit name appears in the numerator of both fractions and the same unit name appears in the denominator of both fractions.

Did you write $\frac{\text{2 tablespoons}}{\text{400 square feet}}$ as your first fraction and $\frac{x\text{ tablespoons}}{\text{1600 square feet}}$ as your second fraction?
Yes ____ No ____

If you answered No, stop and reread the problem to see how these fractions were obtained.

Did you simplify $\frac{2}{400}$ before you solved the proportion?
Yes ____ No ____

If you answered No, go back and complete this step again. Simplifying first will make the calculations easier to perform and will help with accuracy.

Remember to include units for x in your final answer.

If you answered Problem 28 incorrectly, go back and rework the problem using these suggestions.

Name: ______________________________ Date: ______________

Instructor: ___________________________ Topic: ______________

Module 5 Operations on Fractional Expressions
Topic 1 Multiplying and Dividing Fractional Expressions

Vocabulary

common factors • reciprocals • fractional part • inverting the fraction

1. __________ is when the numerator and denominator is interchanged.

2. We multiply in situations that require repeated addition or taking a(n) __________ of something.

3. If the product of two numbers is 1, we say that these two numbers are __________ of each other.

4. Removing a factor of 1 is also referred to as factoring out __________.

Example	Student Practice
1. Find $\frac{3}{7}$ of $\frac{2}{9}$. $\frac{3}{7}$ of $\frac{2}{9}=\frac{3}{7}\cdot\frac{2}{9}=\frac{3\cdot 2}{7\cdot 9}$ $=\frac{3\cdot 2}{7\cdot 3\cdot 3}=\frac{\overset{1}{\cancel{3}}\cdot 2}{7\cdot\underset{1}{\cancel{3}}\cdot 3}=\frac{2}{21}$	**2.** Find $\frac{4}{6}$ of $\frac{3}{5}$.
3. Multiply $\frac{-2}{18}\cdot\frac{9}{11}$. When multiplying positive and negative fractions, we determine the sign of the product and then multiply and simplify. $\frac{-2}{18}\cdot\frac{9}{11}=(-)$ $=-\frac{2\cdot 9}{18\cdot 11}=-\frac{2\cdot 9}{2\cdot 9\cdot 11}$ $=-\frac{\cancel{2}\cdot\cancel{9}}{\cancel{2}\cdot\cancel{9}\cdot 11}=-\frac{1}{11}$	**4.** Multiply $\frac{-14}{25}\cdot\frac{-15}{28}$.

Vocabulary Answers: 1. inverting the fraction 2. fractional part 3. reciprocals 4. common factors.

Example	Student Practice
5. Multiply $12x^3 \cdot \frac{5x^2}{4}$. $\frac{12x^3}{1} \cdot \frac{5x^2}{4} = \frac{4 \cdot 3 \cdot x^3 \cdot 5 \cdot x^2}{1 \cdot 4}$ $= \frac{\not{4} \cdot 3 \cdot 5 \cdot x^3 \cdot x^2}{1 \cdot \not{4}}$ $= \frac{3 \cdot 5 \cdot x^{3+2}}{1} = 15x^5$	**6.** Multiply $\frac{2x^4}{5} \cdot (10x^3)$.
7. Find the area of a triangle with $b = 12$ in. and $h = 7$ in. We evaluate the formula with the given values. $A = \frac{1}{2}bh$ $= \frac{1}{2} \cdot 12 \text{ in.} \cdot 7 \text{ in.}$ $= \frac{1 \cdot 12 \text{ in.} \cdot 7 \text{ in.}}{2 \cdot 1 \cdot 1}$ $= \frac{1 \cdot 2 \cdot 6 \cdot 7 \text{ in.} \cdot \text{ in.}}{2}$ $A = 42 \text{ in.}^2$	**8.** Find the area of a triangle with $b = 10$ ft and $h = 5$ ft.
9. Find the reciprocal. **(a)** $\frac{-7}{8}$ To find the reciprocal, we invert the fraction. $\frac{-7}{8} \to \frac{8}{-7} = -\frac{8}{7}$ **(b)** $6 = \frac{6}{1} \to \frac{1}{6} = \frac{1}{6}$	**10.** Find the reciprocal. **(a)** $-\frac{1}{3}$ **(b)** x

Example	Student Practice
11. Divide $\frac{-4}{11} \div \left(\frac{-3}{5}\right)$. There are 2 negative signs in the division. The number 2 is even, so the answer is positive. $\frac{-4}{11} \div \left(\frac{-3}{5}\right) = \frac{-4}{11} \cdot \left(\frac{5}{-3}\right) = \frac{4 \cdot 5}{11 \cdot 3} = \frac{20}{33}$	**12.** Divide $\frac{7}{13} \div \left(\frac{-15}{17}\right)$.
13. Divide $\frac{7x^4}{20} \div \left(\frac{-14x^2}{45}\right)$. $\frac{7x^4}{20} \div \left(\frac{-14x^2}{45}\right) = \frac{7x^4}{20} \cdot \left(\frac{45}{-14x^2}\right) = -\frac{\cancel{7} \cdot \cancel{5} \cdot 3 \cdot 3 \cdot x^4}{2 \cdot 2 \cdot \cancel{5} \cdot 2 \cdot \cancel{7} \cdot x^2} = -\frac{9x^2}{8}$	**14.** Divide $\frac{10x^8}{27} \div \left(\frac{-24x^5}{18}\right)$.
15. Samuel Jensen has $\frac{9}{40}$ of his income withheld for taxes and retirement. What amount is withheld each week if he earns \$1440 per week? The key phrase is "$\frac{9}{40}$ of his income." The word "of" often indicates multiplication. $\frac{9}{40} \cdot \frac{\$1440}{1} = \$324$ \$324 is withheld for taxes and retirement each week.	**16.** Nancy has a board that is 50 feet long that he wants to cut into 8 equal pieces. How long is each piece?

Extra Practice

1. Multiply. Be sure your answer is simplified.

$$\frac{-1}{6}\cdot\frac{18}{19}$$

2. Multiply. Be sure your answer is simplified.

$$\left(\frac{-3x}{4}\right)\cdot\left(\frac{6}{7x}\right)\cdot\left(\frac{2}{-5x}\right)$$

3. Divide. Be sure your answer is simplified.

$$\left(\frac{-5}{12}\right)\div\frac{25}{36}$$

4. Divide. Be sure your answer is simplified.

$$27x^4\div\frac{9}{4x^3}$$

Concept Check

Explain how you would divide $\frac{-16x^2}{3}$ by $8x$.

Name: ______________________________ Date: ________________
Instructor: ____________________________ Topic: _______________

Module 5 Operations on Fractional Expressions
Topic 2 Least Common Multiples of Algebraic Expressions

Vocabulary
multiples • least common multiple • common factors • build the LCM

1. The smallest of common multiples is called the ____________.

2. We should always factor out ____________ before we multiply the numerators and denominators; otherwise we must simplify the product.

3. A quicker method is to ____________ using the prime factors of each number.

4. To generate a list of ____________ of a number, multiply that number by 1, and then by 2, and then by 3, and so on.

Example	Student Practice
1. **(a)** List the first six multiples of $8x$ and the first six multiples of $12x$. $8x \cdot 1 = 8x,\ 8x \cdot 2 = 16x,\ 8x \cdot 3 = 24x,$ $8x \cdot 4 = 32x,\ 8x \cdot 5 = 40x,\ 8x \cdot 6 = 48x$ $12x \cdot 1 = 12x,\ 12x \cdot 2 = 24x,\ 12x \cdot 3 = 36x,$ $12x \cdot 4 = 48x,\ 12x \cdot 5 = 60x, 12x \cdot 6 = 72x$ **(b)** Which of these multiples are common to both lists? The multiples common to both lists are $24x$ and $48x$.	**2.** **(a)** List the first seven multiples of $6y$ and $10y$. **(b)** Which of these multiples are common to both lists?
3. Find the LCM of 10 and 15. First, we list some multiples of 10: 10, 20, 30, 40, 50, 60. Next, we list some multiples of 15: 15, 30, 45, 60. We see that both 30 and 60 are common multiples. Since 30 is the smaller of these common multiples, we call 30 the least common multiple (LCM).	**4.** Find the LCM of 6 and 9.

Vocabulary Answers: 1. least common multiple 2. common factors 3. build the LCM 4. multiples

Example	Student Practice
5. Find the LCM of 18, 42, and 45. Factor each number, then list the requirements for factorization of LCM, then build the LCM. $18 = 2 \cdot 3 \cdot 3 \rightarrow$ must have a 2 and pair of 3's $\rightarrow \boxed{\text{LCM} = 2 \cdot 3 \cdot 3 \cdot ?}$ $42 = 2 \cdot 3 \cdot 7 \rightarrow$ must have a 2, a 3, and a 7 $\rightarrow \boxed{\text{LCM} = 2 \cdot 3 \cdot 3 \cdot 7 \cdot ?}$ $45 = 3 \cdot 3 \cdot 5 \rightarrow$ must have a pair of 3's and a 5 $\rightarrow \boxed{\text{LCM} = 2 \cdot 3 \cdot 3 \cdot 7 \cdot 5}$ The LCM of 18, 42, and 45 is $2 \cdot 3 \cdot 3 \cdot 7 \cdot 5 = 630$.	**6.** Find the LCM of 14, 24, and 34.
7. Find the LCM of $2x$, x^2, and $6x$. Factor each expression, then list the requirements for factorization of LCM, then build the LCM. $2x = 2 \cdot x \rightarrow$ must have a 2 and an x $\rightarrow \boxed{\text{LCM} = 2 \cdot x \cdot ?}$ $x^2 = x \cdot x \rightarrow$ must have a pair of x's $\rightarrow \boxed{\text{LCM} = 2 \cdot x \cdot x \cdot ?}$ $6x = 2 \cdot 3 \cdot x \rightarrow$ must have a 2, a 3 and an x $\rightarrow \boxed{\text{LCM} = 2 \cdot x \cdot x \cdot 3}$ The LCM of $2x$, x^2, and $6x$ is $2 \cdot x \cdot x \cdot 3 = 6x^2$.	**8.** Find the LCM of $5x$, 20, and $10x^3$.

Example	Student Practice

9. Sonia and Leo are tour guides at a castle. Sonia gives a 40-minute tour of the interior of the castle, and Leo gives a 30-minute tour of the castle grounds. There is a 10-minute break after each tour. If tours start at 8 A.M. what is the next time that both tours will start at the same time?

Read the problem carefully and create a Mathematics Blueprint.

We make a chart to help us develop a plan to solve the problem.

Tours	Number of Minutes after 8 A.M. Tours Start
Interior tours start every 50 minutes (40 min+10-min break).	50, 100, . . .
Grounds tours start every 40 minutes (30 min+10-min break).	40, 80, . . .

First we factor 40 and 50 into a product of prime factors and then find the LCM.

$40 = 2 \cdot 2 \cdot 2 \cdot 5,\ \ 50 = 2 \cdot 5 \cdot 5$

$\text{LCM} = 2 \cdot 2 \cdot 2 \cdot 5 \cdot 5 = 200$

Both tours will start at the same time 200 minutes after 8 A.M.
Next we change minutes to hours and minutes. Since we need to know how many 60's are in 200, we divide.

$200 \text{ minutes} \div 60 \text{ minutes per hour}$

$= 3 \text{ hours and } 20 \text{ minutes after } 8$

$8 + 3 \text{ hours and } 20 \text{ minutes} = 11{:}20 \text{ A.M.}$

At 11:20 A.M. both tours will start at the same time.

10. Refer to example **9** to complete this problem. Sonia's tour of the interior of the castle is reduced to 35 minutes. Determine the next time that both tours will start at the same time.

Extra Practice

1. Find the LCM of 20 and 45.

2. Find the LCM of 3, 6, and 21.

3. Find the LCM $16x$, $70x^2$, and $7x^3$.

4. Two security guards patrol a museum each night. Both security guards begin their patrol from the same point. The security guard patrolling the north wing takes 21 minutes to complete his rounds, while the security guard patrolling the south wing takes 18 minutes to complete his rounds. If both security guards leave the same point at 10:00 P.M., at what time will they cross paths?

Concept Check

Is $=3\times5\times5\times7\times x\times x$ the correct factorization for the LCM of $63x^2$ and $75x^3$? Why or why not?

Name: ______________________________ Date: ________________
Instructor: ____________________________ Topic: ________________

Module 5 Operations on Fractional Expressions
Topic 3 Adding and Subtracting Fractional Expressions

Vocabulary
common denominator • least common denominator • mixed number • equivalent fractions

1. As answers to applications, ___________ are generally easier to understand.

2. To add or subtract fractional expressions with different denominators we find the LCD and write ___________ that have the LCD as the denominator.

3. When fractions have the same denominator, we say that these fractions have a(n) ___________.

4. The ___________ of two fractions is the least common multiple of the two denominators.

Example	**Student Practice**
1. Subtract $\frac{7}{15}-\frac{3}{15}$. $\frac{7}{15}-\frac{3}{15}=\frac{7-3}{15}$ $=\frac{4}{15}$	**2.** Add $\frac{4}{17}+\frac{9}{17}$.
3. Add $\frac{-11}{20}+\left(\frac{-13}{20}\right)$. $\frac{-11}{20}+\left(\frac{-13}{20}\right)=\frac{-11+(-13)}{20}$ $=\frac{-24}{20}$ $=\frac{\cancel{4}(-6)}{\cancel{4}(5)}$ $=-\frac{6}{5}$	**4.** Add $\frac{-3}{10}+\left(\frac{-11}{10}\right)$.

Vocabulary Answers: 1. mixed numbers 2. equivalent fractions 3. common denominator 4. least common denominator

Example	Student Practice
5. Perform the operation indicated. **(a)** $\frac{6}{y}-\frac{2}{y}$ $\frac{6}{y}-\frac{2}{y}=\frac{6-2}{y}=\frac{4}{y}$ **(b)** $\frac{x}{5}+\frac{4}{5}$ $\frac{x}{5}+\frac{4}{5}=\frac{x+4}{5}$	**6.** Perform the operation indicated. **(a)** $\frac{3}{5y}+\frac{4}{5y}$ **(b)** $\frac{x}{8}-\frac{3}{8}$
7. Find the least common denominator (LCD) of the fractions $\frac{1}{12}, \frac{5}{18}$. $12=2\cdot2\cdot3,\ 18=2\cdot3\cdot3$ $\text{LCD}=2\cdot2\cdot3\cdot3=36$ The LCD of $\frac{1}{12}$ and $\frac{5}{18}$ is 36.	**8.** Find the least common denominator of the fractions $\frac{1}{20}, \frac{6}{25}$.
9. Write the equivalent fraction for $\frac{1}{5}$ that has 40 as the denominator. $\frac{1}{5}=\frac{?}{40}$ $\frac{1}{5}=\frac{?}{40}$ $\frac{1\cdot8}{5\cdot8}=\frac{8}{40}$ $\frac{1}{5}=\frac{8}{40}$	**10.** Write the equivalent fractions that have 20 as the denominator. **(a)** $\frac{3}{2}=\frac{?}{20}$ **(b)** $\frac{1}{4}=\frac{?}{20}$

Example	Student Practice
11. Perform the operation indicated. $\frac{-5}{7}+\frac{3}{4}$ Find the LCD of $\frac{-5}{7}$ and $\frac{3}{4}$. $\text{LCD}=28$ Write equivalent fractions. $\frac{-5\cdot 4}{7\cdot 4}=\boxed{\frac{-20}{28}}\quad \frac{3\cdot 7}{4\cdot 7}=\boxed{\frac{21}{28}}$ Add the numerators of the fractions with common denominators. $\frac{-5}{7}+\frac{3}{4}=\boxed{\frac{-20}{28}+\frac{21}{28}}=\frac{1}{28}$	**12.** Perform the operation indicated. $\frac{17}{30}-\frac{7}{18}$
13. Add $\frac{7x}{16}+\frac{3x}{32}$. Find the LCD of $\frac{7x}{16}$ and $\frac{3x}{32}$. $\text{LCD}=32$ Write equivalent fractions. $\frac{7x\cdot 2}{16\cdot 2}=\boxed{\frac{14x}{32}}\quad \frac{3x}{32}=\boxed{\frac{3x}{32}}$ Add fractions with common denominators. $\frac{7x}{16}+\frac{3x}{32}=\boxed{\frac{14x}{32}+\frac{3x}{32}}=\frac{14x+3x}{32}=\frac{17x}{32}$	**14.** Add $\frac{2x}{25}+\frac{x}{5}$.

Example	Student Practice
15. Leila finished $\frac{1}{8}$ of her English term paper before spring break and $\frac{1}{2}$ of the paper during spring break. How much more did she complete during the break than before the break?	**16.** Frank painted $\frac{2}{5}$ of his home red and $\frac{2}{9}$ of his home blue. How much more did he paint red than blue.

The phrase "how much more" indicates that we subtract.

$\frac{1}{2}-\frac{1}{8}$ The LCD is 8.

$\frac{1}{2}=\frac{1\cdot 4}{2\cdot 4}=\boxed{\frac{4}{8}}$ $\frac{1}{8}=\boxed{\frac{1}{8}}$

Subtract. $\frac{1}{2}-\frac{1}{8}=\boxed{\frac{4}{8}-\frac{1}{8}}=\frac{3}{8}$

Extra Practice

1. Perform the operation indicated. Be sure to simplify your answer. $\frac{4}{15}+\left(\frac{-2}{15}\right)$

2. Perform the operation indicated. Be sure to simplify your answer. $\frac{3}{10}+\frac{4}{25}$

3. Perform the operation indicated. Be sure to simplify your answer. $\frac{-3x}{14}-\left(\frac{-8x}{21}\right)$

4. Tran and Alex were collecting canned food as part of their class food drive. Tran collected $\frac{3}{8}$ of the class total and Alex collected $\frac{1}{5}$ of the class total. What part of the total number of cans did Tran and Alex collect together?

Concept Check

(a) What is a common denominator for the fractions $\frac{3x}{20}$ and $\frac{5x}{6}$?

(b) Explain how you would add the fractions.

Name: ______________________________ Date: ______________

Instructor: ____________________________ Topic: ______________

Module 5 Operations on Fractional Expressions
Topic 4 Operations with Mixed Numbers

Vocabulary

carrying • borrow • mixed numbers • improper fractions

1. The process used to __________ with mixed numbers is the opposite of carrying with mixed numbers.

2. Change mixed numbers to __________ before multiplying or dividing.

3. To simplify a mixed number with an improper fraction part, use a process similar to __________ with whole numbers.

4. When adding and subtracting __________, we add or subtract the fractions first and then the whole numbers.

Example	Student Practice
1. Add. $4\frac{1}{8}+3\frac{3}{8}$ $$\begin{array}{r} 4\frac{1}{8} \\ +3\frac{3}{8} \\ \hline 7\frac{4}{8}=7\frac{1}{2} \end{array}$$	**2.** Add. $4\frac{1}{7}+2\frac{5}{7}$
3. Add. $4\frac{2}{3}+2\frac{1}{4}$ The LCD of $\frac{2}{3}$ and $\frac{1}{4}$ is 12. $$\begin{array}{rr} 4\frac{2}{3}\cdot\frac{4}{4}= & 4\frac{8}{12} \\ +2\frac{1}{4}\cdot\frac{3}{3}= & +2\frac{3}{12} \\ \hline & 6\frac{11}{12} \end{array}$$	**4.** Add. $7\frac{1}{4}+5\frac{2}{5}$

Vocabulary Answers: 1. borrow 2. improper fractions 3. carrying 4. mixed numbers

Example	Student Practice
5. Add. $2\frac{5}{7}+6\frac{2}{3}$	**6.** Add. $5\frac{7}{9}+2\frac{3}{4}$

The LCD of $\frac{5}{7}$ and $\frac{2}{3}$ is 21.

$$2\frac{5}{7}\cdot\frac{3}{3}= \quad 2\frac{15}{21}$$
$$+6\frac{2}{3}\cdot\frac{7}{7}=+6\frac{14}{21}$$
$$8\frac{29}{21}=8+1\frac{8}{21}$$
$$=9\frac{8}{21}$$

7. Subtract. $7\frac{4}{15}-2\frac{7}{15}$	**8.** Subtract. $6\frac{3}{20}-4\frac{7}{20}$

We cannot subtract $\frac{4}{15}-\frac{7}{15}$ without borrowing.

$$7\frac{4}{15}=6+1\frac{4}{15}$$
$$=6+\frac{19}{15}$$

$$7\frac{4}{15}= \quad 6\frac{19}{15}$$
$$-2\frac{7}{15}=-2\frac{7}{15}$$
$$4\frac{12}{15}$$

We simplify: $4\frac{12}{15}=4\frac{4}{5}$.

Example	Student Practice
9. Subtract $8-3\frac{1}{4}$. $$\begin{array}{rr} 8= & 7\frac{4}{4} \\ -3\frac{1}{4}= & -3\frac{1}{4} \\ \hline & 4\frac{3}{4} \end{array}$$ When we borrowed 1 from 8, we changed the 1 to $\frac{4}{4}$ so the fraction had the same denominator as $\frac{1}{4}$.	**10.** Subtract $6-2\frac{3}{4}$.
11. Multiply $5\frac{5}{12}\cdot 3\frac{11}{15}$. We change the mixed numbers to improper fractions and then multiply. $$5\frac{5}{12}\cdot 3\frac{11}{15}=\frac{65}{12}\cdot\frac{56}{15}=\frac{\cancel{5}\cdot 13\cdot\cancel{4}\cdot 14}{3\cdot\cancel{4}\cdot\cancel{5}\cdot 3}=\frac{182}{9}\text{ or }20\frac{2}{9}$$	**12.** Multiply $7\frac{4}{5}\cdot 2\frac{6}{7}$.
13. Divide $2\frac{1}{4}\div(-5)$. Recall that to divide we invert the second fraction and multiply. $$2\frac{1}{4}\div(-5)=\frac{9}{4}\div\frac{(-5)}{1}=\frac{9}{4}\cdot\left(-\frac{1}{5}\right)=-\frac{9}{20}$$	**14.** Divide $4\frac{5}{6}\div(-4)$.

Example	Student Practice
15. Ester uses a small piece of painted wood as the base for each centerpiece she makes for banquet tables. She has a long piece of word that measures $13\frac{1}{2}$ feet. She needs to cut it into pieces that are $\frac{1}{2}$ foot long for the centerpiece bases. How many centerpiece bases will she be able to cut from the long piece of wood?	**16.** A recipe for pancakes calls for $2\frac{2}{3}$ cups of flour. If Connor only has a $\frac{1}{3}$-cup measuring utensil, how many times must he fill this utensil to get the desired amount of flour?

Draw a picture.

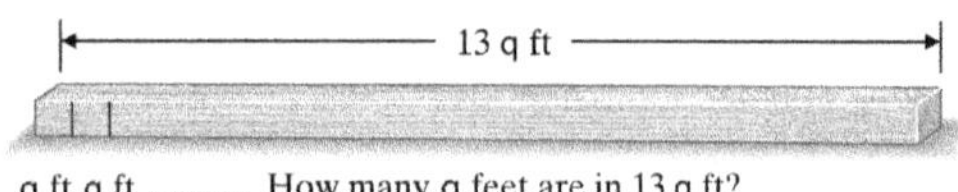

We want to know how many $\frac{1}{2}$s are in $13\frac{1}{2}$, so we must divide $13\frac{1}{2}\div\frac{1}{2}$.

$$13\frac{1}{2}\div\frac{1}{2}=\frac{27}{2}\div\frac{1}{2}=\frac{27}{2}\cdot\frac{2}{1}=27$$

Ester can make 27 centerpiece bases.

Extra Practice

1. Add. Simplify the answer. Express as a mixed number. $5\frac{6}{7}+8\frac{11}{14}$

2. Subtract. Simplify the answer. Express as a mixed number. $14\frac{3}{8}-5\frac{9}{16}$

3. Multiply and simplify your answer. $2\frac{3}{5}\cdot 4\frac{5}{8}$

4. Divide and simplify your answer. $-2\frac{3}{7}\div\frac{3}{14}$

Concept Check

Explain how you would multiply $2\frac{1}{2}\times 3\frac{2}{3}$.

Name: ______________________________ Date: ________________
Instructor: ___________________________ Topic: ________________

Module 5 Operations on Fractional Expressions
Topic 5 Order of Operations and Complex Fractions

Vocabulary
order of operations • grouping symbols • complex fraction • main fraction bar

1. A fraction that contains at least one fraction in the numerator or in the denominator is a(n) ___________.

2. We must perform operations above and then below the ___________ before we divide.

3. Recall that when we work a problem with more than one operation, we must follow the ___________.

4. Although we usually do not write ___________ (parentheses or brackets) around the numerator and denominator of a complex fraction, it is understood that they exist.

Example	Student Practice
1. Simplify $\left(\frac{2}{3}\right)^2 - \frac{2}{9}\cdot\frac{1}{3}$. $\left(\frac{2}{3}\right)^2 - \frac{2}{9}\cdot\frac{1}{3} = \frac{4}{9} - \frac{2}{9}\cdot\frac{1}{3} = \frac{4}{9} - \frac{2}{27}$ $= \frac{4\cdot 3}{9\cdot 3} - \frac{2}{27} = \frac{12}{27} - \frac{2}{27} = \frac{10}{27}$	**2.** Simplify $\left(\frac{2}{5}\right)^3 + \frac{1}{3} \div \frac{7}{9}$.
3. Simplify $\dfrac{(-2)^2+8}{\frac{2}{3}}$. We must follow the order of operations. $\dfrac{\left[(-2)^2+8\right]}{\left(\frac{2}{3}\right)} = \dfrac{12}{\left(\frac{2}{3}\right)} = 12 \div \frac{2}{3}$ $= 12\cdot\frac{3}{2} = \dfrac{\overset{1}{\cancel{2}}\cdot 2\cdot 3\cdot 3}{\underset{1}{\cancel{2}}} = 18$	**4.** Simplify $\dfrac{\frac{3}{4}}{5^2-(-7)}$.

Vocabulary Answers: 1. complex fraction 2. main fraction bar 3.order of operations 4. grouping symbols

Example	Student Practice
5. Simplify $\dfrac{\frac{x^2}{8}}{\frac{x}{4}}$.	**6.** Simplify $\dfrac{\frac{6}{x}}{\frac{2}{x^2}}$.
7. Simplify $\dfrac{\frac{2}{3}+\frac{1}{6}}{\frac{3}{4}-\frac{1}{2}}$.	**8.** Simplify $\dfrac{\frac{3}{4}-\frac{1}{8}}{\frac{7}{9}+\frac{5}{6}}$.

5. Simplify $\dfrac{\frac{x^2}{8}}{\frac{x}{4}}$.

Since the main fraction bar indicates division, we can divide the top fraction by the bottom fraction to simplify.

$$\frac{\frac{x^2}{8}}{\frac{x}{4}}=\frac{x^2}{8}\div\frac{x}{4}=\frac{x^2}{8}\cdot\frac{4}{x}$$

$$=\frac{x^2\cdot \not{4}}{2\cdot\not{4}\cdot x}=\frac{x\cdot x}{2\cdot x}=\frac{x}{2}$$

6. Simplify $\dfrac{\frac{6}{x}}{\frac{2}{x^2}}$.

7. Simplify $\dfrac{\frac{2}{3}+\frac{1}{6}}{\frac{3}{4}-\frac{1}{2}}$.

We write parentheses in the numerator and denominator and follow the order of operations.

$$\frac{\left(\frac{2}{3}+\frac{1}{6}\right)}{\left(\frac{3}{4}-\frac{1}{2}\right)}=\frac{\left(\frac{2\cdot 2}{3\cdot 2}+\frac{1}{6}\right)}{\left(\frac{3}{4}-\frac{1\cdot 2}{2\cdot 2}\right)}=\frac{\left(\frac{4}{6}+\frac{1}{6}\right)}{\left(\frac{3}{4}-\frac{2}{4}\right)}=\frac{\frac{5}{6}}{\frac{1}{4}}$$

Now we divide the top fraction by the bottom fraction.

$$\frac{5}{6}\div\frac{1}{4}=\frac{5}{6}\cdot\frac{4}{1}=\frac{5\cdot\not{2}\cdot 2}{3\cdot\not{2}}=\frac{10}{3}$$

8. Simplify $\dfrac{\frac{3}{4}-\frac{1}{8}}{\frac{7}{9}+\frac{5}{6}}$.

Example	Student Practice
9. A recipe requires $3\frac{2}{3}$ cups of flour to make bread and feed 50 people. How much flour do we need to make bread to feed 120 people?	**10.** To make 3 bracelets you need $21\frac{3}{4}$ inches of yarn. How much yarn do you need to make 28 bracelets?

Since the problem concerns the rate of cups of flour per 50 people, we set up a proportion and solve for the missing number.

$$\frac{3\frac{2}{3}\text{ cups}}{50\text{ people}} = \frac{x\text{ cups}}{120\text{ people}}$$

$$120 \cdot 3\frac{2}{3} = 50x$$

$$120 \cdot \frac{11}{3} = 50x$$

$$\frac{40 \cdot \overset{1}{\cancel{3}} \cdot 11}{\underset{1}{\cancel{3}}} = 50x$$

$$40 \cdot 11 = 50x$$

$$\frac{40 \cdot 11}{50} = \frac{50x}{50}$$

$$\frac{4 \cdot \overset{1}{\cancel{10}} \cdot 11}{5 \cdot \underset{1}{\cancel{10}}} = x$$

$$\frac{44}{5} = x \text{ or } x = 8\frac{4}{5}$$

We need $8\frac{4}{5}$ cups of flour.

Extra Practice

1. Simplify $\left(\frac{3}{5}\right)^3 \cdot \left(\frac{1}{3}\right)^3$.

2. Simplify $\left(\frac{2}{5}-\frac{3}{10}\right)\left(\frac{2}{5}+\frac{3}{10}\right)$.

3. Simplify $\dfrac{\frac{x^2}{3}}{\frac{x}{18}}$.

4. Simplify $\dfrac{\frac{16}{30}-\frac{2}{15}}{\frac{14}{40}+\frac{9}{20}}$.

Concept Check

Explain how you would simplify $\dfrac{1+2\times 3}{\frac{1}{2}}$.

Name: ______________________________ Date: ________________

Instructor: ____________________________ Topic: ______________

Module 5 Operations on Fractional Expressions
Topic 6 Solving Applied Problems Involving Fractions

Vocabulary

fractions • improper fractions

1. We must change mixed numbers to ___________ before we perform division.

2. For applied problems involving ____________, we may need to draw a picture to help us determine which operation to use.

Example	Student Practice
1. Jason planted a rectangular rose garden in the center of his 26-foot by 20-foot backyard. Around the garden there is a sidewalk that is $3\frac{1}{2}$ feet wide. The garden and sidewalk take up the entire 26-foot by 20-foot yard. What are the dimensions of the rose garden? Read the problem carefully and create a Mathematics Blueprint. We find the length and width of the rose garden. $L=26-\left(3\frac{1}{2}+3\frac{1}{2}\right)$ $L=26-7=19$ We find the width of the rose garden. $W=20-\left(3\frac{1}{2}+3\frac{1}{2}\right)$ $W=20-7=13$ The dimensions of the garden are 19 feet by 13 feet.	**2.** Leona wants a rectangular coy pond in the middle of her 32-foot by 28-foot backyard. She wants around the pond to be a rock formation that is $2\frac{1}{4}$ feet wide. The pond and rocks will take up the entire 32-foot by 28-foot yard. **(a)** What are the dimensions of the coy pond. **(b)** How much will it cost to put a fence around the coy pond if the fencing costs $\$3\frac{1}{2}$ per linear foot?

Vocabulary Answers: 1. improper fractions 2. fractions

Example	Student Practice
3. Marian is planning to build a fence on her farm. She determines that she must make 115 wooden fence posts that are each $3\frac{3}{4}$ feet in length. The wood to make the fence posts is sold in 20-foot lengths. How many 20-foot pieces of wood must Marian purchase so that she can make 115 fence posts?	**4.** Martin wishes to make 92 shelves that are each $3\frac{3}{8}$ feet in length. The wood to make the shelves is sold in 10-foot lengths. How many 10-foot pieces of wood must Martin purchase so that he can make 92 shelves?

Read the problem carefully and create a Mathematics Blueprint.

We must divide to find out how many $3\frac{3}{4}$-foot sections are in 20 feet.

$$20 \div 3\frac{3}{4} = 20 \div \frac{15}{4} = 20 \cdot \frac{4}{15}$$

$$= \frac{4 \cdot \cancel{5} \cdot 4}{\cancel{5} \cdot 3} = \frac{16}{3} \text{ or } 5\frac{1}{3}$$

5 posts can be cut from each 20-foot piece of wood, with some wood left over.

Now we must find how many of the 20-foot pieces are needed. We must find how many groups of 5 are in 115. We divide $115 \div 5$.

$$115 \div 5 = \frac{115}{5} = \frac{\cancel{5} \cdot 23}{\cancel{5}} = 23$$

Marian must purchase 23 pieces of wood.

Check the answer by rounding $3\frac{3}{4}$ to 4 and reworking the problem.

Extra Practice

1. Amy and Alex have 55 feet of ribbon. How much ribbon will be left if they use $\frac{1}{5}$ of the ribbon to decorate their raffle table and they use $\frac{5}{7}$ of the ribbon to hang nametags?

2. Allison and Tyler are planning a cookout. They invited 27 people (including themselves) to the cookout. If they estimate that each person can eat $\frac{1}{3}$ pound of meat, $\frac{3}{5}$ pound of potato salad, and $\frac{1}{2}$ pound of fruit, how much meat, salad, and fruit must they order?

3. Plastic tubing is sold in 40-foot bundles. To complete an order, Kieran needs to cut 48 sections of plastic tubing that are each $3\frac{5}{8}$ feet long. How many bundles of plastic tubing will Kieran need to purchase to complete the order?

4. The instructions for making a concrete driveway require $13\frac{1}{3}$ parts cement, 40 parts sand, and 90 parts aggregate. How many parts cement are required to make 8 concrete driveways?

Concept Check

Choose the correct operation you must use to answer each of the following questions: Add, Subtract, Multiply, or Divide. You do not need to calculate the answer.

(a) Jason ran $2\frac{1}{3}$ miles, and Lester ran $2\frac{7}{8}$ miles. How much farther did Lester run than Jason?

(b) Beatrice earns \$780 per week and has $\frac{1}{13}$ of her paycheck placed in a savings account. How much money does she put in her savings each week?

(c) Samuel has 14 pounds of candy and must place it in $\frac{2}{3}$-pound bags. How many bags can he fill?

Name: ______________________________ Date: ________________
Instructor: ____________________________ Topic: ________________

Module 5 Operations on Fractional Expressions
Topic 7 The Multiplication Principle of Equality

Vocabulary
equal • multiply • reciprocal • simplification

1. To solve an equation of the form $\frac{x}{a} = b$, we can ____________ both sides of the equation by the same nonzero number.

2. If both sides of an equation are multiplied by the same nonzero number, the results on both sides are ____________ in value.

3. It is important that we remember to perform any necessary ____________ of an equation before we find the solution.

4. When you multiply a fraction by its ____________, the product is 1.

Example	Student Practice
1. Solve $\frac{x}{-4} = 28$.	**2.** Solve $\frac{y}{-3} = -20$.
Since we are dividing the variable x by -4, we can undo the division and get x alone by multiplying by -4.	
$\frac{x}{-4} = 28$	
$\frac{-4 \cdot x}{-4} = 28 \cdot (-4)$	
Simplify: $\frac{-4x}{-4} = x$ and $28 \cdot (-4) = -112$.	
$x = -112$	
Be sure that you check your solution.	

Vocabulary Answers: 1. multiply 2. equal 3. simplification 4. reciprocal

Example	Student Practice
3. Solve $\frac{x}{2^3}=\frac{1}{2}+\frac{1}{4}$.	**4.** Solve $\frac{y}{4^2}=\frac{1}{4}+\frac{1}{8}$.

We simplify each side of the equation first and then we find the solution.

$$\frac{x}{2^3}=\frac{1}{2}+\frac{1}{4}$$

Simplify: $2^3=8$

$$\frac{x}{8}=\frac{1}{2}+\frac{1}{4}$$

Add: $\frac{1}{2}+\frac{1}{4}=\frac{2}{4}+\frac{1}{4}=\frac{3}{4}$

$$\frac{x}{8}=\frac{3}{4}$$

We undo the division by multiplying both sides by 8.

$$\frac{8\cdot x}{8}=\frac{3}{4}\cdot 8$$

Multiply to find the solution: $8\cdot\frac{3}{4}=6$

$$x=6$$

We leave the check for the student.

Example	Student Practice
5. Solve for the variable and check your solution. $-\frac{3}{4}x=12$	**6.** Solve for the variable and check your solution. $-\frac{5}{7}y=-10$

$$-\frac{3}{4}x=12$$

Multiply both sides of the equation by $-\frac{4}{3}$ because $\left(-\frac{4}{3}\right)\left(-\frac{3}{4}\right)=1$.

$$\left(-\frac{4}{3}\right)\left(-\frac{3}{4}\right)x=12\left(-\frac{4}{3}\right)$$

$$1x=-\frac{4\cdot \cancel{3}\cdot 4}{\cancel{3}}$$

$$x=-16$$

Check.

$$-\frac{3}{4}x=12$$

Replace x with -16.

$$\left(-\frac{3}{4}\right)(-16)\stackrel{?}{=}12$$

$$12=12$$

Extra Practice

1. Solve and check your solution. $\frac{x}{8}=22$

2. Solve and check your solution.
$\frac{x}{-9}=-27+12$

3. Solve and check your solution. $\frac{1}{8}x=-14$

4. Solve and check your solution. $\frac{5}{7}x=25$

Concept Check

To solve the equation $\frac{x}{-5}=6$, Amy multiplied both sides of the equation by 5 to obtain $x=30$. Is this correct? Why or why not?

MATH COACH

Mastering the skills you need to do well on the test.

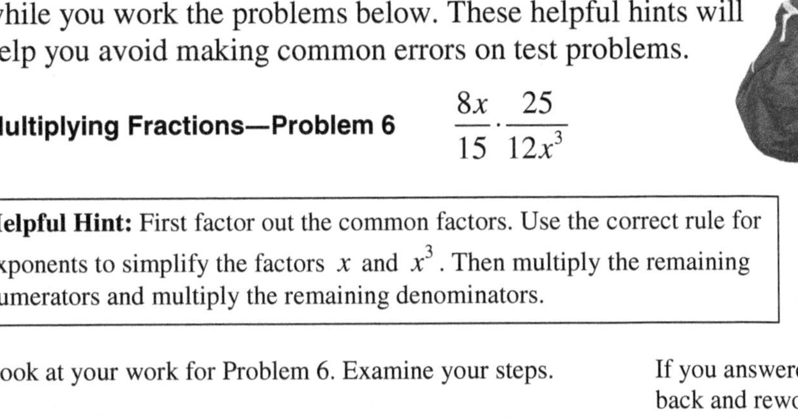

Watch the MATH COACH videos in MyMathLab® or on YouTube™ while you work the problems below. These helpful hints will help you avoid making common errors on test problems.

Multiplying Fractions—Problem 6 $\frac{8x}{15} \cdot \frac{25}{12x^3}$

> **Helpful Hint:** First factor out the common factors. Use the correct rule for exponents to simplify the factors x and x^3. Then multiply the remaining numerators and multiply the remaining denominators.

Look at your work for Problem 6. Examine your steps.

Did you simplify the fractions before multiplying?
Yes ____ No ____

If you answered No, stop and complete these calculations.

Did you divide 8 and 12 by 2, and 25 and 15 by 5?
Yes ____ No ____

If you answered Yes, make sure you did not stop with 20 in the numerator and 18 in the denominator. The fraction can be simplified further.

Did you use the quotient rule to simplify?
Yes ____ No ____

If you answered No, go back and perform this step.

If you answered Problem 6 incorrectly, go back and rework the problem using these suggestions.

Adding Fractions with Different Denominators—Problem 15 $\frac{2}{21} + \frac{5}{9}$

> **Helpful Hint:** Make sure you understand how to find the LCD. Then rewrite all fractions with the LCD as the denominator. Remember, you do not add the denominators of fractions.

Did you factor 21 into $3 \cdot 7$, and 9 into $3 \cdot 3$?
Yes ____ No ____

If you answered No, stop and complete this step.

Did you get $3 \cdot 3 \cdot 7$ or 63 as the LCD? Yes ____ No ____

If you answered No, review how to find the LCD of 21 and 9.

Did you multiply $\frac{2}{21}$ by $\frac{3}{3}$ and $\frac{5}{9}$ by $\frac{7}{7}$ to Rewrite the fractions? Yes ____ No ____

If you answered No, review how to write equivalent fractions using 63 as the LCD. Now go back and rework the problem using these suggestions.

Following the Order of Operations with Fractions—Problem 19 $\left(\frac{2}{3}\right)^2+\frac{1}{2}\cdot\frac{1}{4}$

Helpful Hint: Write out the rule for the order of operations and refer to it as you complete each step. Then be sure to write down each step. Skipping steps often leads to errors.

Did you square both 2 and 3 in $\left(\frac{2}{3}\right)^2$ as your first step?

Yes ____ No ____

Did you multiply $\frac{1}{2}\cdot\frac{1}{4}$ as your next step?

Yes ____ No ____

If you answered No to either question, go back and make these corrections.

Did you rewrite $\frac{4}{9}$ and $\frac{1}{8}$ with the LCD 72 as the denominator and then add? Yes ____ No ____

If you answered No, stop and consider why this step must occur before adding the final two fractions.

If you answered Problem 22 incorrectly, go back and rework the problem using these suggestions.

Solving Applied Problems Involving Fractions—Problem 28 Anna wishes to build 2 bookcases, each with 5 shelves. Each shelf is $3\frac{1}{2}$ feet long. The wood for the shelves is sold in 10-foot boards. How many boards does Anna need to buy for the shelves?

Helpful Hint: Use the Mathematics Blueprint for Problem Solving to help organize your work. Be sure to change mixed numbers to improper fractions before doing any other steps. Then change any division to multiplication by inverting the second fraction.

Did you realize that the problem requires you to perform the calculation 10 divided by $3\frac{1}{2}$? Yes ____ No ____

If you answered No, draw a diagram to help you better understand the problem.

Did you change $3\frac{1}{2}$ to $\frac{7}{2}$ and then rewrite the division as $10\cdot\frac{2}{7}$? Yes ____ No ____

If you answered No to either step, stop and perform these calculations.

Did you think that your calculations were now complete?
Yes ____ No ____

If you answered Yes, go back and read the problem again. Your answer refers to the number of shelves that can be made out of one 10-foot board. You still need to find out how many 10-foot boards are needed to make 10 shelves, each with a length of $3\frac{1}{2}$ feet. Make sure that you answer the question asked in the problem.

Now go back and rework the problem using these suggestions.

Name: ______________________________ Date: ________________
Instructor: ___________________________ Topic: ________________

Module 6 Decimals and Percents
Topic 1 Understanding Decimal Fractions

Vocabulary
decimal point • place-value chart • decimal fraction • rounding

1. The "." in the decimal .3 is called a ___________.

2. The rule for ___________ decimals is similar to the rule for whole numbers.

3. A ___________ is helpful in understanding the meaning of decimal fractions as well as how to write decimal fractions in different forms.

4. A ___________ is a fraction whose denominator is a power of 10.

Example	Student Practice
1. Write a word name for the decimal 0.561. The place value of the last digit is the last word in the word name. The last digit is 1 and is in the thousandths place. We do not include 0 as part of the word name. thousandths place ↓ 0.56[1] Five hundred sixty-one thousandths	**2.** Write a word name for each decimal. **(a)** 0.734 **(b)** 6.23
3. Write the word name for a check written to Shandell Strong for \$126.87. One hundred twenty-six and $\frac{87}{100}$ Dollars	**4.** Write the word name for a check written to Spin-to-Win Sprinklers for \$343.51. ______________________ Dollars

Vocabulary Answers: 1. decimal point 2. rounding 3. place-value chart 4. decimal fraction

Example	Student Practice
5. Write 0.86132 using fractional notation. Do not simplify. We do not need to write 0 as part of the fraction. $0.86132=\frac{86,132}{100,000}$ Notice that the decimal has 5 decimal places and the power of ten in the denominator has 5 zeros.	**6.** Write 1.4928 using fractional notation. Do not simplify.
7. Write $7\frac{56}{1000}$ as a decimal. Since there are 3 zeros, we move the decimal point 3 places to the left. $7\frac{56}{1000}=7\frac{\overleftarrow{056}}{\cancel{1000}}=7.056$ Note that we had to insert a 0 before 56 so we could move the decimal point 3 places to the left.	**8.** Write $1\frac{23}{1000}$ as a decimal.
9. Replace the ? with $<$ or $>$. 0.24 ? 0.244 Add a zero to 0.24 so that both decimal parts have the same number of digits. 0.240 ? 0.244 The tenths digits and the hundredths digits are equal. The thousandths digits differ. Since $0<4$, $0.240<0.244$.	**10.** Replace the ? with $<$ or $>$. 0.236 ? 0.23

Example	Student Practice
11. Round 237.8435 to the nearest hundredth. The round-off place digit is in the hundredths place. 204.8[4]35 The digit to the right of the round-off place digit is less than 5, so we do not change the round-off place digit. Drop all digits to the right of the round-off place digit. 237.84 237.8435 rounded to the nearest hundredth is 237.84.	**12.** Round 34.05482 to the nearest thousandth.
13. Round to the nearest hundredth. Alex and Lisa used 204.9954 kilowatt-hours of electricity in their house in June. We locate the digit in the hundredths place. 204.9[9]54 Since the digit to the right of 9 is 5, we increase 9 to 10 by changing 9 to 0. Then we must increase the 9 in the tenths place by 1, followed by increasing 4 to 5. 205.00 Note that we must include the two zeros after the decimal point because we were asked to round to the nearest hundredth. Thus, 204.9954 rounds to 205.00 kilowatt-hours.	**14.** Round to the nearest hundredth. In November, Cassie and Roger used 150.9971 kilowatt-hours of electricity in their house.

Extra Practice

1. Write a word name for the decimal 3.48.

2. Write the fraction $19\frac{243}{1000}$ as a decimal.

3. Replace the ? with < or >. 0.35 ? 0.355

4. Round 3.1415 to the nearest thousandth.

Concept Check

Explain how you know how many zeros to put in the denominator of your answer when you write 8.6711 as a fraction.

Name: ______________________________ Date: ________________
Instructor: ____________________________ Topic: ________________

Module 6 Decimals and Percents
Topic 2 Adding and Subtracting Decimal Expressions

Vocabulary
line up • evaluate • combine • estimate

1. To ___________ an expression, replace the variable with the given number and simplify.

2. The first step to add or subtract decimals is to write the numbers vertically and ___________ the decimal points.

3. We can ___________ a sum or difference of decimals by rounding each decimal to the nearest whole number.

4. To ___________ like terms we add coefficients of the like terms and the variable part stays the same.

Example	Student Practice
1. Add $40+8.77+0.9$. We can write any whole number as a decimal by placing a decimal point at the end of the number: $40 = 40.$ Line up the decimal points and add zeros so that each number has the same number of decimal places. $\begin{array}{r} \overset{1}{4}0.00 \\ 8.77 \\ +\,0.90 \\ \hline 49.67 \end{array}$	**2.** Add $28+3.87+0.2$.
3. Subtract $19.02-8.6$. $\begin{array}{r} 1\overset{8}{\not 9}.\overset{10}{\not 0}2 \\ -8.60 \\ \hline 10.42 \end{array}$	**4.** Subtract $32.06-4.2$.

Vocabulary Answers: 1. evaluate 2. line up 3. estimate 4. combine

Example	Student Practice
5. Perform the operation indicated. $-9.79-(-0.68)$ To subtract, we add the opposite of the second number. $-9.79-(-0.68)$ $-9.79+(0.68)$ Next, to add numbers with different signs, we keep the sign of the larger absolute value and subtract. -9.79 $\underline{0.68}$ -9.11 The answer is negative since $\lvert -9.79 \rvert$ is larger than $\lvert 0.68 \rvert$.	**6.** Perform the operation indicated. $-2.64-(-10.23)$
7. Combine like terms. $11.2x+3.6x-7.1y$ $11.2x$ and $3.6x$ are like terms, so we line up the decimal points and add them. $11.2x$ $\underline{+3.6x}$ $14.8x$ We have $11.2x+3.6x-7.1y$ $=14.8x-7.1y.$ We cannot combine $14.8x$ and $7.1y$ since they are not like terms.	**8.** Combine like terms. $2.3b+7.4b-8.2a$

Example	Student Practice
9. Evaluate $x + 3.12$ for $x = 0.11$. We replace the variable with 0.11. $x + 3.12 = 0.11 + 3.12$ Next we line up decimal points and add. $\begin{array}{r} 3.12 \\ +\,0.11 \\ \hline 3.23 \end{array}$	**10.** Evaluate $y + 5.89$ for $y = -7.21$.
11. Julie runs on her treadmill every day, She wants to run approximately 25 miles each week to prepare for a track race. She logged the distance she ran each day this week on the following chart. Estimate the total number of miles Julie ran this week. **Monday** 2.13 mi; **Tuesday** 2.79 mi; **Wednesday** 2.9 mi; **Thursday** 3.11 mi; **Friday** 3.8 mi; **Saturday** 4.12 mi; **Sunday** 4.9 mi We round each decimal to the nearest whole number and then add. $2 + 3 + 3 + 3 + 4 + 4 + 5 = 24$ miles Julie ran approximately 24 miles.	**12.** Antonio has kept track of how much he spends at the Laundromat for his last 5 trips. The total amount of each trip is as follows: \$9.75, \$11.25, \$10.50, \$10.75, \$8.75. Estimate the total amount he spent at the Laundromat over the past 5 trips.

Example	Student Practice

13. The table compares some of the top players' average season statistics.

MVP	Points	Rebounds
1. Allen Iverson	31.1	3.1
2. Tim Duncan	25.5	12.7
3. Kobe Bryant	28.3	6.3
4. LeBron James	29.7	7.3

Source: Orange County Register

(a) How many more points did Iverson average than Bryant?

$$\begin{array}{lr} \text{Allen Iverson's points} & \overset{2}{\cancel{3}}\overset{\overset{10}{\cancel{0}}}{\cancel{1}}.\overset{11}{\cancel{1}} \\ \text{Kobe Bryant's points} & \underline{-28.3} \\ & 2.8 \end{array}$$

(b) Find the total of the average number of rebounds made by Iverson, Duncan, and James.

Iverson	3.1
Duncan	12.7
James	7.3
	23.1

14. Use the table in example **13** to answer the following.

(a) How many more rebounds did Duncan average than James?

(b) Find the total of the average number of points made by all four players listed on the chart.

Extra Practice

1. Add $7.334 + 21.04$.

2. Subtract $-12.6 - (-6.2)$.

3. Combine like terms. $10.02x - 7.39x + 3.4y$

4. Evaluate $x - 0.97$ for $x = 13.23$.

Concept Check

Explain how you would evaluate $x - 3.1$ for $x = 0.866$.

Name: ______________________________ Date: ________________
Instructor: ____________________________ Topic: _______________

Module 6 Decimals and Percents
Topic 3 Multiplying and Dividing Decimal Expressions

Vocabulary
repeating decimals • positive • negative • dividend

1. To divide a decimal by a whole number, we place the decimal point in the quotient directly above the decimal point in the ___________.

2. Decimals that have a digit, or group of digits, that repeats are called ___________.

3. To multiply or divide positive and negative decimals, the sign of the answer will be ___________ if the problem has an odd number of negative signs.

4. To multiply or divide positive and negative decimals, the sign of the answer will be ___________ if the problem has an even number of negative signs.

Example	Student Practice
1. Multiply 5.33×7.2. We write the multiplication just as we would if there were no decimal points. We need 3 decimal places since there are 2 in the first factor and 1 in the second. $\begin{array}{r} 5.33 \\ \times 7.2 \\ \hline 1066 \\ 3731 \\ \hline 38.376 \end{array}$	**2.** Multiply 34.1×2.78.
3. Multiply $(-2)(4.51)$. The number of negative signs, 1, is odd so the product is negative. $\begin{array}{r} 4.51 \\ \times(-2) \\ \hline -9.02 \end{array}$	**4.** Multiply $(-5)(7.93)$.

Vocabulary Answers: 1. dividend 2. repeating decimals 3. negative 4. positive

Example	Student Practice
5. Multiply 0.2345×1000. Since 1000 has three zeros, we move the decimal point to the right three places. $0.2345 \times 1000 = 234.5$	**6.** Multiply $(0.4256)(10^6)$.
7. Divide $2.3 \div 5$. Place the decimal point directly above the decimal point in the dividend and divide as if there were no decimal point. Add zeros as needed. $2.3 \div 5 \rightarrow 5\overline{)2.30}$ = 0.46 $\underline{2\,0}$ 30 $\underline{30}$ 0	**8.** Divide $3.6 \div 8$.
9. Divide $-0.185 \div 13$. Round your answer to the nearest thousandth. We must divide one place beyond the thousandths place, to the ten thousandths place, so we can round to the nearest thousandth. $13\overline{)-0.1850}$ = $-.0142$ $\underline{13}$ 55 $\underline{52}$ 30 $\underline{26}$ 4 $-0.185 \div 13 \approx -0.014$	**10.** Divide $-0.8469 \div 17$. Round your answer to the nearest thousandth.

Example	Student Practice
11. Divide $6.93 \div 2.2$. Since the divisor, 2.2, is not a whole number, multiply the dividend and quotient by 10 and the divisor becomes a whole number. $\frac{(6.93)(10)}{(2.2)(10)} = \frac{69.3}{22} = 69.3 \div 22$ $\begin{array}{r} 3.15 \\ 22\overline{)69.30} \\ \underline{66} \\ 33 \\ \underline{22} \\ 110 \\ \underline{110} \\ 0 \end{array}$ $6.93 \div 2.2 = 3.15$	**12.** Divide $10.92 \div 2.4$.
13. Divide $0.7 \div 1.5$. The divisor is not a whole number, so move the decimal point one place to the right. $\begin{array}{r} 0.466 \\ 15\overline{)7.000} \\ \underline{6\,0} \\ 1\,00 \\ \underline{90} \\ 100 \\ \underline{90} \\ 10 \end{array}$ Thus, $0.7 \div 1.5 = 0.4\overline{6}$ because if we continued dividing, the 6 would repeat.	**14.** Divide $1.3 \div 1.6$.

Example	Student Practice
15. Write as a decimal. $5\frac{7}{11}$	**16.** Write as a decimal. $3\frac{7}{12}$

$5\frac{7}{11}$ means $5+\frac{7}{11}$, so divide 7 by 11.

$$\begin{array}{r} 0.6363 \\ 11\overline{)7.0000} \\ \underline{66} \\ 40 \\ \underline{33} \\ 70 \\ \underline{66} \\ 40 \\ \underline{33} \end{array}$$

Since the pattern of 40 minus 33 repeats, $\frac{7}{11}=0.\overline{63}$ and $5\frac{7}{11}=5.\overline{63}$.

Extra Practice

1. Multiply $(15.23)(-3)$.

2. Multiply 2.3681×1000.

3. Divide $13.5525\div4.17$.

4. Write $11\frac{4}{15}$ as a decimal. Round to the nearest hundredth when necessary. If a repeating decimal is obtained, use proper notation such as $0.\overline{3}$.

Concept Check

Marc multiplied 0.097×0.5 and obtained the answer 0.485. Is Marc's answer correct? Why or why not?

Name: ______________________________ Date: ________________
Instructor: ____________________________ Topic: ________________

Module 6 Decimals and Percents
Topic 4 Estimating with Percents

Vocabulary
percents • last digit • 15% • 20%

1. To estimate 10% of a whole number, delete the ___________.

2. ___________ can be described as ratios whose denominators are 100.

3. To estimate ___________, double 10%.

4. To estimate ___________, add 10% and 5%

Example	Student Practice
1. For the number 59,040, estimate the following.	**2.** For the number 74,605, estimate the following.
(a) 10%	**(a)** 10%
To estimate 10% of 59,040, first we round to the nearest thousand, then we delete the last digit.	
$59{,}040 \to 59{,}000$	
$59{,}00\not{0}$	
10% of $59{,}040 \approx 5900$	
(b) 1%	**(b)** 1%
To estimate 1% of 59,040, first we round to the nearest thousand, then we delete the last two digits.	
$59{,}040 \to 59{,}000$	
$59{,}0\not{0}\not{0}$	
1% of $59{,}040 \approx 590$	

Vocabulary Answers: 1. last digit 2. percents 3. 20% 4. 15%

Example	Student Practice
3. For the number 1205, estimate the following. **(a)** 5% of 1205 We round 1205 to the nearest hundred: 1200. To find 5%, we find $\frac{1}{2}\times 10\%$ of the number. $\frac{1}{2}\times(10\% \text{ of } 1200)=\frac{1}{2}\times 120=60$ **(b)** 15% of 1205 To find 15%, we add $(10\% \text{ of } 1200)+(5\% \text{ of } 1200)$. $15\% \text{ of } 1200=120+60=180$ **(c)** 6% of 1205 To find 6% we add: $(5\% \text{ of } 1200)+(1\% \text{ of } 1200)$. $6\% \text{ of } 1200=60+12=72$	**4.** For the number 3206, estimate the following. **(a)** 10% of 3206 **(b)** 20% of 3206 **(c)** 11% of 3206
5. Loren would like to leave a 15% tip for her dinner, If the total bill at a restaurant is \$20.76, estimate the tip Loren should leave. We round \$20.76 to the nearest ten: \$20. $15\% \text{ of } 20=(10\% \text{ of } 20)+(5\% \text{ of } 20)$ $=2+1$ $=\$3$	**6.** Doug would like to leave a 20% tip for his breakfast. If the total bill at a restaurant is \$11.04, estimate the tip that Doug should leave.

Example	Student Practice
7. Everything in a store is on sale for 30% off the original price. The discount is calculated at the cash register at the time of purchase. Josh buys 2 shirts priced at \$19.95 each, one pair of pants priced at \$28.00, and a pair of shoes priced at \$39.99	**8.** Josh bought 3 shirts, 3 pairs of pants, and 2 pairs of shoes from the same store in example **7**.
(a) Round the original price of each item to the nearest ten and then estimate the total cost of the items before the discount. Shirt: $\$19.95 \rightarrow \$20$ Pants: $\$28 \rightarrow \$30$ Shoes: $\$39.99 \rightarrow \$40$ We add these amounts: $\$20+\$20+\$30+\$40=\$110$.	**(a)** Round the original price of each item to the nearest ten and then estimate the total cost of the items before the discount.
(b) Estimate the amount of the 30% discount. Round the discount to the nearest ten. 10% of $\$110=\$11$ 30% of $\$110=\$11+\$11+\$11$ $=\$33$ The estimated discount rounded to the nearest ten is \$30.	**(b)** Estimate the amount of the 30% discount. Round the discount to the nearest ten.
(c) Estimate the total cost of the items after the discount is taken. We subtract the total estimated cost minus the estimated discount. $\$110-\$30=\$80$ The estimated cost of the items after the discount is \$80.	**(c)** Estimate the total cost of the items after the discount is taken.

Extra Practice

1. For the number 407, estimate 30%.

2. For the number 3006, estimate 8%.

3. For the number 420,070, estimate 4%.

4. Matt and Andrea paid the realtor 7% commission on the sale price of their home. If they sold their home for $205,000, estimate how much commission was paid to the realtor.

Concept Check

We can estimate 35% of 200 by finding 3 times 10% of 200, then adding $\frac{1}{2}$ times 10% of 200. Explain two other ways you can estimate 35% of 200.

Name: ______________________________ Date: ________________
Instructor: ____________________________ Topic: _______________

Chapter 6 Decimals and Percents
Topic 5 Percents

Vocabulary
percents • left • right • whole

1. To write a decimal as a percent, move the decimal point two places to the ___________.

2. ___________ can be described as ratios whose denominators are 100.

3. Decimals, fractions, or percents are used to describe parts of a ___________.

4. To write a percent as a decimal, move the decimal point two places to the ___________.

Example	Student Practice
1. State using percents. 13 out of 100 radios are defective. $\frac{13}{100} = 13\%$ 13% of the radios are defective.	**2.** State using percents. 3 out of 100 calculators are defective.
3. Last year's attendance at a school's winter formal was 100 students. This year the attendance was 121. Write this year's attendance as a percent of last year's. We must write this year's attendance (121) as a percent of last year's (100). $\frac{\text{This year's attendance} \rightarrow 121}{\text{Last year's attendance} \rightarrow 100} = 121\%$ This year's attendance at the formal was 121% of last year's. Note there is 121 parts out of 100 parts. This means we have more than one whole amount and thus more than 100%.	**4.** Last year, attendance at a town meeting was 100 residents. This year the attendance was 134. Write this year's attendance as a percent of last year's.

Vocabulary Answers: 1. right 2. percents 3. whole 4. left

Example	Student Practice

5. There are 100 milliliters (mL) of a solution in a container, Sara takes 0.3 mL of the solution. What percentage of the solution does Sara take?

$\frac{0.3}{100} = 0.3\%$ of the solution

6. There are 100 mL of solution in a container. Evan takes 0.5 mL of the solution. What percentage of the solution does Evan take?

7. (a) Write 3.8% as a decimal.

We move left on the chart, so the decimal point moves 2 places left.

Decimal ⟵ Percent

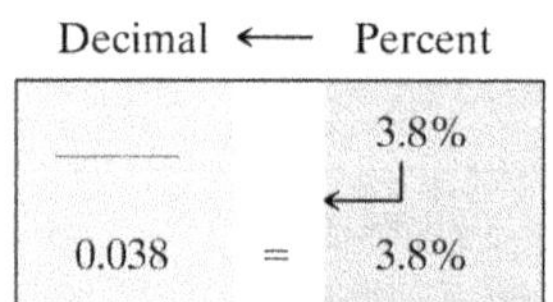

(b) Write 0.009 as a percent.

We move right on the chart, so the decimal point moves 2 places right.

Decimal ⟶ Percent

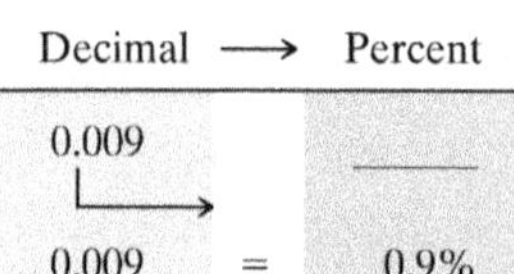

8. (a) Write 1.5% as a decimal.

(b) Write 0.008 as a percent.

9. Complete the table of equivalent notations.

Decimal Form	Percent Form
0.457	
	58.2%
	0.6%
2.9	

Decimal Form	Percent Form
0.457	45.7%
0.582	58.2%
0.006	0.6%
2.9	290%

10. Complete the table of equivalent notations.

Decimal Form	Percent Form
0.423	
	64.4%
	0.3%
90.1	

Example	Student Practice
11. **(a)** Write $\frac{211}{500}$ as a percent.	**12.** **(a)** Write $\frac{67}{125}$ as a percent.

Compute: $211 \div 500 = 0.422$, then move the decimal point 2 places to the right.

Fraction ⟶ Decimal ⟶ Percent

$\frac{211}{500}$	⟶	0.422		$\frac{?}{\ }$
$\frac{211}{500}$	⟶	0.422	⟶	42.2%

(b) Write 42.2% as a fraction.

Move the decimal point 2 places to the left. Note that $0.422 = \frac{422}{1000} = \frac{211}{500}$.

Fraction ⟵ Decimal ⟵ Percent

$\frac{?}{\ }$		0.422	⟵	42.2%
$\frac{211}{500}$	⟵	0.422	⟵	42.2%

12. **(b)** Write 34.8% as a fraction.

13. Write $\frac{5}{9}$ as a percent. Round to the nearest hundredth of a percent.

First we change $\frac{5}{9}$ to a decimal:

$5 \div 9 = 0.55555\ldots$. We must carry out the division at least five places beyond the decimal point so that we can move the decimal point to the right two places, and we then round to the nearest hundredth of a percent.

$0.55555\ldots = 55.555\ldots\% \approx 55.56\%$

Remember that if you are not asked to round, $\frac{5}{9} = 55.\bar{5}\%$.

14. Write $\frac{2}{5}\%$ as a fraction.

Extra Practice

1. Write 245% as a decimal.

2. Write 0.07 as a percent.

3. Write $5\frac{1}{3}$ as a percent. Round to the nearest hundredth of a percent.

4. Write $\frac{1}{8}$% as a fraction.

Concept Check

Explain how you would change 0.43% to a decimal, then to a fraction.

Name: ______________________________ Date: ________________
Instructor: ___________________________ Topic: ________________

Module 6 Decimals and Percents
Topic 6 Solving Percent Problems Using Equations

Vocabulary

base • amount • percent • of

1. We can write the relationship $\frac{\text{amount}}{\text{base}} = \text{percent}$ as amount= ____________ ×base.

2. The part being compared to the base is called the ____________.

3. When we translate the statement into symbols, we replace "____________" with ×.

4. The entire quantity is called the ____________.

Example	**Student Practice**
1. Translate into an equation and solve.	**2.** Translate into an equation and solve.
(a) What is 25% of 40? $n = 25\% \times 40$ $n = 0.25 \times 40$ $n = 10$	**(a)** What is 30% of 30?
(b) 10 is 25% of what number? $10 = 25\% \times n$ $10 = 0.25 \times n$ $\frac{10}{0.25} = n$ $40 = n$	**(b)** 9 is 30% of what number?
(c) 10 is what percent of 40? $10 = n\% \times 40$ $\frac{10}{40} = n\%$ $0.25 = n\%$ $25 = n$	**(c)** 9 is what percent of 30?

Vocabulary Answers: 1. percent 2. amount 3. degrees 4. base

Example	Student Practice
3. Translate into an equation and solve. 50 is what percent of 40? We should expect to get more than 100% since 50 is more than the base 40. 50 is what percent of 40? $50 = n\% \times 40$ $\frac{50}{40} = n\%$ $1.25 = n\%$ $125 = n$ 50 is 125% of 40.	**4.** Translate into an equation and solve. 80 is what percent of 50?
5. Find 55% of 36. $n = 55\% \times 36$ $= 0.55 \times 36$ $= 19.8$ 19.8 is 55% of 36.	**6.** Find 48% of 87.
7. Marilyn has 850 out of 1000 points possible in her English class. What percent of the total points does Marilyn have? We must find the percent, so we write the statement that represents the percent situation. 850 is what percent of 1000? $850 = n\% \times 1000$ $\frac{850}{1000} = n\%$ $0.85 = n\%$ $85 = n$ Marilyn has 85% of the total points.	**8.** Hank earned 65 out of 80 points on a test. What percentage of total points did he earn?

Example	Student Practice
9. Sean's bill for his dinner at the Spaghetti House was \$19.75. How much should he leave for a 15% tip? Round this amount to the nearest cent. We must find the amount, that is, the part of the base of \$19.75. What is 15% of \$19.75? $n = 15\% \times \$19.75$ $= 0.15 \times \$19.75$ $= \$2.9625$ $\approx \$2.96$ The tip is \$2.96	**10.** Johanna left a \$3 tip for her dinner, which cost \$17.89. What percent of the total bill did Johanna leave for a tip? Round your answer to the nearest hundredth of a percent.
11. Sergio stayed in a luxury hotel on a Saturday night and paid \$230 for that night. If the rate on Saturday night is 15% higher than it is on Sunday night, how much will Sergio pay for the room on Sunday night? Read the problem carefully and create a Mathematics Blueprint. Let x = the room rate on Sunday. The room rate on Sunday plus 15% of the Sunday rate equals the room rate on Saturday. $100\%x + 15\%x = \$230$ $115\%x = 230$ $1.15x = 230$ $x = \frac{230}{1.15}$ $x = 200$ Sergio will pay a rate of \$200 to stay Sunday night.	**12.** Refer to example **11** to answer the following. If the rate on Saturday night is 25% higher than it is on Sunday night, how much will Sergio pay to stay Sunday?

Extra Practice

1. Translate into an equation and solve. What is 83% of 155?

2. Translate into an equation and solve. 240 is 80% of what number?

3. Translate into an equation and solve. 280 is what percent of 70?

4. Julie's bill for dinner at the Seafood Hut was \$34.75. How much should she leave for a 15% tip? Round the amount to the nearest cent.

Concept Check

The owner of M&R Windows determined that 0.8% of the products ordered from the manufacturer are defective. Explain how you would determine how many windows the owner should expect to be defective in a shipment from the manufacturer of 375 windows.

Name: ______________________ Date: ______________

Instructor: ______________________ Topic: ______________

Module 6 Decimals and Percents
Topic 7 Solving Percent Problems Using Proportions

Vocabulary

base • percent • amount • variables

1. The __________ is the part being compared to the whole.

2. The letters *a*, *b*, and *p* to represent amount, base, and percent are called __________.

3. The __________ is the entire quantity or total involved.

4. Usually, the easiest part to identify is the __________.

Example	Student Practice
1. Identify the percent number p.	**2.** Identify the percent number p.
(a) Find 15% of 360.	**(a)** Find 20% of 369.
The value of p is 15.	
(b) 28% of what is 25?	**(b)** 34% of what is 54?
The value of p is 28.	
(c) What percent of 18 is 4.5?	**(c)** What percent of 15 is 3.5?
The value of p is unknown.	
3. Identify the base b and the amount a.	**4.** Identify the base b and the amount a.
(a) 25% of 520 is 130.	**(a)** 20% of 42 is 8.4.
The base is the entire quantity. $b = 520$. The amount is the part compared to the whole, $a = 130$.	
(b) 19 is 50% of what?	**(b)** 35 is 40% of what?
The amount 19 is the part of the base. The base is unknown.	

Vocabulary Answers: 1. amount 2. variables 3. base 4. percent

Example	Student Practice
5. Find the percent p, base b, and amount a. **(a)** What is 77% of 210? The amount is unknown. The value of p is 77. The base usually follows the word "of." Here, $b = 210$. **(b)** What percent of 21 is 17? The value of p is not known. The base usually follows the word "of." Here, $b = 21$. The amount is 17. $a = 17$.	**6.** Find the percent p, base b, and amount a. **(a)** What is 23% of 524? **(b)** What percent of 4 is 37?
7. Find 260% of 40. The percent $p = 260$. The number that is the base usually appears after the word "of." The base $b = 40$. The amount is unknown. We use the variable *a*. Thus, $\frac{a}{b} = \frac{p}{100}$ becomes $\frac{a}{40} = \frac{260}{100}$. If we simplify the fraction on the right-hand side, we have the following. $\frac{a}{40} = \frac{13}{5}$ $5a = (40)(13)$ $5a = 520$ $\frac{5a}{5} = \frac{520}{5}$ $a = 104$ Thus 260% of 40 is 104.	**8.** Find 540% of 50.

Example	Student Practice
9. 65% of what is 195?	**10.** 76% of what is 228?
11. 19 is what percent of 95?	**12.** 175 is what percent of 140?

9. 65% of what is 195?

The percent $p = 65$. The base is unknown. We use the variable b. The amount a is 195. Thus,

$\frac{a}{b} = \frac{p}{100}$ becomes $\frac{195}{b} = \frac{65}{100}$.

If we simplify the fraction on the right-hand side, we have the following.

$$\frac{195}{b} = \frac{13}{20}$$
$$(20)(195) = 13b$$
$$3900 = 13b$$
$$\frac{3900}{13} = \frac{13b}{13}$$
$$300 = b$$

Thus 65% of 300 is 195.

10. 76% of what is 228?

11. 19 is what percent of 95?

The percent is unknown. We use the variable p. The base $b = 95$. The amount $a = 19$. Thus,

$\frac{a}{b} = \frac{p}{100}$ becomes $\frac{19}{95} = \frac{p}{100}$.

Cross-multiplying, we have the following.

$$(100)(19) = 95p$$
$$1900 = 95p$$
$$\frac{1900}{95} = \frac{95p}{95}$$
$$20 = p$$

12. 175 is what percent of 140?

Example	Student Practice
13. Sonia has \$29.75 deducted from her weekly salary of \$425 for a retirement plan. What percent of Sonia's salary is withheld for the retirement plan?	**14.** Vincent has \$84 deducted from his bi-weekly salary of \$1050 for taxes. What percent of Vincent's salary is withheld for taxes?

We must find the percent p. The base $b = 425$. The amount $a = 29.75$. Thus,

$$\frac{a}{b} = \frac{p}{100} \text{ becomes } \frac{29.75}{425} = \frac{p}{100}.$$

When we cross-multiply, we obtain the following.

$$100(29.75) = 425p$$
$$2975 = 425p$$
$$\frac{2975}{425} = \frac{425p}{425}$$
$$7 = p$$

We see that 7% of Sonia's salary is deducted for the retirement plan.

Extra Practice

1. Identify the percent p, base b, and amount a. Do not solve for the unknown.
   What is 88% of 198?

2. Find 0.02% of 950.

3. 1800 is 225% of what?

4. 96 is what percent of 240?

Concept Check

In the following percent proportion, what can you say about the percent number if the value of the amount is larger than the base? $\frac{\text{amount}}{\text{base}} = \frac{\text{percent number}}{100}$

Name: ______________________________ Date: ________________
Instructor: ___________________________ Topic: ________________

Module 6 Decimals and Percents
Topic 8 Solving Applied Problems Involving Percents

Vocabulary
commission • commission rate • interest • principal • interest rate • time

1. ____________ is the percent used in computing the interest.

2. ____________ is the amount deposited or borrowed.

3. ____________ is the money earned or paid for the use of money.

4. ____________ is when a person's earnings are a certain percentage of the sales he/she makes.

Example	Student Practice
1. Alex is a car salesman and earns a commission rate of 9% of the price of each car he sells. If he earned \$3150 commission this month, what were his total sales for the month? $\text{commision} = \text{commision rate} \times \text{total sales}$ $\$3150 = 9\% \times n$ $\$3150 = 0.09 \times n$ $\frac{\$3150}{0.09} = \frac{0.09n}{0.09}$ $\$35,000 = n$ Alex's total sales were \$35,000. We will estimate to check our answer. We round 9% to 10% and then verify that 10% of his total sales (\$35,000) is approximately his commission (\$3150). 10% of $\$35,000 = \$3500$, which is close to his commission of \$3150.	**2.** Leslie is a real estate agent with a 7% commission rate. If Leslie sells a home for \$274,000, what amount of commission will she earn?

Vocabulary Answers: 1. interest rate 2. principal 3. interest 4. commission

Example	Student Practice
3. The enrollment at Laird Elementary School was 450 students in 2011. In 2012 the enrollment decreased by 36 students. What was the percent decrease? Decrease of 36 students is what percent of 450? decrease = percent decrease × original amount $36 = n\% \times 450$ $\frac{36}{450} = n\%$ $0.08 = n\%$ $8 = n$ The enrollment decreased by 8%.	**4.** A suit is on sale for $75.25 off the original price of $215. By what percent is the price of the suit reduced?
5. Arnold earned $26,000 a year and received a 6% raise. How much is his new yearly salary? First, we find the amount of the raise. percent increase × original amount = increase (raise) $6\% \times 26{,}000 = 0.06 \times 26{,}000 = \$1560$ Now we find his new yearly salary. original salary + raise = new salary $\$26{,}000 + 1560 = \$27{,}560$ Arnold's new salary is $27,560. Check. 6% is a little more than $\frac{1}{2}$ of 10%. Use this fact to check the answer.	**6.** Penny earned $37,000 a year and received a 4% raise. How much is her new yearly salary?

Example	Student Practice
7. An advertisement states that all items in a department store are reduced 30% off the original list price. What is the sale price of a flat-screen television set with a list price of \$2700? First, we find the amount of the discount. $\text{percent decrease} \times \text{original amount}$ $= \text{decrease}$ $30\% \times 2700 = 0.30 \times 2700 = \$810$ Next we find the sale price. $\text{original price} - \text{discount} = \text{sale price}$ $\$2700 - \$810 = \$1890$ The sale price is \$1890.	**8.** 1300 people voted in the yearly town election last year. This year the number of voters dropped by 23%. How many people voted in the town election this year?
9. Larsen borrowed \$9400 from the bank at a simple interest rate of 13%. **(a)** Find the interest on the loan for 1 year. $P = \text{principal} = \$9400$ $R = \text{rate} = 13\%$ $T = \text{time} = 1 \text{ year}$ $I = P \times R \times T$ $I = \$9400 \times 0.13 \times 1 = \$1222$ The interest for 1 year is \$1222. **(b)** How much does Larsen pay back at the end of the year when he pays off the loan? $\$9400 + 1222 = \$10,622$	**10.** Julia put \$1500 in a savings account that pays 4% simple interest per year. **(a)** How much interest will Julia earn in 2 years? **(b)** How much money will be in Julia's savings account at the end of two years?

Example	Student Practice
11. Find the interest on a loan of \$2500 that is borrowed at a simple interest rate of 9% for 3 months.	**12.** Find the interest on a loan of \$2000 that is borrowed at a simple interest rate of 13% for 6 months.

We must change 3 months to years since the formula requires that the time be in years: $T = 3 \text{ months} = \frac{3}{12} = \frac{1}{4}$ year.

$I = P \times R \times T$

$I = 2500 \times 0.09 \times \frac{1}{4} = 225 \times \frac{1}{4} = 56.25$

The interest for 3 months is \$56.25.

Extra Practice

1. You must pay annual property tax of 1.25% of the value of your home. If your home is worth \$140,000, how much must you pay in taxes each year?

2. Bill works as a phone solicitor and is paid an 8% commission on the amount of sales he makes. If Bill earned \$280 in commissions last week, what were his total sales for the week?

3. Elizabeth's salary last year was \$43,500. If she gets a 5.5% pay raise, what is her new salary?

4. Marta borrows \$7500 at a simple annual interest rate of 8%. Six months later, she repays the loan. How much interest does she pay on the loan?

Concept Check

Explain how to find simple interest on a loan of \$6500 at an annual rate of 9% for a period of 4 months.

MATH COACH

Mastering the skills you need to do well on the test.

Watch the MATH COACH videos in MyMathLab® or on YouTube™ while you work the problems below. These helpful hints will help you avoid making common errors on test problems.

Subtracting Decimal Expression—Problem 9

Perform the operation indicated. $18.8 - 6.23$

Helpful Hint: Add zeros at the end of each number, if necessary, so that the same number of digits appear to the right of each decimal point. Remember to line up the decimal points.

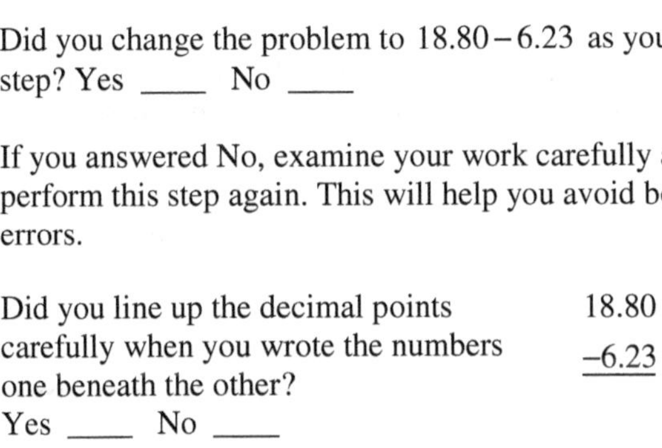

Did you change the problem to $18.80 - 6.23$ as your first step? Yes ____ No ____

If you answered No, examine your work carefully and perform this step again. This will help you avoid borrowing errors.

Did you line up the decimal points carefully when you wrote the numbers one beneath the other?
Yes ____ No ____

$$\begin{array}{r} 18.80 \\ \underline{-6.23} \end{array}$$

If you answered No, rewrite the problem on paper. Write out your steps and show your borrowing.

If you answered Problem 9 incorrectly, go back and rework the problem using these suggestions.

Dividing a Decimal by a Decimal—Problem 13

Perform the operation indicated. $15.75 \div 3.5$

Helpful Hint: First determine how many decimal places you must move the decimal point to the right in the divisor to make it a whole number. Then move the decimal point the *same number of places* to the right in the dividend. Be sure to place the decimal point in the quotient directly above the decimal point in the dividend.

Did you change 3.5 to 35, and 15.75 to 157.5 first?
Yes ____ No ____

If you answered No, stop and make this correction.

Did you remember to write the decimal point in the quotient directly above the decimal point in the dividend?
Yes ____ No ____

If you answered No, stop and make this correction.

When you performed the last step of your division, did you multiply 5×35 to obtain 175? Yes ____ No ____

If you answered No, examine your work and check each step for calculation errors.

Now go back and rework the problem using these suggestions.

Changing Between Fractions, Decimals, and Percents—Problem 21

Fill in the blanks. Complete the table of equivalent notations.

Fraction Form	Decimal Form	Percent Form
(a) ________	**(b)** ________	**(c)** 5%

Helpful Hint: For this problem, it is easier to convert the percent to a decimal first. To change a percent to a decimal, move the decimal point two places to the **left** and remove the % symbol. To change a decimal to a fraction, write the decimal part over a denominator that has a 1 and the same number of zeros as the number of decimal places.

Did you remember to write 5% as 5.0%, then move the decimal point two places to the left? Yes ____ No ____

If you answered No, go back and complete this step again. Remember that when you change a percent to a decimal, you are dividing that number by 100. That is why the decimal point moves two places to the left.

To change 0.05 to a fraction, did you determine that the denominator of the fraction must have a 1 and 2 zeros? Then did you write 5 as the numerator of the fraction?
Yes ____ No ____

If you answered No, reread the Helpful Hint regarding changing a decimal to a fraction.

Did you reduce your fraction to lowest terms?
Yes ____ No ____

If you answered Problem 21 incorrectly, go back and rework the problem using these suggestions.

Solving Applied Problems Involving Percents—Problem 28

A computer is reduced 24% from the original price of $3300.

(a) How much is the computer reduced in price?
(b) What is the sale price?

Helpful Hint: Write a simple percent statement in your own words that describes the situation in the applied problem.

For part **(a)**, were you able to write a statement such as "The discount is 24% of the original price of $3300?"
Yes ____ No ____

If you answered No, reread the problem and see if you can write a similar statement.

Were you able to write the equation $n = 0.24 \times 3300$ to find the answer to part **(a)**? Yes ____ No ____

If you answered No, check your calculations to be sure that you correctly changed 24% to a decimal.

For part **(b)**, once you know how much the computer is reduced in price, did you subtract as follows?
original price − discount = sale price
Yes ____ No ____

Remember to include the dollar sign ($) as the units for both answers.

Now go back and rework the problem using these suggestions.

Name: ______________________________ Date: ________________
Instructor: ___________________________ Topic: ________________

Module 7 Measurement and Geometric Figures
Topic 1 Using Unit Fractions with U.S. and Metric Units

Vocabulary
unit fraction • basic unit • left • right

1. In general, to change from a larger to a smaller metric unit, we move the decimal point to the ____________.

2. A(n) ____________ is a fraction that shows the relationship between units and is equal to 1.

3. The _____________ of length in the metric system is the meter.

4. In general, to change from a smaller to a larger metric unit, we move the decimal point to the ____________.

Example	Student Practice
1. Convert 35 yards to feet. We write the relationship between feet and yards as a unit fraction. Since $3\text{ ft} = 1\text{ yd}$, we have the unit fraction $\frac{3\text{ ft}}{1\text{ yd}}$. $35\text{ yd} = \underline{\ ?\ }\text{ ft}$ Now multiply by the unit fraction and then divide out the units "yd." $35\text{ yd}\times\frac{3\text{ ft}}{1\text{ yd}}$ $= 35\ \cancel{\text{yd}}\times\frac{3\text{ ft}}{1\ \cancel{\text{yd}}}$ $= 35\times 3\text{ ft} = 105\text{ ft}$	**2.** Convert 360 minutes to hours.

Vocabulary Answers: 1. right 2. unit fraction 3. mathematics blueprint 4. left

Example	Student Practice
3. Convert 560 quarts to gallons. We write the relationship between quarts and gallons: $4\text{ qt} = 1\text{ gal}$. We want to end up with gallons, so we write 1 gal in the numerator of the unit fraction: $\frac{1\text{ gal}}{4\text{ qt}}$. $560\text{ qt} = \underline{?}\text{ gal}$ Multiply by the appropriate unit fraction. Then divide out the units "qt." $560\text{ qt} \times \frac{1\text{ gal}}{4\text{ qt}}$ $= 560\ \cancel{\text{qt}} \times \frac{1\text{ gal}}{4\ \cancel{\text{qt}}}$ $= 560 \times \frac{1}{4}\text{ gal} = \frac{560\text{ gal}}{4} = 140\text{ gal}$	**4.** Convert 128 ounces to pounds.
5. The all-night garage charges \$1.50 per hour for parking both day and night. A businessman left his car there for $2\frac{1}{4}$ days. How much was he charged? Read the problem carefully and create a Mathematics Blueprint. We change $\frac{1}{4}$ to a decimal and change days to hours. $2\frac{1}{4}\text{ days} = 2.25\ \cancel{\text{days}} \times \frac{24\text{ hr}}{1\ \cancel{\text{day}}} = 54\text{ hr}$ Now find the total charge for parking, $54\ \cancel{\text{hr}} \times \frac{\$1.50}{1\ \cancel{\text{hr}}} = \$81$. Check by verifying that is your answer in the desired units.	**6.** A businesswoman parked her car at a garage for $1\frac{1}{4}$ days. The garage charges \$1.25 per hour. How much did she pay to park the car?

Example	Student Practice
7. Answer parts **(a)** and **(b)**.	**8.** Answer parts **(a)** and **(b)**.
(a) Change 7 kilometers to meters. To go from kilometer to meter (basic unit), we move 3 places to the right on the prefix chart, so we move the decimal point 3 places to the right. 7 km = 7.000 m = 7000 m	**(a)** Change 5 meters to centimeters.
(b) Change 30 liters to centiliters. To go from liter (basic unit) to centiliter, we move 2 places to the right on the prefix chart. Thus we move the decimal point 2 places to the right. 30 L = 30.00 cL = 3000 cL	**(b)** Change 40 centigrams to milligrams.
9. Answer parts **(a)** and **(b)**.	**10.** Answer parts **(a)** and **(b)**.
(a) Change 7 centigrams to grams. To go from centigrams to grams, we move 2 places to the left on the prefix chart. Thus we move the decimal point 2 places to the left. 7 cg = 0.07 g = 0.07 g	**(a)** Change 5 milliliters to liters.
(b) Change 56 millimeters to kilometers. To go from millimeters to kilometers, we move the decimal point 6 places to the left. 56 mm = 0.000056 km = 0.000056 km	**(b)** Change 49 centimeters to kilometers.

Example	Student Practice
11. A special cleaning fluid used to rinse test tubes in a chemistry lab costs \$40.00 per liter. What is the cost per milliliter?	**12.** A purified acid costs \$120 per liter. What does it cost per milliliter?

Read the problem carefully and create a Mathematics Blueprint.

Change liters to milliliters.

$1\text{ L} = 1000\text{ mL}$

Replace 1 L with 1000 mL.

$$\frac{\$40}{1\text{ L}} = \frac{\$40}{1000\text{ mL}}$$
$$= \$0.04 \text{ per mL}$$

Check your answer. A milliliter is a very small part of a liter. Therefore it should cost much less for 1 milliliter of fluid than it does for 1 liter. \$0.04 is much smaller than \$40.00, so our answer seems reasonable.

Extra Practice

1. Convert. 5280 feet = ? miles

2. Convert. 1410 minutes = ? hours

3. Andrew walked 3.2 kilometers from his house to his friend's house. How many meters did he walk?

4. Fill in the blanks with the correct values.

0.01 L = ? kL = ? mL

Concept Check

Explain how you would convert 240 ounces to pounds.

Name: ______________________________ Date: ________________

Instructor: ____________________________ Topic: ________________

Module 7 Measurement and Geometric Figures
Topic 2 Converting Between the U.S. and Metric Systems

Vocabulary

equivalent values • unit fraction • Celsius scale • Fahrenheit system

1. To convert from one unit to another we multiply by a(n) ____________ that is equivalent to 1.

2. In the metric system, temperature is measured on the ____________.

3. In the ____________ water boils at $212°$ $(212°F)$ and freezes at $32°$ $(32°F)$.

4. To convert between U.S. and metric units, it is necessary to know ____________.

Example	Student Practice
1. Answer parts **(a)** through **(d)**.	**2.** Answer parts **(a)** through **(d)**.
(a) Convert 26 m to yd. $26 \not{m} \times \frac{1.09 \text{ yd}}{1 \not{m}} = 28.34 \text{ yd}$	**(a)** Convert 9 ft to m.
(b) Convert 1.9 km to mi. $1.9 \not{km} \times \frac{0.62 \text{ mi}}{1 \not{km}} = 1.178 \text{ mi}$	**(b)** Convert 17 qt to L.
(c) Convert 2.5 L to qt. $2.5 \not{L} \times \frac{1.06 \text{ qt}}{1 \not{L}} = 2.65 \text{ qt}$	**(c)** Convert 22 L to gal.
(d) Convert 5.6 lb to kg. $5.6 \not{lb} \times \frac{0.454 \text{ kg}}{1 \not{lb}} = 2.5424 \text{ kg}$	**(d)** Convert 5 oz to g.

Vocabulary Answers: 1. unit fraction 2. Celsius scale 3. mathematics blueprint 4. equivalent values

Example	Student Practice
3. Convert 235 cm to feet. Round your answer to the nearest hundredth of a foot. Our first unit fraction converts centimeters to inches. Our second unit fraction converts inches to feet. $235\ \cancel{cm} \times \frac{0.394\ \cancel{in.}}{1\ \cancel{cm}} \times \frac{1\ \text{ft}}{12\ \cancel{in.}} = \frac{92.59}{12}\ \text{ft}$ $= 7.7158\overline{3}$ Round to the nearest hundredth we have 7.72 ft.	**4.** Convert 194 cm to feet. Round your answer to the nearest hundredth of a foot.
5. Convert 100 km/hr to mi/hr. We multiply by the unit fraction that relates mi to km. $\frac{100\ \cancel{km}}{\text{hr}} \times \frac{0.62\ \text{mi}}{1\ \cancel{km}} = 62\ \text{mi/hr}$ Thus 100 km/hr is approximately equal to 62 mi/hr.	**6.** Convert 86 km/hr to mi/hr.
7. A camera film that is 35 mm wide is how many inches wide? First convert from millimeters to centimeters by moving the decimal point in the number 35 one place to the left, $35\ \text{mm} = 3.5\ \text{cm}$. Then convert to inches using a unit fraction. $3.5\ \cancel{cm} \times \frac{0.394\ \text{in.}}{1\ \cancel{cm}} = 1.379\ \text{in.}$	**8.** The sheriff's department uses 22-mm revolvers. If such a gun fires a bullet 22 mm wide, how many inches wide is the bullet? (Round to the nearest hundredth.)

Example	Student Practice
9. Convert $176°F$ to Celsius temperature. Use the formula that gives us Celsius degrees. $C = \frac{5 \times F - 160}{9}$ $= \frac{5 \times 176 - 160}{9}$ First multiply, then subtract. $\frac{880 - 160}{9} = \frac{720}{9} = 80$ The temperature is $80°C$.	**10.** Convert $185°F$ to Celsius temperature.
11. Hester is planning a visit from his home in Rhode Island to Brazil. He checked the weather report for the part of Brazil where he will visit and finds that the temperature during the day is $37°C$. If the temperature in Rhode Island is currently $87°F$, what is the difference between the higher and lower temperatures in degrees Fahrenheit? Use the formula that gives us Fahrenheit degrees. $F = 1.8 \times C + 32$ $= 1.8 \times 37 + 32$ $= 66.6 + 32$ $= 98.6$ It is $98.6°F$ in Brazil. Now find the difference in Fahrenheit temperatures. $98.6° - 87° = 11.6°F$	**12.** On a cold winter day in London, Drew notices that the temperature reads $2°C$. She calls home to Los Angeles, California, and finds out that the temperature is $75°F$. What is the difference between the higher and lower temperatures in degrees Fahrenheit?

Extra Practice

1. Convert 3.2 mi to km. Round your answer to the nearest hundredth if necessary.

2. Convert 14.5 gal to L. Round your answer to the nearest hundredth if necessary.

3. Convert 45 km/hr to mi/hr. Round your answer to the nearest hundredth if necessary.

4. Convert 77°F to Celsius.

Concept Check

Explain how you would convert 50 km/hr to mi/hr.

Name: ______________________________ Date: ________________
Instructor: ____________________________ Topic: ________________

Module 7 Measurement and Geometric Figures
Topic 3 Angles

Vocabulary
geometry • point • line • line segment • ray • angle • sides • vertex
degrees • right angle • perpendicular • straight angle • acute angle • obtuse angle
supplementary angles • complementary angles • vertical angles • adjacent angles
parallel line • transversal • alternate interior angles • corresponding angles

1. A portion of a line called a(n) ____________ has a beginning and an end.

2. Two angles that are opposite of each other are called ____________.

3. Two angles that have a sum of 180° are called ____________.

4. Two angles that share a common side are called ____________.

Example	Student Practice
1. Use the figures below to answer the following.	**2.** Use the figure below to answer the following.

Example

1. Use the figures below to answer the following.

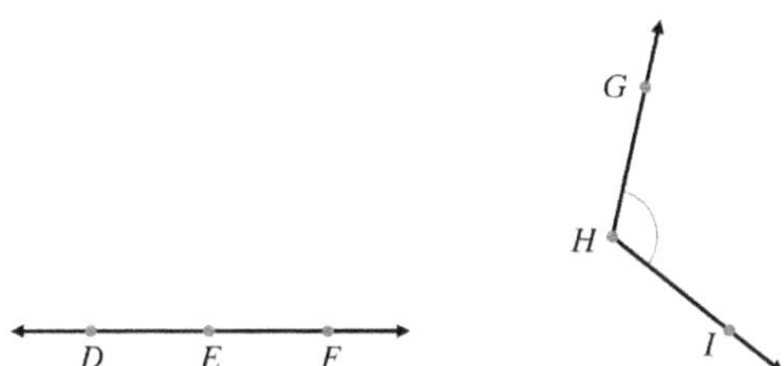

(a) State the measure of angle DEF.

Since $\angle DEF$ is a straight angle, it measures $180°$.

(b) State the three ways we can name the obtuse angle.

We can name the obtuse angle as $\angle GHI$, $\angle$IHG, or $\angle H$. Be sure the letter representing the vertex is the middle letter.

Student Practice

2. Use the figure below to answer the following.

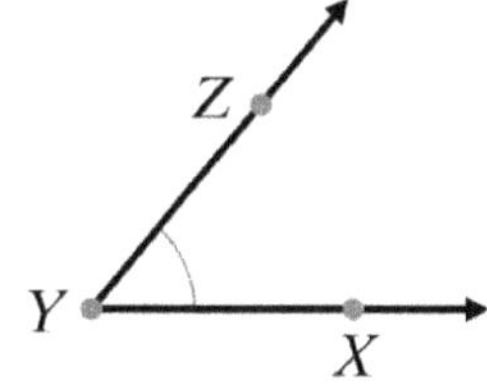

(a) Using a single letter, name the angle in the figure.

(b) State the three ways we can name the acute angle.

Vocabulary Answers: 1. line segment 2. vertical angles 3. supplementary angles 4. adjacent angles

Example	Student Practice
3. The measure of $\angle J$ is $31°$. Find the supplement of $\angle J$. If we let $\angle S =$ the supplement of $\angle J$, then we have the following. The supplement of $\angle J$ plus $\angle J = 180°$. Translate this to symbols. $\angle S + \angle J = 180°$ Now replace $\angle J$ with $31°$ and finally, solve for $\angle S$. $\angle S + \angle J = 180°$ $\angle S + 31° = 180°$ $+ \quad -31° \quad -31°$ $\angle S = 149°$ The supplement of $\angle J$ measure $149°$.	**4.** The measure of $\angle A$ is $24°$. Find the complement of $\angle A$.
5. If $\angle a = 80°$, find the measure of $\angle b$. 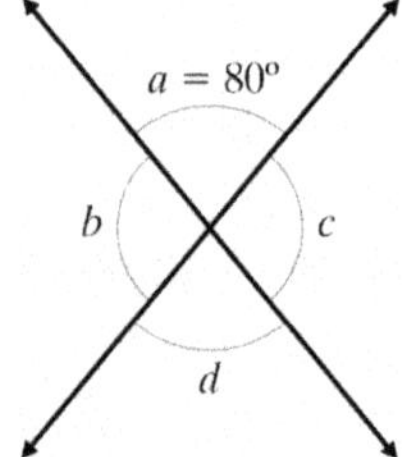 Since $\angle a$ and $\angle b$ are adjacent angles of intersecting lines, we know that they are supplementary angles. Thus, $\angle a + \angle b = 180°$. $\angle a + \angle b = 180°$ $80° + \angle b = 180°$ $+ \quad -80° \quad -80°$ $\angle b = 100°$	**6.** If $\angle a = 70°$, find the measure of $\angle d$. 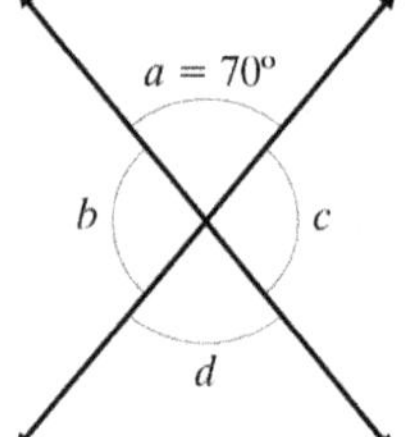

Example	Student Practice

7. In the following figure $a \parallel b$, and the measure of $\angle z$ is $56°$. Find the measure of $\angle y$, $\angle x$, $\angle w$, and $\angle v$.

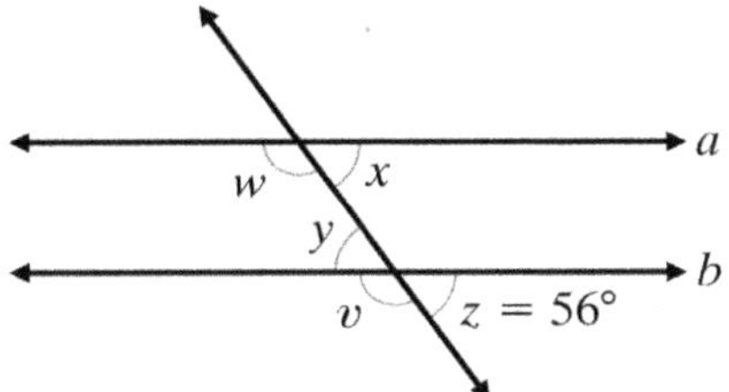

The vertical angles z and y have the same measure.

$\angle z = \angle y = 56°$

The alternate interior angles y and x have the same measure.

$\angle y = \angle x = 56°$

$\angle w$ is the supplement of $\angle x$ and $\angle x = 56°$. The sum of supplementary angles is $180°$.

$\angle w + \angle x = 180°$

Substitute $56°$ for $\angle x$ and solve for $\angle w$.

$$\begin{array}{lrcrcr} & \angle w & + & 56° & = & 180° \\ + & & - & 56° & & -56° \\ \hline & & & \angle w & = & 124° \end{array}$$

The corresponding angles w and v have the same measure.

$\angle w = \angle v = 124°$

8. In the following figure $m \parallel n$, and the measure of $\angle a = 65°$. Find the measures of $\angle b$, $\angle c$, $\angle e$, and $\angle d$.

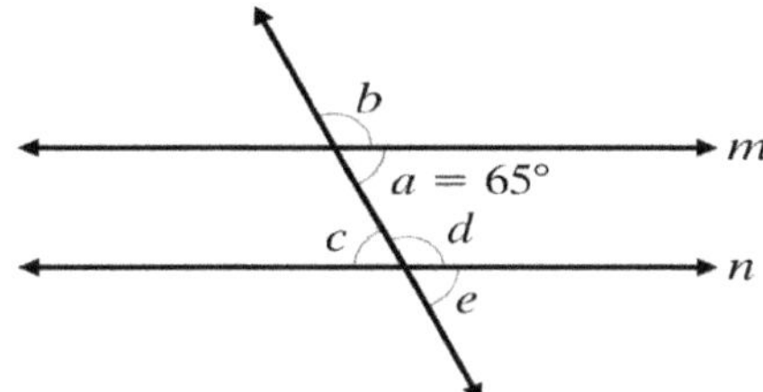

Extra Practice

1. In the following figure, $p \parallel q$ and $\angle a = 68°$, find the measure of $\angle g$.

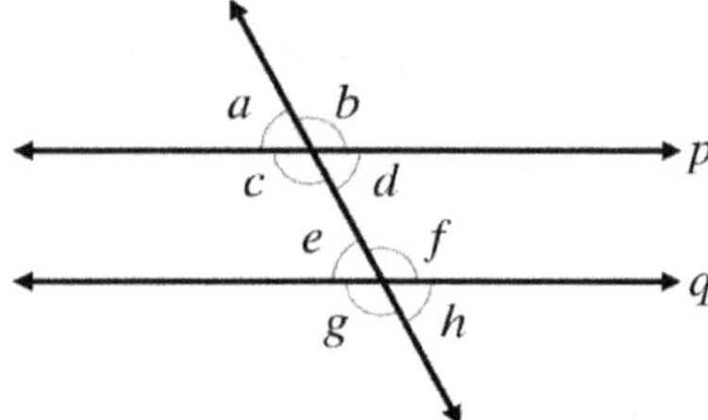

2. Use the figure in extra practice **1** to answer the following. If $\angle a = 68°$, find the measure of $\angle a + \angle b$.

3. In the following figure, find the measure of $\angle DBE + \angle EBC$.

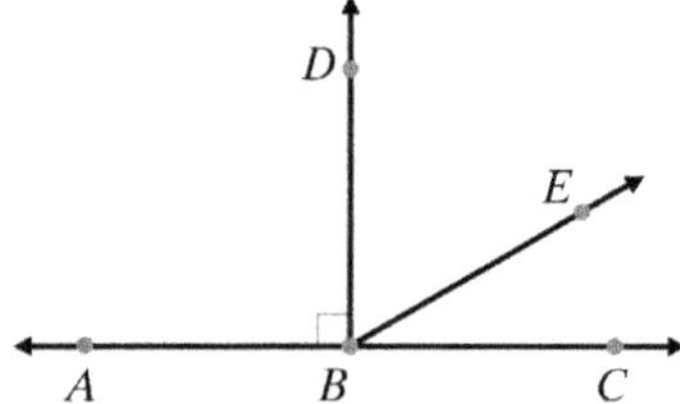

4. Use the figure in extra practice **3** to answer the following. If $\angle ABE = 143°$ find the measure of $\angle EBC$.

Concept Check

In the figure shown below, explain what the relationship is between angle a and angle d. If you know the measure of angle a, how can you find angle d?

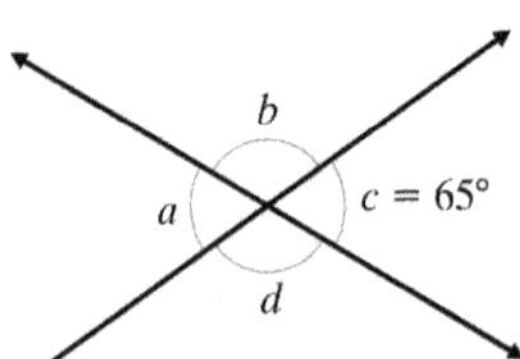

Name: ______________________________ Date: ________________

Instructor: ____________________________ Topic: ________________

Module 7 Measurement and Geometric Figures
Topic 4 The Circle

Vocabulary

pi • center • circle • radius • diameter • circumference • area of a circle

1. The distance around a circle is called its ____________.

2. The ____________ of a circle is the length of a line segment from the center to a point on the circle.

3. The ____________ is the product of π times the radius squared.

4. A(n) ____________ is a figure for which all points on the figure are at an equal distance from a given point.

Example	Student Practice
1. Find the circumference of a circle if the diameter is 9.2 meters. Use $\pi \approx 3.14$. Since we are given the diameter, we use the formula for circumference that includes the diameter. Write the formula. $C = \pi d$ Now substitute the values given. $C = (3.14)(9.2 \text{ m})$ $C \approx 28.888 \text{ m}$ Since π is approximately equal to 3.14, our answer is an approximate value.	**2.** Find the circumference of a circle if the diameter is 8.3 meters. Use $\pi \approx 3.14$.

Vocabulary Answers: 1. circumference 2. radius 3. area of a circle 4. circle

Example	Student Practice

3. The larger of two bicycles has a 26-inch wheel diameter, and the smaller bicycle has a 24-inch wheel diameter.

(a) If the wheels on the larger bicycle complete 12 revolutions, what distance does the larger bicycle travel?

The distance traveled by the larger bicycle during 1 revolution is equal to the measure of the circumference. Multiply the circumference by 12 to find the distance traveled in 12 revolutions.

$$\begin{aligned} C &= \pi d \\ &= (12)(\pi d) \\ &= (12)(3.14)(26 \text{ in.}) \\ &= (12)(81.64 \text{ in.}) \\ &= 979.68 \text{ in.} \end{aligned}$$

(b) How many revolutions must each wheel on the smaller bicycle complete to travel the same distance as the larger bicycle in part **(a)**?

First, find the distance traveled for 1 revolution (the circumference).

$$C = \pi d = (3.14)(24 \text{ in.}) = 75.36 \text{ in.}$$

The smaller bike travels 75.36 inches per revolution. Divide 979.68 in. by 75.36 in. to find the number of revolutions the smaller bike must make to travel the same distance as the larger bike, $979.68 \div 75.36 = 13$.

The smaller bike must complete 13 revolutions.

4. The larger of two bicycles has a 28-inch wheel diameter, and the smaller bicycle has a 22-inch wheel diameter.

(a) If the wheels on the larger bicycle complete 11 revolutions, what distance does the larger bicycle travel?

(b) How many revolutions must each wheel on the smaller bicycle complete to travel the same distance as the larger bicycle in part **(a)**?

Example	Student Practice
5. Lester wants to buy a circular braided rug that is 8 feet in diameter. Find the cost of the rug at \$35 per square yard. Read the problem carefully and create a Mathematics Blueprint. The diameter is 8 feet, so the radius is 4 feet. $r=\frac{8}{2}=4\text{ ft}$ Now find the area. $A=\pi r^2=(3.14)(4\text{ ft})^2$ Square the radius first and then multiply. $A=(3.14)(4\text{ ft})^2=(3.14)(16\text{ ft}^2)$ $\approx 50.24\text{ ft}^2$ Change square feet to square yards. Since $1\text{ yd}=3\text{ ft}$, $(1\text{ yd})^2=(3\text{ ft})^2$. That is, $1\text{ yd}^2=9\text{ ft}^2$. $50.24\ \cancel{\text{ft}^2}\times\frac{1\text{ yd}^2}{9\ \cancel{\text{ft}^2}}\approx 5.58\text{ yd}^2$ (rounded to the nearest hundredth) Finally, find the cost. $\frac{\$35}{1\ \cancel{\text{yd}^2}}\times 5.58\ \cancel{\text{yd}^2}=\$195.30$ The rug cost \$195.30.	**6.** Jen wants to buy a circular crocheted tablecloth that is 80 inches in diameter. Find the cost of the tablecloth at \$6 per square foot.

Extra Practice

1. Find the circumference of the circle. Use $\pi \approx 3.14$. Round your answer to the nearest tenth.

   radius = 8.5 in.

2. A wheel makes 5 revolutions. Determine how far the bicycle travels in inches. Use $\pi \approx 3.14$. Round your answer to the nearest tenth.

   The diameter of the wheel is 42 in.

3. Find the area of the circle. Use $\pi \approx 3.14$. Round your answer to the nearest tenth.

   diameter = 3.5 in.

4. A circular, decorative window in a church has a diameter of 72 inches. Due to age, the window needs to have the insulating strip that surrounds the window replaced. How many feet of insulating strip are needed? Use $\pi \approx 3.14$. Round your answer to the nearest hundredth.

Concept Check

Jasmine must determine the area of a semicircle with a diameter of 10 inches. Explain how you would find the area of the semicircle.

Name: ______________________________ Date: ________________
Instructor: ___________________________ Topic: ________________

Module 7 Measurement and Geometric Figures
Topic 5 Volume

Vocabulary

volume of a cylinder • volume of a sphere • volume of a cone
volume of a pyramid • volume

1. The ____________ is obtained by multiplying the area of the base B by the height h and dividing by 3.

2. The ____________ is 4 times π times the radius cubed divided by 3.

3. The ____________ is π times the radius squared times the height divided by 3.

4. The ____________ is the area of its circular base, πr^2, times the height, h.

Example	**Student Practice**
1. Find the volume of a tin can with radius 7.62 centimeters and height 15.24 centimeters. Round your answer to the nearest hundredth. $V = \pi r^2 h = (3.14)(7.62\text{ cm})^2(15.24\text{ cm})$ $V = 2778.59\text{ cm}^3$ The volume of the tin can is approximately 2778.59 cubic centimeters.	**2.** Find the volume of a tin can with radius 9.45 centimeters and height 13.52 centimeters. Round your answer to the nearest hundredth.
3. How much air is needed to fully inflate a soccer ball if the radius of the inner lining is 5 inches? Round your answer to the nearest hundredth. $V = \frac{4\pi r^3}{3} = \frac{4(3.14)(5\text{ in.})^3}{3}$ $V = 523.33\text{ in.}^3$ Approximately 523.33 cubic inches of air is needed to fully inflate the ball.	**4.** How much air is needed to fully inflate a beach ball if the radius of the inner lining is 9 inches? Round your answer to the nearest hundredth.

Vocabulary Answers: 1. volume of a pyramid 2. volume of a sphere 3. volume of a cone 4. volume of a cylinder

Example

5. A cylindrical thermos has a layer of insulation around its sides. The radius R to the outer edge of the insulation is 15 centimeters, and the radius r to the inner edge is 13 centimeters. The thermos is 27 centimeters tall. Use $\pi \approx 3.14$.

(a) What volume of coffee can the thermos hold?

We draw a picture of the thermos.

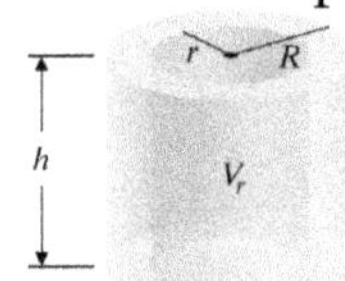

We find the volume of the inner shaded region (V_r) to determine how much coffee the thermos will hold. The radius is 13 cm and the height is 27 cm.

$V_r = \pi r^2 h = (3.14)(13 \text{ cm})^2 (27 \text{ cm})$

$V_r = 14{,}327.82 \text{ cm}^3$

(b) What is the volume of the insulated region?

We find the volume of the entire cylinder (V_R) minus the volume of the inner region (V_r), $V_R - V_r$. Start by finding V_R.

$V_R = \pi r^2 h = (3.14)(15 \text{ cm})^2 (27 \text{ cm})$

$V_R = 19{,}075.5 \text{ cm}^3$

Now subtract to find the volume of the insulated region.

$V_R - V_r = 19{,}075.5 - 14{,}327.82$

$V_R - V_r = 4747.69$ cubic centimeters

Student Practice

6. A cylindrical thermos has a layer of insulation around its sides. The radius R to the outer edge of the insulation is 14 centimeters, and the radius r to the inner edge is 11 centimeters. The thermos is 25 centimeters tall. Use $\pi \approx 3.14$.

(a) What volume of coffee can the thermos hold?

(b) What is the volume of the insulated region?

Example	**Student Practice**

7. Find the volume of a cone of radius 7 meters and height 9 meters. Round to the nearest tenth.

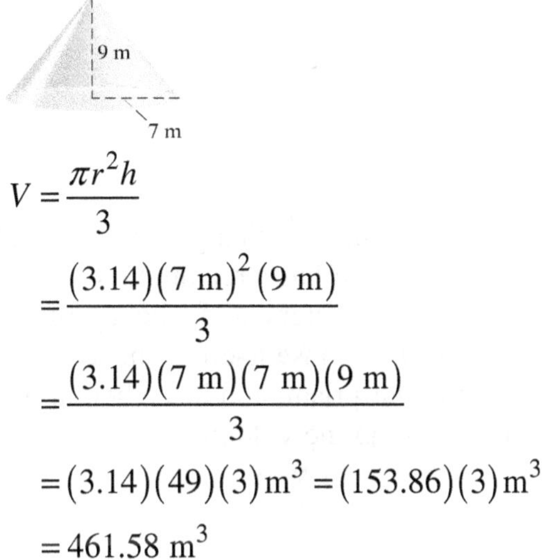

$$V = \frac{\pi r^2 h}{3}$$

$$= \frac{(3.14)(7\text{ m})^2(9\text{ m})}{3}$$

$$= \frac{(3.14)(7\text{ m})(7\text{ m})(9\text{ m})}{3}$$

$$= (3.14)(49)(3)\text{m}^3 = (153.86)(3)\text{m}^3$$

$$= 461.58\text{ m}^3$$

$V \approx 461.6\text{ m}^3$ rounded to the nearest tenth.

8. Find the volume of a cone of radius 9 meters and height 18 meters. Round to the nearest tenth.

9. Find the volume of a pyramid with $\text{height} = 6$ meters, length of $\text{base} = 7$ meters, width of $\text{base} = 5$ meters.

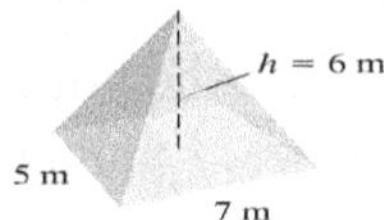

The base is a rectangle. Find its area, $\text{Area of base} = (7\text{ m})(5\text{ m}) = 35\text{ m}^2$.

Substitute the area of the base, 35 m^2, and the height of 6 m in the formula for the volume of the pyramid.

$$V = \frac{Bh}{3} = \frac{(35\text{ m}^2)(6\text{ m})}{3} = (35)(2)\text{ m}^3$$

$$= 70\text{ m}^3$$

10. Find the volume of a pyramid with $\text{height} = 12$ meters, length of $\text{base} = 9$ meters, width of $\text{base} = 6$ meters.

Extra Practice

1. Find the volume of a pyramid with a height of 3.5 ft and a square base of 1.5 ft on a side. Round your answer to the nearest tenth.

2. Determine the amount of air required to fill a rubber ball with a radius of 7.5 centimeters. Use $\pi \approx 3.14$. Round your answer to the nearest tenth.

3. A box is 12 inches long, 8 inches wide and 3 inches tall. It is topped by a cylinder with a diameter of 6 inches and a height of 15 inches. Find the combined volume of the box and cylinder. Use $\pi \approx 3.14$.

4. A house is built with a pyramid-shaped roof. The roof has a rectangular base of 8 meters by 6 meters, and a height of 5 meters. The house itself is shaped like a rectangular prism, and has a height of 9 meters. Find the volume of the house.

Concept Check

After you find the volume of a tin can with a radius of 6 inches and a height of 10 inches to be 1130.40 in.^3, rounded to the nearest hundredth, you decide that you must find the volume of a similar can that has all the same measurements, except the height is 5 inches. Explain how you could do this without using the formula.

Name: ______________________________ Date: ________________
Instructor: ____________________________ Topic: ________________

Module 7 Measurement and Geometric Figures
Topic 6 Similar Geometric Figures

Vocabulary
similar triangles • corresponding angles • corresponding sides
perimeters • geometric figures

1. The length of ___________ of similar triangles have the same ratio.

2. The ___________ of similar triangles have the same ratios as the corresponding sides.

3. The ___________ of similar triangles are equal.

4. Two triangles with the same shape but not necessarily the same size are called ___________.

Example	**Student Practice**
1. The two triangles below are similar. Find the length of side n. Round to the nearest tenth. 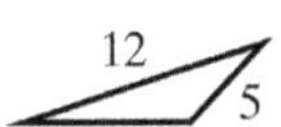 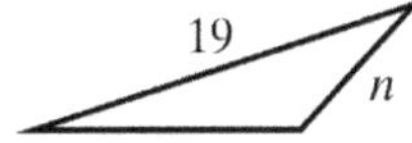 The ratio of 12 to 19 is the same as the ratio of 5 to n, $\frac{12}{19}=\frac{5}{n}$. Now cross-multiply and simplify. $12n=(19)(5)$ $12n=95$ Next, divide both sides by 12 and round to the nearest tenth. $\frac{12n}{12}=\frac{95}{12}$ $n=7.91\overline{6}$ $n\approx 7.9$	**2.** The two triangles are similar. Find the length of side n. Round to the nearest tenth. 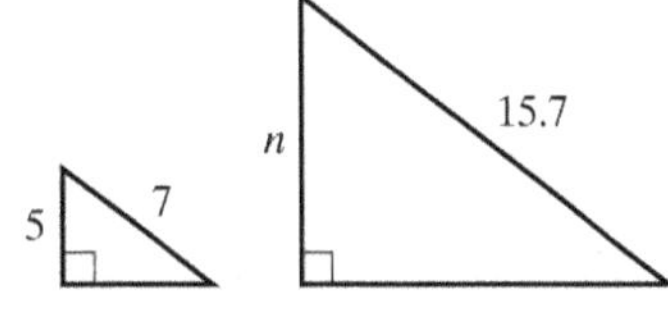

Vocabulary Answers: 1. corresponding sides 2. perimeters 3. corresponding angles 4. similar triangles

Example	**Student Practice**
3. Two triangles are similar. The smaller triangle has sides 5 yards, 7 yards, and 10 yards. The 7-yard side on the smaller triangle corresponds to a side of 21 yards on the larger triangle. What is the perimeter of the larger triangle?	**4.** Two triangles are similar. The smaller triangle has sides 7 yards, 11 yards, and 15 yards. The 15-yard side on the smaller triangle corresponds to the side of 60 yards on the larger triangle. What is the perimeter of the larger triangle?

We draw the two triangles.

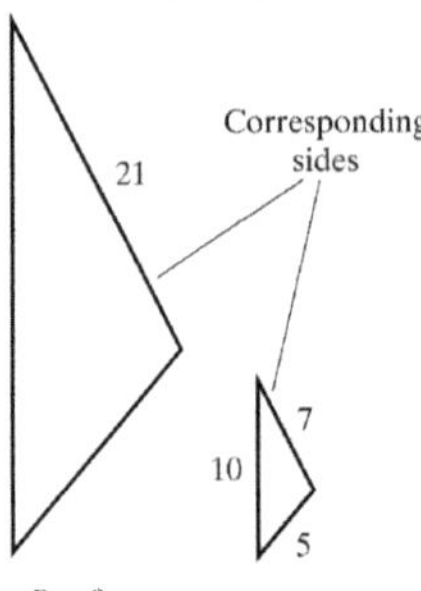

The perimeters of similar triangles have the same ratios as the corresponding sides. Therefore, we begin by finding the perimeter of the smaller triangle.

$$5 \text{ yd} + 7 \text{ yd} + 10 \text{ yd} = 22 \text{ yd}$$

We can now write equal ratios. Let $P =$ the unknown perimeter. We set up the ratios as $\frac{\text{smaller triangle}}{\text{larger triangle}}$. Make sure to write the terms of the second ratio in the same order.

$$\frac{22}{P} = \frac{7}{21}$$
$$7P = (21)(22)$$
$$7P = 462$$
$$\frac{7P}{7} = \frac{462}{7}$$
$$P = 66$$

The perimeter of the larger triangle is 66 yards.

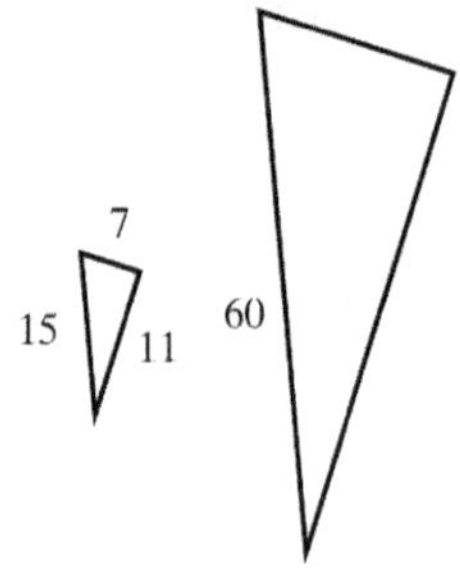

Example	**Student Practice**

5. A flagpole casts a shadow of 36 feet. At the same time, a tree that is 3 feet tall has a shadow of 5 feet. How tall is the flagpole?

The shadows cast by the sun shining on vertical objects at the same time of day form similar triangles. We draw a picture and organize our information.

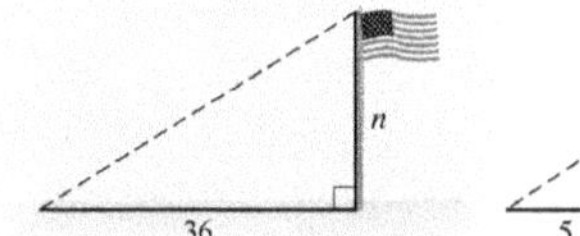

Let $n =$ the height of the flagpole. Thus we can say that n is to 3 as 36 is to 5.

$$\frac{n}{3}=\frac{36}{5}$$
$$5n=(3)(36)$$
$$5n=108$$

Now divide both sides by 5, $n=21.6$.

6. A flagpole casts a shadow of 30 feet. At the same time a tree that is 7.2 feet tall has a shadow of 8 feet. How tall is the flagpole?

7. The two rectangles shown below are similar because the corresponding sides of the two rectangles have the same ratio. Find the width of the larger rectangle.

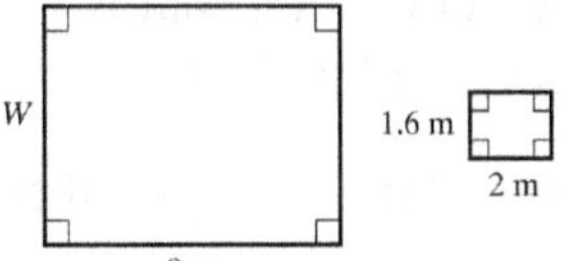

Let $W =$ the width of the larger rectangle.

$$\frac{W}{1.6}=\frac{9}{2}$$
$$2W=(1.6)(9)$$
$$2W=14.4$$

Now divide both sides by 2, $W=7.2$.

8. The two rectangles shown below are similar because the corresponding sides of the two rectangles have the same ratio. Find the width of the larger rectangle.

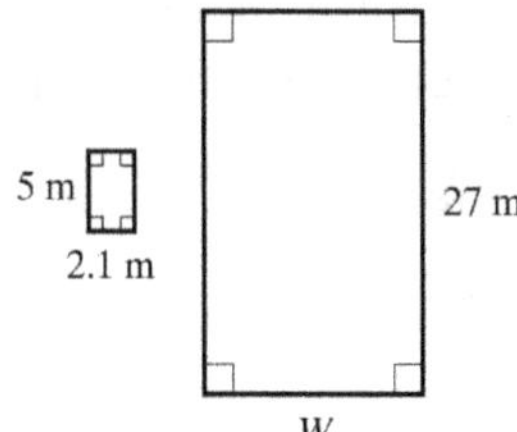

Extra Practice

1. For the pair of similar triangles below, find the missing side n. Round your answer to the nearest tenth if necessary.

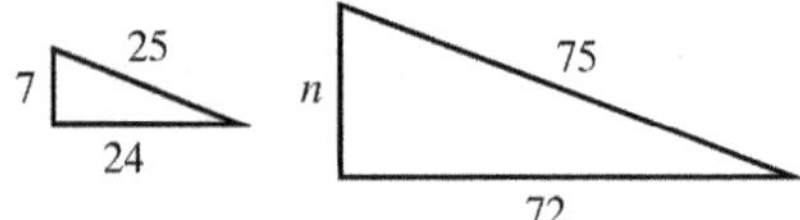

2. The pair of rectangles below are similar. Find the missing side n. Round to the nearest tenth if necessary.

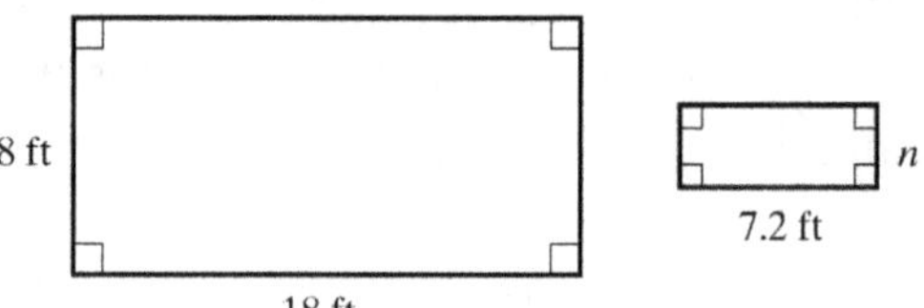

3. A tree casts a shadow of 6 feet. At the same time, a building that is 50 feet tall has a shadow that is 30 feet. How tall is the tree?

4. Keenan is rock climbing in Colorado. He is 6 ft tall and his shadow measures 9 ft long. The rock he wants to climb casts a shadow of 540 ft. How tall is the rock he wants to climb?

Concept Check

A tree that is 13.5 feet tall casts a shadow of 3 feet. At the same time, a building that is 45 ft tall casts a shadow of 10 feet. Using the properties of similar triangles, the height of the building is found to be 45 feet. The shadow cast by the building later in the day is $\frac{1}{2}$ of the length that it was earlier. Explain how you would find the shadow cast by the tree at the same time that afternoon.

MATH COACH

Mastering the skills you need to do well on the test.

Watch the MATH COACH videos in MyMathLab® or on YouTube™ while you work the problems below. These helpful hints will help you avoid making common errors on test problems.

Converting Units of Measurement—Problems 3(a) and (b)
Sharon purchased 2 kilograms of hamburger meat.
(a) How many grams did she purchase?
(b) How many pounds did she purchase?

> **Helpful Hint:** Recall that 1 kilogram equals 1000 grams, and 1 kilogram is approximately 2.2 pounds.

For Part (a), did you understand that you have to move the decimal point 3 places? Yes ____ No ____

Did you remember to move the decimal point to the right?
Yes ____ No ____

If you answered No to either of these questions, remember that the number of zeros in 1000 tells you how many places to move the decimal point. When converting from a larger to a smaller unit, you move the decimal point to the right.

For Part (b), did you use a unit fraction with kilograms in the denominator and pounds in the numerator?
Yes ____ No ____

Did you write $2\text{ kg}\times\left(\frac{2.2\text{ lb}}{1\text{ kg}}\right)$?

Yes ____ No ____

If you answered No to either of these questions, consider why this is the correct unit fraction and perform this step. Be careful with your calculations.

If you answered Problem 3(a) or 3(b) incorrectly, go back and rework the problems using these suggestions.

Solving Applied Problems Involving Right Triangles—Problem 21 A ladder is placed against the side of a building. The top of the ladder is 14 feet from the ground. The base of the ladder is 11 feet from the building. What is the approximate length of the ladder? Round to the nearest tenth.

> **Helpful Hint:** It is wise to first draw a picture and label each side of the figure. This will help you recognize that you must use the Pythagorean Theorem: $c^2 = a^2 + b^2$. Make sure you know which sides are the legs and which side is the hypotenuse: $\text{Hypotenuse}^2 = \text{leg}^2 + (\text{other leg})^2$.

When you substituted values, did you get $c^2 = 11^2 + 14^2$?
Yes ____ No ____

If you answered No, reread the problem and consider how to substitute these values correctly into the Pythagorean Theorem formula.

Did you simplify the equation to get $c^2 = 317$?
Yes ____ No ____
Next, did you approximate the square root of 317 and round to the nearest tenth? Yes ____ No ____

If you answered No to either of these questions, check your calculations carefully. You can use the square root function on your calculator to approximate square roots. Remember to include the units in your final answer.

Now go back and rework this problem using these suggestions.

Finding the Area of Circles—Problem 24 Find the area of a circle whose radius is 1.2 centimeters. Round you answer to the nearest tenth.

Helpful Hint: Try to memorize all the formulas and how to use them correctly. Take the extra time to write down the formula. This will help you to avoid errors in calculations.

Did you write down the formula $A = \pi r^2$?
Yes ____ No ____

If you answered No, go back and complete this step.

Did you perform the calculation $3.14 \times (1.2)$ to get your final answer?
Yes ____ No ____

If you answered Yes, stop and study the formula carefully. Notice that you must square the radius, 1.2 *before you multiply*. Remember to include the units in your final answer.

If you answered Problem 24 incorrectly, go back and rework the problem using these suggestions.

Finding the Volume of Pyramids—Problem 29 Find the volume of a pyramid with height 12 meters, length of base 10 meters, and width of base 7 meters.

Helpful Hint: It is important that you understand how to use this formula, $V = \frac{Bh}{3}$. The B refers to the area of the rectangular base, and the h refers to the height. You must find the area of the base first and substitute this value for B in the formula.

Did you perform the calculations $10 \text{ m} \times 7 \text{ m}$ as your first step to find B ?
Yes ____ No ____

If you answered No, stop and make this correction to find the area of the base, B .

Next, did you write down $V = \frac{70 \text{ m}^2 (12 \text{ m})}{3}$?

Yes ____ No ____

Did you get a final unit of m^3 ?
Yes ____ No ____

If you answered No to either of these questions, substitute the values carefully and remember that the units must be multiplied too.

Now go back and rework the problem using these suggestions.

Name: ______________________________ Date: ________________

Instructor: ____________________________ Topic: ________________

Module 8 Introduction to Polynomials
Topic 1 Adding and Subtracting Polynomials

Vocabulary

add polynomials • distributive property • opposite • subtract polynomials

1. To remove parentheses from some expressions such as $(-1)(2x+6)$, use the ___________ to multiply each term inside the parentheses by -1.

2. To ___________, combine like terms.

3. To ___________, change the sign of each term in the second polynomial and then add.

4. When a negative sign precedes parentheses, we find the ___________ of the expression by changing the sign of each term inside the parentheses.

Example	Student Practice
1. Identify the terms of the polynomial. $xy^2-7y^2-2x+5y$ We include the sign in front of the term as part of the term. Polynomial: $xy^2-7y^2-2x+5y$ Terms: $+xy^2,-7y^2,-2x,+5y$	**2.** Identify the terms of the polynomial. $a^2b+5b^2-6a-3b$
3. Perform the operations indicated. $(-4x^2+5x-2)+(3x^2+4)$ We must combine like terms. Rearrange terms so that like terms are grouped together, then add like terms. $(-4x^2+5x-2)+(3x^2+4)$ $=-4x^2+3x^2+5x-2+4$ $=-1x^2+5x+2$ $=-x^2+5x+2$	**4.** Perform the operations indicated. $(5a^2-8a+2)+(-4a^2-3)$

Vocabulary Answers: 1. distributive property 2. add polynomials 3. subtract polynomials 4. opposite

Example	Student Practice
5. Simplify $-(-2a+5b-7c)$. Since there is a $-$ in front of the parentheses, we change the sign of each term. $-(-2a+5b-7c)=2a-5b+7c$	**6.** Simplify $-(3x+7y-4z)$.
7. Perform the operations indicated. $(3x^2+5x-7)-(6x^2-8x-1)$ A $-$ sign in front of the parentheses indicates we are subtracting. We change the signs of the terms in the second polynomial, and then add. $(3x^2+5x-7)-(6x^2-8x-1)$ $=3x^2+5x-7+(-6x^2)+8x+1$ Simplify by combining like terms: $3x^2-6x^2=-3x^2$; $5x+8x=+13x$; $-7+1=-6$. $3x^2+5x-7+(-6x^2)+8x+1$ $=-3x^2+13x-6$	**8.** Perform the operations indicated. $(x^2+5x-2)-(4x^2-5x+9)$

Name: ______________________________ Date: ________________
Instructor: ___________________________ Topic: ________________

Module 8 Introduction to Polynomials
Topic 1 Adding and Subtracting Polynomials

Vocabulary
add polynomials • distributive property • opposite • subtract polynomials

1. To remove parentheses from some expressions such as $(-1)(2x+6)$, use the ____________ to multiply each term inside the parentheses by -1.

2. To ____________, combine like terms.

3. To ____________, change the sign of each term in the second polynomial and then add.

4. When a negative sign precedes parentheses, we find the ____________ of the expression by changing the sign of each term inside the parentheses.

Example	Student Practice
1. Identify the terms of the polynomial. $xy^2-7y^2-2x+5y$ We include the sign in front of the term as part of the term. Polynomial: $xy^2-7y^2-2x+5y$ Terms: $+xy^2,-7y^2,-2x,+5y$	**2.** Identify the terms of the polynomial. $a^2b+5b^2-6a-3b$
3. Perform the operations indicated. $(-4x^2+5x-2)+(3x^2+4)$ We must combine like terms. Rearrange terms so that like terms are grouped together, then add like terms. $(-4x^2+5x-2)+(3x^2+4)$ $=-4x^2+3x^2+5x-2+4$ $=-1x^2+5x+2$ $=-x^2+5x+2$	**4.** Perform the operations indicated. $(5a^2-8a+2)+(-4a^2-3)$

Vocabulary Answers: 1. distributive property 2. add polynomials 3. subtract polynomials 4. opposite

Example	Student Practice
5. Simplify $-(-2a+5b-7c)$. Since there is a $-$ in front of the parentheses, we change the sign of each term. $-(-2a+5b-7c)=2a-5b+7c$	**6.** Simplify $-(3x+7y-4z)$.
7. Perform the operations indicated. $(3x^2+5x-7)-(6x^2-8x-1)$ A $-$ sign in front of the parentheses indicates we are subtracting. We change the signs of the terms in the second polynomial, and then add. $(3x^2+5x-7)-(6x^2-8x-1)$ $=3x^2+5x-7+(-6x^2)+8x+1$ Simplify by combining like terms: $3x^2-6x^2=-3x^2$; $5x+8x=+13x$; $-7+1=-6$. $3x^2+5x-7+(-6x^2)+8x+1$ $=-3x^2+13x-6$	**8.** Perform the operations indicated. $(x^2+5x-2)-(4x^2-5x+9)$

Example	Student Practice
9. Perform the operations indicated.	**10.** Perform the operations indicated.
$6x-3(-4x^2+3)-(-2x^2+x-5)$	$3a-(7a^2+2a)-6(a^2+9a-4)$
First we multiply (-3) times the binomial $(-4x^2+3)$ to remove parentheses. $-3(-4x^2)=+12x^2$ and $-3(+3)=-9$	
$6x-3(-4x^2+3)-(-2x^2+x-5)$ $=6x+12x^2-9-(-2x^2+x-5)$	
We remove parentheses and change the sign of each term inside the parentheses.	
$6x+12x^2-9-(-2x^2+x-5)$ $=6x+12x^2-9+2x^2-x+5$	
We combine like terms: $6x-x=5x$; $12x^2+2x^2=+14x^2$; $-9+5=-4$.	
$6x+12x^2-9+2x^2-x+5$ $=5x+14x^2-4$	
We write the polynomial so that the powers of x decrease as we read from left to right.	
$5x+14x^2-4=14x^2+5x-4$	

Extra Practice

1. Identify the terms of the polynomial.

$$5a^4 - 3a^3 + 2a^2 - 7a + 1$$

2. Perform the operations indicated.

$$\left(3y^2 - 2y - 5\right) + \left(4y^2 - 6y + 4\right)$$

3. Simplify $-\left(-4a^4 - 6a^2 + 9\right)$.

4. Perform the operations indicated.

$$7x - 3(x+4) - (-5x - 9) - (4x + 3)$$

Concept Check

Mitchell subtracted two polynomials as follows.

$$\left(-6x^2 + 3x - 1\right) - \left(4x^2 + 2x - 7\right)$$
$$= -6x^2 + 3x - 1 - 4x^2 + 2x - 7$$
$$= -10x^2 + 5x - 8$$

Did Mitchell complete the problem correctly? Why or why not?

Name: ______________________________ Date: ________________

Instructor: ___________________________ Topic: ________________

Module 8 Introduction to Polynomials
Topic 2 Multiplying Polynomials

Vocabulary

negative • FOIL method • distributive property

1. You can use the ___________ only when you multiply a binomial times a binomial.

2. When we multiply a binomial times a trinomial, we use the ___________ twice since we must multiply each term of the binomial times the trinomial.

3. When multiplying by a ___________ monomial, it is a good idea to check the product, verifying that the sign of each term changes.

Example	**Student Practice**
1. Multiply $-4x(2x-6y-7)$. We multiply each term by $-4x$. $-4x(2x-6y-7)$ $=-4x(2x)-4x(-6y)-4x(-7)$ $=-8x^2+24xy+28x$	**2.** Multiply $-5a(4a-8b+2)$.
3. Multiply $(3x^2+x-6)(-7x^3)$. We move the monomial to the left side. $-7x^3(3x^2+x-6)$ We multiply each term by $-7x^3$. $-7x^3(3x^2+x-6)$ $=-21x^5-7x^4+42x^3$ Since we are multiplying by a negative monomial, check the sign of each term in the product to be sure it changed.	**4.** Multiply $(4z^2-8z+3)(-5z^5)$.

Vocabulary Answers: 1. FOIL method 2. distributive property 3. negative

Example	Student Practice
5. Multiply $(2x+3)(3x^2+5x-1)$. We multiply $2x$ times $3x^2+5x-1$ and then +3 times $3x^2+5x-1$. $(2x+3)(3x^2+5x-1)$ $=2x(3x^2+5x-1)+3(3x^2+5x-1)$ $=2x\cdot 3x^2+2x\cdot 5x+2x(-1)$ $+3\cdot 3x^2+3\cdot 5x+3(-1)$ We multiply. $=6x^3+10x^2-2x+9x^2+15x-3$ We combine like terms. $=6x^3+19x^2+13x-3$	**6.** Multiply $(4a+7)(2a^2-a+6)$.
7. Use the distributive property to multiply $(x+1)(x+4)$. We multiply each term of $(x+1)$ times the binomial $(x+4)$. Then we use the distributive property again; multiply $x(x+4)$ and $(+1)(x+4)$. $(x+1)(x+4)=x(x+4)+1(x+4)$ $=x\cdot x+x\cdot 4+1\cdot x+1\cdot 4$ $=x^2+4x+1x+4$ Combine like terms. $(x+1)(x+4)=x^2+5x+4$	**8.** Use the distributive property to multiply $(x+5)(x+4)$.

Example	Student Practice
9. Multiply $(x+4)(x+3)$. Multiply the First terms, x and x, $x \cdot x = x^2$. Multiply the Outer terms, x and 3, $x \cdot 3 = 3x$. Multiply the Inner terms, 4 and x, $4 \cdot x = 4x$. Multiply the Last terms, 4 and 3, $4 \cdot 3 = 12$. Add the results and combine like terms. $(x+4)(x+3) = x^2 + 3x + 4x + 12$ $= x^2 + 7x + 12$	**10.** Multiply $(y+5)(y+1)$.
11. Multiply $(y-1)(y-7)$. Pay special attention to the signs of the terms when we multiply. Multiply the First terms, y and y, $y \cdot y = y^2$. Multiply the Outer terms, y and -7, $y \cdot (-7) = -7y$. Multiply the Inner terms, -1 and y, $(-1) \cdot y = -1y$. Multiply the Last terms, -1 and -7, $(-1)(-7) = +7$. Add the results and combine like terms. $(y-1)(y-7) = y^2 - 8y + 7$	**12.** Multiply $(x-2)(x-7)$.

Example	Student Practice
13. Multiply $(2x+3)(x-1)$.	**14.** Multiply $(x+3)(7x-4)$.
Be sure to check the sign of each term.	
Multiply the First terms, $2x$ and x, $2x \cdot x = 2x^2$.	
Multiply the Outer terms, $2x$ and -1, $2x \cdot (-1) = -2x$.	
Multiply the Inner terms, 3 and x, $3 \cdot x = +3x$.	
Multiply the Last terms, 3 and -1, $3 \cdot (-1) = -3$.	
Add the results and combine like terms.	
$(2x+3)(x-1) = 2x^2 + x - 3$	

Extra Practice

1. Use the distributive property to multiply $(2x-1)(x^2+3x+1)$.

2. Use FOIL to multiply $(x+7)(x-5)$.

3. Use FOIL to multiply $(-x-9)(-2x+9)$.

4. Simplify.
$-2x(x^2+3x+1)+(x-3)(x+4)$

Concept Check

Multiply each of the following.

1. $(x+1)(x+2)$ **2.** $(x-1)(x-2)$

(a) Explain why the middle terms in each product of 1 and 2 have opposite signs.
(b) Explain why the last terms in each product of 1 and 2 have the same sign.

Name: ______________________________ Date: ________________
Instructor: ___________________________ Topic: _______________

Module 8 Introduction to Polynomials
Topic 3 Translate from English to Algebra

Vocabulary
more than • sum of • increased by • added to • greater than • plus • minus
decreased by • less than • subtracted from • smaller than • fewer than • of
diminished by • difference between • double • twice • product of • times
divided by • quotient of

1. "The ____________ a number and three" translates to $x+3$.

2. "The ____________ a number and four" translates to $x-4$.

3. "The ____________ a two and a number" translates to $2x$.

4. "The ____________ a number and five" translates to $\frac{x}{5}$.

Example	Student Practice
1. Write each English phrase as an algebraic expression.	**2.** Write each English phrase as an algebraic expression.
(a) A quantity is increased by five. $x+5$	**(a)** Five less than a quantity
(b) Double the value. $2x$	**(b)** Triple the discount.
(c) One-third of the weight $\frac{x}{3}$ or $\frac{1}{3}x$	**(c)** One-fifth of the height
(d) Seven less than a number $x-7$ Note that the variable or expression that follows the words "less than" always comes first.	**(d)** Ten more than a number

Vocabulary Answers: 1. sum of 2. difference between 3. product of 4. quotient of

Example	Student Practice
3. Write each English phrase as an algebraic expression. **(a)** Seven more than double a number $2x+7$ **(b)** The value of the number is increased by seven and then doubled. Note that the word "then" tells us to add x and 7 before doubling. $2(x+7)$ Note that this is not the same as $2x+7$. **(c)** One-half of the sum of a number and 3 $\frac{1}{2}(x+3)$	**4.** Write each English phrase as an algebraic expression. **(a)** Eight more than triple a number **(b)** The value of the number is increased by eight and then tripled. **(c)** One-fourth of the sum of a number and 6
5. Use a variable and an algebraic expression to describe two quantities in the English sentence "Mike's salary is \$2000 more than Fred's salary." The two quantities that are being compared are Mike's and Fred's salaries. Since Mike's salary is being compared to Fred's salary, we let the variable represent Fred's salary. The choice of the letter f helps us to remember that the variable represents Fred's salary. Let f = Fred's salary. Then, $f + \$2000$ = Mike's salary, since Mike's salary is \$2000 more than Fred's.	**6.** Use a variable and an algebraic expression to describe two quantities in the English sentence "Tom's car has 12,500 more miles on it than Chuck's truck."

Example	Student Practice
7. The length of a rectangle is 3 meters shorter than twice the width. Use a variable and an algebraic expression to describe the length and the width. Draw a picture of the rectangle and label the length and width. The length of the rectangle is being compared to the width. Use the letter w for width. Let $w =$ the width. Express the length in terms of the width. The length is 3 meters shorter than twice the width. Then $2w - 3 =$ the length. A picture of the rectangle is shown. 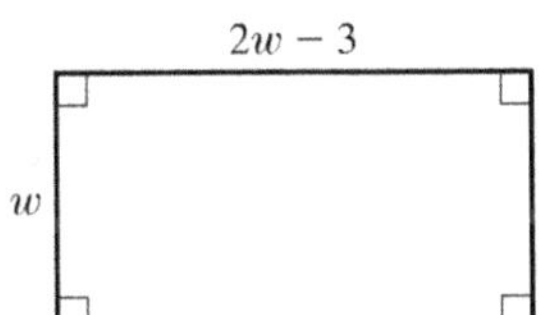	**8.** The length of a rectangle is 7 inches shorter than triple the width. Use a variable and an algebraic expression to describe the length and the width. Draw a picture of the rectangle and label the length and width.
9. The first angle of a triangle is triple the second angle. The third angle of the triangle is $12°$ more than the second angle. Describe each angle algebraically. Draw a diagram of the triangle and label its parts. Since the first and third angles are described in terms of the second angle, we let the variable represent the number of degrees in the second angle. Let $s =$ the number of degrees in the second angle. Then $3s =$ the number of degrees in the first angle and $s + 12 =$ the number of degrees in the third angle. 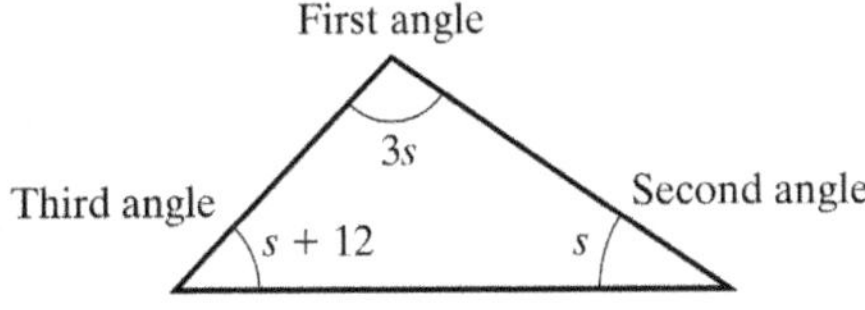	**10.** The first angle of a triangle is five times the second angle. The third angle of the triangle is $40°$ more than the second angle. Describe each angle algebraically. Draw a diagram of the triangle and label its parts.

Extra Practice

1. Write an algebraic expression for the quantity. Let x represent the unknown value.

   a value increased by 12

2. Write an algebraic expression for the quantity. Let x represent the unknown value.

   one-fourth of a number, decreased by one

3. Write an algebraic expression for the quantities being compared.

   The value of Alicia's car is $2300 more than the value of Allison's car.

4. Write an algebraic expression for the quantities being compared.

   Patricia owns 17 more comic books than Scott, Walter owns three times as many as Scott, and Adrienne owns five fewer than four times as many as Scott.

Concept Check

Explain how you would decide whether to use $\frac{1}{3}(x+7)$ or $\frac{1}{3}x+7$ as an algebraic expression for the phrase "one-third of the sum of a number and seven."

Name: ______________________________ Date: ________________

Instructor: ____________________________ Topic: ________________

Module 8 Introduction to Polynomials
Topic 4 Factoring Using the Greatest Common Factor

Vocabulary

common factors • greatest common factor • divide • factor an expression

1. To factor out the GCF we can multiply the expression by the GCF and then ____________ the expression by the GCF.

2. We call 1, 2, 3, and 6 ____________ of 6 and 12.

3. To ____________, we find an equivalent expression written as a product.

4. We call the largest common factor the ____________.

Example	Student Practice
1. Find the GCF of 8 and 16. Think of a factor, the largest factor, that will divide into both 8 and 16. The largest common factor is 8, therefore the GCF of 8 and 16 is 8.	**2.** Find the GCF of 10 and 5.
3. Find the GCF of 8, 20, and 28. First we write each number as a product of prime factors in exponent form. Then we identify the common factors and use the smallest power that appears on these factors to find our GCF. $8 = 2\cdot2\cdot2 \quad = 2^3$ $20 = 2\cdot2\cdot \quad 5 \quad = 2^2\cdot5$ $28 = 2\cdot2\cdot \quad 7 = 2^2\cdot7$ Notice that $2\cdot2$ or 2^2 is common to 8, 20, 28. The greatest common factor of 8, 20, 28 is 4.	**4.** Find the GCF of 15, 20, and 30.

Vocabulary Answers: 1. Divide 2. common factors 3. factor an expression 4. greatest common factor

Example	Student Practice
5. Find the GCF. **(a)** $15x^2y+18x^3$ Rewrite each term using prime factors written in exponent form, then identify the common prime factors and use the smallest power that appears on each of these factors. $15x^2y=3^1\cdot5\cdot x^2\cdot y$ $18x^3=3^2\cdot2\cdot x^3$ The GCF of $15x^2y+18x^3$ is $3x^2$. **(b)** $a^4bc+a^2b^2$ $a^4bc=a^4\cdot b^1\cdot c$ $a^2b^2=a^2\cdot b^2$ The GCF of $a^4bc+a^2b^2$ is a^2b.	**6.** Find the GCF. **(a)** $6x^5y^2+10x^2$ **(b)** $x^3y^2z+y^7z^4$
7. Factor $12x+15$ and check your solution. 3 is the greatest common factor of $12x+15$. Write each term as a product and factor out the GCF, 3. $12x+15=3\cdot4x+3\cdot5$ $=3(4x+5)$ Check. We multiply using the distributive property. $3(4x+5)=3\cdot4x+3\cdot5$ $=12x+15$	**8.** Factor $20x+35$ and check your solution.

Example	Student Practice
9. Factor $8x-12y+16$ and check your solution. 4 is the greatest common factor of $8x-12y+16$. Write each term as a product and factor out the GCF, 4. $8x-12y+16=4\cdot 2x-4\cdot 3y+4\cdot 4$ $=4(2x-3y+4)$ We should always check our answer using the distributive property. $4(2x-3y+4)=8x-12y+16$	**10.** Factor $12x+24y-30$ and check your solution.
11. Factor $5xy^2+15xy^3$. We find the GCF of the expression. $5xy^2=\quad 5\cdot x\cdot y^2$ $15xy^3=3\cdot 5\cdot x\cdot y^3$ $\downarrow\ \downarrow\ \downarrow$ The GCF $=\ 5\cdot x\cdot y^2=5xy^2$ We factor out the GCF from the expression. $\text{GCF}\left(\frac{5xy^2}{\text{GCF}}+\frac{15xy^3}{\text{GCF}}\right)$ $5xy^2+15xy^3=5xy^2\left(\frac{5xy^2}{5xy^2}+\frac{15xy^3}{5xy^2}\right)$ $=5xy^2(1+3y)$ The check is left to the student.	**12.** Factor $14x^2y+35x^4y^3$.

Extra Practice

1. Find the GCF for the expression.
$x^4y^5 - x^3y^6$

2. Factor $3x-9$. Check by multiplying.

3. Factor $30x-12y+18$. Check by multiplying.

4. Factor $27x^3y^2-34a^3b^3$. Check by multiplying.

Concept Check

For the expression $12xy+16x$

(a) Is xy part of the GCF? Why or why not?

(b) State the GCF.

(c) Factor $12xy+16x$.

MATH COACH

Mastering the skills you need to do well on the test.

Watch the MATH COACH videos in MyMathLab® or on YouTube™ while you work the problems below. These helpful hints will help you avoid making common errors on test problems.

Subtracting Polynomials—Problem 5

Perform the operations indicated. $(-7p-2)-(3p+4)$

> **Helpful Hint:** Be careful when subtracting expressions. Students often forget to change the sign of *each term* in the second polynomial. Take extra time and check your work to be sure you did not make this error.

After removing parentheses, did you get $-3p+4$ as the last two terms? Yes ____ No ____

If you answered Yes, you forgot to change the sign of every term in the second polynomial. Stop now and make this correction.

Did you get $-7p-3p=-10p$ and $-2-4=-6$ after removing parentheses? Yes ____ No ____

If you answered No, consider grouping like terms together before subtracting. Then go back and complete this step again.

If you answered any of Problem 5 incorrectly, go back and rework the problem using these suggestions.

Multiplying Binomials Using FOIL—Problem 13 Multiply $(x+3)(x-2)$.

> **Helpful Hint:** To help with accuracy, try to draw arrows when following the FOIL method. Pay particular attention to the sign of each number to avoid errors.

Using FOIL, did you get $\mathbf{F}=x^2$, $\mathbf{O}=-2x$, $\mathbf{I}=+3x$, and $\mathbf{L}=-6$? Yes ____ No ____

If you answered No, check to see if you made a sign error or used the FOIL method incorrectly.

Is the middle term of your answer equal to $+x$?
Yes ____ No ____

If you answered No, look at your work for adding the inner and outer terms. It may help with accuracy to write out the step $-2x+3x=x$.

Now go back and rework the problem using these suggestions.

Writing Variable Expressions When Comparing Two or More Quantities—Problem 17 The width of a piece of wood is 3 inches shorter than the length. Define the variable expressions for the length and width of the piece of wood.

Helpful Hint: It is a good idea to let the variable represent the quantity to which things are being compared. When writing expressions, remember that math symbols are not always written in the same order as they are read in the statement.

Did you let the variable represent the length?

Yes ____ No ____

If you answered No, stop now and make this correction.

Did you write the variable expression: width $= 3 - L$?
Yes ____ No ____

If you answered Yes, go back and reread the problem carefully. Notice that the phrase "is shorter than the length" means 3 taken away from L.

Remember that the question asks for the variable expressions for both length and width, so you will need to write both of these answers.

If you answered any of Problem 17 incorrectly, go back and rework the problem using these suggestions.

Factoring Out the GCF from the Polynomial—Problem 25 Factor $2x^2y - 6xy^2$.

Helpful Hint:

- Write each number as a product using *only* prime factors. Then align all common prime factors.
- When forming the GCF, choose factors that are common to every term and use the smallest power on these common factors.
- After factoring, double-check to see if the product is factored completely.

Did you factor each term as follows?

$$2x^2y = 2^1 \cdot \quad x^2 \cdot y^1$$
$$6xy^2 = 2^1 \cdot 3^1 \cdot x^1 \cdot y^2$$

Notice that common factors are aligned. ↑ ↑ ↑

Yes ____ No ____

If you answered No, go back and perform this step again.

Did you factor $2x^2y - 6xy^2$ into $2\left(x^2y - 3xy^2\right)$?

Yes ____ No ____

If you answered Yes, the expression is *not factored completely*. To avoid this situation, you must find the correct GCF. Align common factors as shown in the first step. Then you will see that 2, x, and y are common to both terms and therefore part of the GCF.

Did you choose $2x^2y^2$ as the GCF?
Yes ____ No ____

If you answered Yes, check your work again. You *did not use* the *smallest* power on each common factor: 2^1, x^1, and y^1.

Now go back and rework the problem using these suggestions.

Name: ______________________________ Date: ________________
Instructor: ______________________________ Topic: ________________

Module 9 Equations, Inequalities, and Applications
Topic 1 Addition and Multiplication Principles of Equality

Vocabulary
identity • equivalent equations • addition principle • opposite in sign • terminating decimal • multiplicative inverse • additive inverse • multiplication principle • division principle

1. A number that is opposite in sign to another number is called its ___________.

2. "When checking the solution to an equation, if the same value appears on both sides of the equals sign at the last step, the equation is called a(n)______________."

3. The ______________ states that dividing both sides of an equation by the same nonzero number results in an equivalent equation.

4. A ___________ is a decimal with a definite number of digits.

Example	Student Practice
1. Solve for x. $x+16=20$ Use the addition principle to add -16 to both sides and simplify. $x+16=20$ $x+16+(-16)=20+(-16)$ $x+0=4$ $x=4$ Substitute the value found for x into the original equation to verify the solution. $x+16=20$ $4+16\stackrel{?}{=}20$ $20=20$ The solution checks.	**2.** Solve for x. $x+8=42$

Vocabulary Answers: 1. additive inverse 2. identity 3. multiplication principle 4. multiplicative inverse

Example	Student Practice
3. Is 10 the solution to the equation $-15+2=x-3$? If not, find the solution. Substitute 10 for x in the equation. $-15+2=x-3$ $-15+2\stackrel{?}{=}10-3$ $13\neq 7$ The values are not equal. Thus, 10 is not the solution. Solve the original equation to find the solution. Start by simplifying. $-15+2=x-3$ $-13=x-3$ Add 3 to both sides. $-13=x-3$ $-13+3=x-3+3$ $-10=x$ The check is left to the student.	**4.** Is 26 the solution to the equation $x-12=28-4$? If not, find the solution.
5. Find the value of x that satisfies the equation. $\frac{1}{5}+x=-\frac{1}{10}+\frac{1}{2}$ To be combined, the fractions must have common denominators. Rewrite each fraction as an equivalent fraction with a denominator of 10 and simplify. $\frac{1}{5}\cdot\frac{2}{2}+x=-\frac{1}{10}+\frac{1}{2}\cdot\frac{5}{5}$ $\frac{2}{10}+x=-\frac{1}{10}+\frac{5}{10}$ $\frac{2}{10}+x=\frac{4}{10}$ Add $-\frac{2}{10}$ to each side.	**6.** Find the value of x that satisfies the equation. $\frac{1}{3}+x=-\frac{1}{21}+\frac{1}{7}$

Example	Student Practice
(Continued) $\frac{2}{10}+\left(-\frac{2}{10}\right)+x=\frac{4}{10}+\left(-\frac{2}{10}\right)$ $x=\frac{2}{10}=\frac{1}{5}$ The check is left to the student.	
7. Solve for x. $\frac{1}{3}x=-15$ We know that $(3)\left(\frac{1}{3}\right)=1$. Multiply each side of the equation by 3 to isolate x. $3\left(\frac{1}{3}x\right)=3(-15)$ $\left(\frac{3}{1}\right)\left(\frac{1}{3}\right)x=-45$ $x=-45$ Check. $\frac{1}{3}(-45)\stackrel{?}{=}-15$ $-15=-15$ The solution checks.	**8.** Solve for x. $-\frac{1}{7}x=-3$
9. Solve for x. $5x=125$ Divide both sides by 5. $\frac{5x}{5}=\frac{125}{5}$ $x=25$ Check. $5x=125$ $5(25)\stackrel{?}{=}125$ $125=125$ The solution checks.	**10.** Solve for x. $6x=42$

Example	Student Practice
11. Solve for x. $-78=5x-8x$ Combine the like terms on the right side. $-78=5x-8x$ $-78=-3x$ Divide both sides by the coefficient -3. $\frac{-78}{-3}=\frac{-3x}{-3}$ $26=x$ The check is left to the student.	**12.** Solve for x. $24=4x-8x$

Extra Practice

1. Solve for x. Check your answers.

$1.7+4.2+x=19.23-9.8$

2. Is 28 the solution to the equation $-17+x-4=12-8+3$? If it is not, find the solution.

3. Solve for x. Check your answers.

$-5=\frac{x}{5}$

4. Is 7 the solution to the equation $\frac{3}{7}x=3$? If it is not, find the correct solution.

Concept Check

Explain how you would check to verify whether $x=3.8$ is the solution to $-1.3+1.6+3x=-6.7+4x+3.2$.

Name: ______________________________ Date: ________________
Instructor: ___________________________ Topic: ________________

Module 9 Equations, Inequalities, and Applications
Topic 2 Using Both Principles of Equality Together

Vocabulary

distributive property • like terms • multiplication principle • addition principle

1. To solve an equation of the form $ax^2 + bx = c$, we must use both the addition principle and the ____________.

2. If the variable appears on both sides of the equation, apply the ____________ to the variable term to collect the variable terms on one side of the equation.

3. If an equation contains parentheses, first apply the ____________ to remove the parentheses.

4. Where it is possible, first collect __________ on one or both sides of the equation.

Example	Student Practice
1. Solve for x. $5x+3=18$ Use the addition principle to add -3 to both sides of the equation and simplify. $5x+3=18$ $5x+3+(-3)=18+(-3)$ $5x=15$ Use the division principle to divide both sides by 5 and simplify. $5x=15$ $\frac{5x}{5}=\frac{15}{5}$ $x=3$ Check. $5(3)+3\stackrel{?}{=}18$ $15+3\stackrel{?}{=}18$ $18=18$	**2.** Solve for x. $6x-8=4$

Vocabulary Answers: 1. multiplication principle 2. addition principle 3. distributive property 4. like terms

Example	Student Practice
3. Solve for x and check your solution. $9x+3=7x-2$ Add $7x$ to both sides of the equation to get all variable terms on one side and combine like terms. $9x+3=7x-2$ $9x+(-7x)+3=7x+(-7x)-2$ $2x+3=-2$ Add -3 to both sides and simplify. $2x+3=-2$ $2x+3+(-3)=-2+(-3)$ $2x=-5$ Divide both sides by 2 and simplify. $2x=-5$ $\frac{2x}{2}=-\frac{5}{2}$ $x=-\frac{5}{2}$ Check. $9x+3=7x-2$ $9\left(-\frac{5}{2}\right)+3\stackrel{?}{=}7\left(-\frac{5}{2}\right)-2$ $-\frac{45}{2}+3\stackrel{?}{=}-\frac{35}{2}-2$ $-\frac{45}{2}+\frac{6}{2}\stackrel{?}{=}-\frac{35}{2}-\frac{4}{2}$ $-\frac{39}{2}=-\frac{39}{2}$ The solution checks.	**4.** Solve for x and check your solution. $4x+6=x+4$

Example	Student Practice
5. Solve for x and check your solution. $5x+26-6=9x+12x=25$ Combine like terms. Then add $(-5x)$ to both sides. $5x+26-6=9x+12x$ $5x+20=21x$ $5x+(-5x)+20=21x+(-5x)$ $20=16x$ Divide both sides by 16. $\frac{20}{16}=\frac{16x}{16}$ $\frac{5}{4}=x$ The check is left for the student.	**6.** Solve for x and check your solution. $3x+22-5=8x-x$
7. Solve for x and check your solution. $4(x+1)-3(x-3)=25$ Multiply by 4 and -3 to remove the parentheses. Then combine like terms. Be careful of the signs. $4(x+1)-3(x-3)=25$ $4x+4-3x+9=25$ $x+13=25$ Subtract 13 from both sides to isolate the variable. $x+13=25$ $x+13-13=25-13$ $x=12$ The check is left to the student.	**8.** Solve for x and check your solution. $-6(x-3)=4(2x+5)$

Example	Student Practice
9. Solve for x and check your solution. $0.3(1.2x-3.6)=4.2x-16.44$	**10.** Solve for z and check your solution. $0.6(z-2)=0.4(z+7)$

Remove the parentheses.

$$0.3(1.2x-3.6)=4.2x-16.44$$
$$0.36x-1.08=4.2x-16.44$$

Solve.

$$0.36x-1.08=4.2x-16.44$$
$$0.36x+(-0.36x)-1.08=4.2x+(-0.36x)-16.44$$
$$-1.08+16.44=3.84x-16.44+16.44$$
$$15.32=3.84x$$
$$\frac{15.36}{3.84}=\frac{3.84x}{3.84}$$
$$4=x$$

The check is left to the student.

Extra Practice

1. Solve for x and check your solution.
$3x+15=30$

2. Solve for x and check your solution.
$2x-3=x+9$

3. Solve for x and check your solution.
$18-3=16x-4x+12$

4. Solve for z and check your solution.
$3(2z-3)-3=3z-2(2z-1)$

Concept Check

Explain how you would solve the equation $3(x-2)+2=2(x-4)$.

Name: ______________________________ Date: ________________

Instructor: ______________________________ Topic: ________________

Module 9 Equations, Inequalities, and Applications
Topic 3 Solving Equations with Fractions

Vocabulary

LCD • multiplication principle • infinite number of solutions • no solution • decimal

1. A(n) ____________ is a fraction written a special way.

2. If there is no value of x for which an equation is true, the equation has ____________.

3. To eliminate the fractions in an equation, multiply each term by the ____________.

Example	Student Practice
1. Solve for x. $\frac{1}{4}x-\frac{2}{3}=\frac{5}{12}x$	**2.** Solve for x. $\frac{1}{18}x+\frac{2}{3}=\frac{5}{6}x$

To eliminate the fractions, multiply both sides by the LCD, 12, and apply the distributive property to simplify.

$$\frac{1}{4}x-\frac{2}{3}=\frac{5}{12}x$$

$$12\left(\frac{1}{4}x-\frac{2}{3}\right)=12\left(\frac{5}{12}x\right)$$

$$\left(\frac{12}{1}\right)\left(\frac{1}{4}\right)x-\left(\frac{12}{1}\right)\left(\frac{2}{3}\right)=\left(\frac{12}{1}\right)\left(\frac{5}{12}\right)x$$

$$3x-8=5x$$

Add $-3x$ to both sides, then divide by 2.

$$3x-8=5x$$

$$3x+(-3x)-8=5x+(-3x)$$

$$-8=2x$$

$$-\frac{8}{2}=\frac{2x}{2}$$

$$-4=x$$

The check is left to the student.

Vocabulary Answers: 1. decimal 2. no solution 3. LCD

Example

3. Solve for x and check your solution.

$$\frac{x}{3}+3=\frac{x}{5}-\frac{1}{3}$$

Multiply each term by the LCD, 15.

$$15\left(\frac{x}{3}\right)+15(3)=15\left(\frac{x}{5}\right)-15\left(\frac{1}{3}\right)$$
$$5x+45=3x-5$$

Subtract $3x$ and 45 from both sides.

$$5x+45=3x-5$$
$$5x-3x+45-45=3x-3x-5-45$$
$$2x=-50$$

Divide by 2.

$$\frac{2x}{2}=\frac{-50}{2}$$
$$x=-25$$

Check.

$$\frac{(-25)}{3}+3\stackrel{?}{=}\frac{(-25)}{5}-\frac{1}{3}$$
$$-\frac{25}{3}+\frac{9}{3}\stackrel{?}{=}-\frac{5}{1}-\frac{1}{3}$$
$$-\frac{16}{3}\stackrel{?}{=}-\frac{15}{3}-\frac{1}{3}$$
$$-\frac{16}{3}=-\frac{16}{3}$$

The solution checks.

Student Practice

4. Solve for x and check your solution.

$$\frac{x}{2}+2=\frac{x}{3}-\frac{1}{2}$$

Example	Student Practice
5. Solve for x and check your solution. $\frac{1}{3}(x-2)=\frac{1}{5}(x+4)+2$ Remove the parentheses. $\frac{x}{3}-\frac{2}{3}=\frac{x}{5}+\frac{4}{5}+2$ Multiply all terms by the LCD, 15. Then, solve the resulting equation. $15\left(\frac{x}{3}\right)-15\left(\frac{2}{3}\right)=15\left(\frac{x}{5}\right)+15\left(\frac{4}{5}\right)+15(2)$ $5x-10=3x+12+30$ $5x-10=3x+42$ $2x=52$ $x=26$ The check is left to the student.	**6.** Solve for x and check your solution. $\frac{1}{4}(x-4)=\frac{1}{6}(x+2)+8$
7. Solve for x. $0.2(1-8x)+1.1=-5(0.4x-0.3)$ Remove parentheses. Then, multiply each term by 10 to get integer coefficients and solve. $0.2-1.6x+1.1=-2.0x+1.5$ $2-16x+11=-20x+15$ $4x+13=15$ $4x=2$ $x=\frac{1}{2}$ or 0.5 The decimal form of the solution should only be given if it is a terminating decimal. The check is left to the student.	**8.** Solve for x. $0.3(1-6x)+1.3=3(-0.8x+0.6)$

Extra Practice

1. Solve for p. Check your solution.

$$\frac{1}{3}p = p - 1$$

2. Solve for x. Check your solution.

$$\frac{x+2}{3} = \frac{x}{18} + \frac{1}{9}$$

3. Solve for x. Check your solution.

$$\frac{3}{4}(2x+6) - 3 = 3(x+3)$$

4. Solve for x. Check your solution.

$$0.4(x-6) = 0.7(2x+3) + 0.5$$

Concept Check

Explain how you would solve $\frac{x+5}{6} = \frac{x}{2} + \frac{3}{4}$.

Name: ______________________________ Date: ________________
Instructor: ___________________________ Topic: ________________

Module 9 Equations, Inequalities, and Applications
Topic 4 Formulas

Vocabulary

formulas • distance • rate • time • straight line

1. The formula $d = rt$ is used to find the ____________ when we know the rate and time.

2. ____________ are equations with one or more variables that describe real-life situations.

3. A(n) ____________ can be described by an equation of the form $Ax + By = C$, where A, B, and C are real numbers and A and B are not both zero.

Example	**Student Practice**
1. American Airlines recently scheduled a nonstop flight in a new Boeing 777 from Chicago to London. The approximate air distance traveled on the flight was 3975 miles. The average speed of the aircraft on the trip was 530 miles per hour. How many hours did it take the Boeing 777 to fly this trip? (Source:www.aa.com) Use the distance formula, $d = rt$ and substitute the known values. $d = rt$ $3975 = 530t$ Divide both sides by 530 to solve for t. $3975 = 530t$ $\frac{3975}{530} = \frac{530t}{530}$ $7.5 = t$ It took the jet about 7.5 hours to fly this trip from Chicago to London.	**2.** An airline is planning a route from Charlotte, NC, to Los Angeles, CA. The distance between the two cities is approximately 2120 miles and the jet can do the trip in 5 hours. Find the average rate of speed for the plane.

Vocabulary Answers: 1. distance 2. formulas 3. straight line

Example	Student Practice
3. Solve for t. $d = rt$ Divide both sides of the equation by the coefficient of t, which is r, to isolate t. $d = rt$ $\frac{d}{r} = \frac{rt}{r}$ $\frac{d}{r} = t$	**4.** Solve for r. $C = 2\pi r$
5. Solve for y. $3x - 2y = 6$ To isolate the term containing y, subtract $3x$ from both sides. $3x - 2y = 6$ $3x - 3x - 2y = 6 - 3x$ $-2y = 6 - 3x$ Divide both sides by -2. $-2y = 6 - 3x$ $\frac{-2y}{-2} = \frac{6-3x}{-2}$ $y = \frac{6-3x}{-2}$ Rewrite the right hand side as two fractions, and simplify. $y = \frac{6-3x}{-2}$ $y = \frac{6}{-2} + \frac{-3x}{-2}$ $y = \frac{3x}{2} - 3$ This is known as the slope-intercept form of the equation of a line.	**6.** Solve for y. $-x + 4y = -12$

Example	Student Practice
7. A trapezoid is a four-sided figure with two parallel sides. In some houses, two of the sides of the roof are in the shape of the trapezoid. If the parallel sides are a and b, and the altitude is h, the area is given by $A=\frac{h}{2}(a+b)$. Solve the equation for a.	**8.** Solve the equation $M=\frac{2}{5}(4y+2x)$ for x.

Remove the parentheses by multiplying.

$$A=\frac{h}{2}(a+b)$$
$$A=\frac{ha}{2}+\frac{hb}{2}$$

Multiply all terms by the LCD, 2, and simplify.

$$A=\frac{ha}{2}+\frac{hb}{2}$$
$$2(A)=2\left(\frac{ha}{2}\right)+2\left(\frac{hb}{2}\right)$$
$$2A=ha+hb$$

Isolate the a term by subtracting hb from both sides.

$$2A=ha+hb$$
$$2A-hb=ha+hb-hb$$
$$2A-hb=ha$$

Divide by h, the coefficient of a.

$$2A-hb=ha$$
$$\frac{2A-hb}{h}=\frac{ha}{h}$$
$$\frac{2A-hb}{h}=a$$

Extra Practice

1. The perimeter of a square is given by $P = 4s$. Solve for s.

2. The slope-intercept form of line is represented by $y = mx + b$. Solve for x.

3. The volume of a rectangular prism is given by $V = LWH$.

(a) Solve for W.

(b) Use this result to find W when $V = 616 \text{ ft}^3$, $L = 8$ ft, and $H = 11$ ft.

4. The equation of a line is $3x + 7y = 13$.

(a) Solve for y.

(b) Use this result to find y when $x = 2$.

Concept Check

Explain how you would solve for x in the equation $y = \frac{3}{8}x - 9$.

Name: ______________________________ Date: ________________
Instructor: ______________________________ Topic: ________________

Module 9 Equations, Inequalities, and Applications
Topic 5 Using Equations to Solve Word Problems

Vocabulary
understand the problem • write an equation • solve and state the answer • check

1. When solving word problems, the last step is to ____________, which involves checking the solution in the original equation and determining if the answer is reasonable.

2. When solving word problems, the third step is to ____________, which involves solving the equation to determine the answer to the problem.

3. When solving word problems, the second step is to ____________, which involves looking for key words to help you to translate the words into algebraic symbols and expressions.

Example	**Student Practice**
1. Two-thirds of a number is eighty-four. What is the number? Understand the problem. Draw a sketch. Let $x =$ the unknown number. Write an equation. The word "of" translates to multiplication and the word "is" translates to equals. $\frac{2}{3}x = 84$ Solve and state the answer. $\frac{2}{3}x = 84$ $2x = 252$ $x = 126$ The check is left to the student.	**2.** Four-fifths of a number is sixty-eight. What is the number?

Vocabulary Answers: 1. check 2. solve and state the answer 3. write an equation

Example	Student Practice
3. Five more than six times a quantity is three hundred five. Find the number. Understand the problem. Read the problem carefully. You may not need to draw a sketch. Let $x =$ the unknown quantity. Write an equation. "More than" translates to addition, "times" translates to multiplication, and "is" translates to equals. $5+6x=305$ Solve and state the answer. $6x+5=305$ $6x+5-5=305-5$ $6x=300$ $\frac{6x}{6}=\frac{300}{6}$ $x=50$ The quantity, or number, is 50. Check. Is five more than six times 50 three hundred five? $6(50)+5\stackrel{?}{=}305$ $300+5\stackrel{?}{=}305$ $305=305$ The answer checks.	**4.** Nineteen more than four times a quantity is two hundred seventy-nine. Find the number.

Example	Student Practice
5. The smallest angle of an isosceles triangle measures $24°$. The other two angles are larger. What are the measurements of the other two angles? Let $x =$ the measure in degrees of each of the larger angles. Draw a sketch. 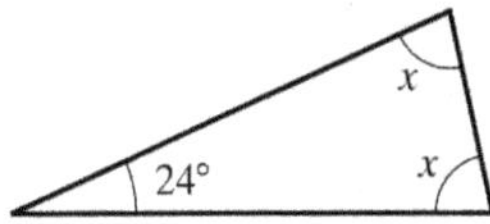 Write an equation showing the sum of all three angles is $180°$, solve for x. $24° + x + x = 180°$ $x = 78°$	**6.** The largest angle of an isosceles triangle measures $116°$. The other two angles are smaller. What are the measurements of the other two angles?
7. Two people travel in separate cars. They each travel a distance of 330 miles on an interstate highway. To maximize fuel economy, Fred travels at exactly 50 mph. Sam travels at exactly 55 mph. How much time did the trip take each person? Read the problem carefully and create a Mathematics Blueprint. Write an equation using the formula distance $= (\text{rate})(\text{time})$ or $d = rt$. Substitute the known values into the formula and solve for t. $d = rt$ and $d = rt$ $330 = 50t_f$ and $330 = 55t_s$ $6.6 = t_f$ and $6.6 = 6t_s$ It took Fred 6.6 hours. It took Sam 6 hours. The check is left to the student.	**8.** Two people travel in separate cars. They each travel a distance of 780 miles on an interstate highway. To maximize fuel economy, John travels at exactly 65 mph. Yuri travels at exactly 60 mph. How much time did the trip take each person?

Extra Practice

1. What number minus 312 gives 234? Check your solution.

2. A number is tripled and then increased by 27. The result is 72. What is the original number? Check your solution.

3. The local health food store has six times as many energy drinks as energy bars. There are 108 energy drinks in stock. How many energy bars are in stock? Check to see if your answer is reasonable.

4. Amy's cell phone company charges $10.50 per month for 200 minutes of use and $0.15 for each additional minute. Last month Amy's cell phone bill was $55.50. How many additional minutes was she charged for? Check to see if your answer is reasonable.

Concept Check

Explain how you would set up an equation to solve the following problem.
Phil purchased two shirts for $23 each and then purchased several pairs of socks. The socks were priced at $0.75 per pair. How many pairs of socks did he purchase if the total cost was $60.25?

Name: ______________________________ Date: ________________
Instructor: ____________________________ Topic: ________________

Module 9 Equations, Inequalities, and Applications
Topic 6 Solving Word Problems Involving Money and Percents

Vocabulary
percent • simple interest • compound interest • interest

1. ___________ is a charge for borrowing money or an income from investing money.

2. ___________ is computed by multiplying the amount of money borrowed or invested times the rate of interest times the period of time over which it is borrowed or invested.

3. Applied situations often require finding a ___________ of an unknown number.

Example	Student Practice
1. A business executive rented a car. The Supreme Car Rental Agency charged \$39 per day and \$0.28 per mile. The executive rented the car for two days and the total rental cost was computed to be \$176. How many miles did the executive drive the rented car? Understand the problem. It is known that it costs \$176 to rent the car for two days. It is necessary to find the number of miles the car was driven. Let $m=$ the number of miles driven in the rented car. Write an equation. Use the relationship for calculating the total cost. per-day cost + mileage cost = total cost $(39)(2) \quad + \quad (0.28)m \quad = \quad 176$ Solve and state the answer. $78+0.28m=176$ $0.28m=98$ $m=350$ The executive drove 350 miles. The check is left to the student.	**2.** A business woman rented a room at a motel. The motel charged \$52 per day and \$2.50 per hour of internet use. The business woman stayed in the room for four days and the total charge was computed to be \$248. How many hours of internet use did the business woman accrue?

Vocabulary Answers: 1. interest 2. simple interest 3. percent

Example	Student Practice
3. A sofa was marked with the following sign: "The price of this sofa has been reduced by 23%. You save \$138 if you buy now." What was the original price of the sofa? Understand the problem. Let $s=$ the original price of the sofa. Then $0.23s=$ the amount of the price reduction, which is \$138. Write an equation and solve. $0.23s=138$ $\frac{0.23s}{0.23}=\frac{138}{0.23}$ $s=600$ The original price of the sofa was \$600. The check is left to the student.	**4.** A refrigerator was marked with the following sign: "The price of this refrigerator has been reduced by 33%. You save \$264 if you buy now." What was the original price of the refrigerator?
5. A woman invested an amount of money in two accounts for one year. She invested some at 8% simple interest and the rest at 6% simple interest. Her total amount invested was \$1250. At the end of the year she had earned \$86 in interest. How much money had she invested in each account? The simple interest formula is $I = prt$. Let $x=$ amount invested at 8%. Then $\$1250-x=$ amount invested at 6%. Write an equation and solve for x. $0.08x+0.06(1250-x)=86$ $x=550$ The amount invested at 8% is \$550. The amount invested at 6% is \$700.	**6.** A man invested an amount of money in two accounts for one year. He invested some at 6% simple interest and the rest at 5% simple interest. His total amount invested was \$3500. At the end of the year he had earned \$195 in interest. How much money had he invested in each account?

Example	Student Practice
7. When Bob got out of math class, he had to make a long-distance call. He had exactly enough dimes and quarters to make a phone call that would cost \$2.55. He had one fewer quarter than he had dimes. How many coins of each type did he have?	**8.** When Rebecca got out of chemistry class, she went to the vending machines for a snack. She had exactly enough nickels and dimes to get a combination of items costing \$2.75. She had four fewer dimes than she had nickels. How many coins of each type did she have?

Let $d =$ the number of dimes. Then $d-1=$ the number of quarters. The total value of the coins was \$2.55. Each dime is worth \$0.10 and each quarter is worth \$0.25. So, d dimes are worth $0.10d$ dollars and $(d-1)$ quarters are worth $0.25(d-1)$ dollars. Now write an equation for the total value, and solve.

$$\begin{aligned} 0.10d+0.25(d-1)&=2.55 \\ 0.10d+0.25d-0.25&=2.55 \\ 0.35d-0.25&=2.55 \\ 0.35d&=2.80 \\ d&=8 \end{aligned}$$

Find the number of quarters Bob has, $d-1=8-1=7$. Thus, Bob has eight dimes and seven quarters.

Check the answer. Bob has $8-7=1$ less quarter than he has dimes. Check that eight dimes and seven quarters are worth \$2.55.

$$\begin{aligned} 8(\$0.10)+7(\$0.25)&\stackrel{?}{=}\$2.55 \\ \$0.80+\$1.75&\stackrel{?}{=}\$2.55 \\ \$2.55&=\$2.55 \end{aligned}$$

The solution checks.

Extra Practice

1. Angelina received a pay raise this year. The raise was 5% of last year's salary. This year, Angelina earned $17,640. What was her salary before the raise?

2. Find the simple interest on $6,500 borrowed at 15% for one year.

3. Randall is due to receive a 7% raise, which in dollars will be $3,780 per year. What is his current salary?

4. Mr. Finch keeps money in his pillowcase. Right now, he has equal numbers of five, ten, and twenty-dollar bills, with no other denominations. He has exactly $1505. How many bills does he have all together?

Concept Check

Explain how you would set up an equation to solve the following problem.
Robert has $2.55 in change consisting of nickels, dimes, and quarters. He has twice as many dimes as quarters. He has one more nickel than he has quarters. How many of each coin does he have?

Name: ______________________________ Date: ______________

Instructor: ___________________________ Topic: ______________

Module 9 Equations, Inequalities, and Applications
Topic 7 Solving Inequalities in One Variable

Vocabulary

inequalities • is less than • is greater than • solution • solution set • graph

1. The set of all numbers that make the inequality true is called a(n) ___________.

2. Comparisons of values, such as one value being greater than or less than another value, are called ___________.

3. A(n) ___________ is a visual representation of the solution set.

4. One number ___________ another if it is to the right of the other on the number line.

Example	Student Practice
1. In each statement, replace the question mark with the symbol < or >.	**2.** In each statement, replace the question mark with the symbol < or >.
(a) $3 ? -1$ $3 > -1$ because 3 is to the right of -1 on the number line.	**(a)** $-4 ? -10$
(b) $-2 ? 1$ $-2 < 1$ because -2 is to the left of 1.	**(b)** $2 ? -2$
(c) $-3 ? -4$ $-3 > -4$ because -3 is to the right of -4.	**(c)** $-1 ? 4$
(d) $0 ? 3$ $0 < 3$ because 0 is to the left of 3.	**(d)** $0 ? -7$
(e) $-3 ? 0$ $-3 < 0$ because -3 is left of 0.	**(e)** $5 ? 8$

Vocabulary Answers: 1. solution set 2. inequality 3. graph 4. is greater than

Example

3. State each mathematical relationship in words and then graph it.

(a) $x<-2$

We state "x is less than -2."

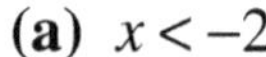

$x<-2$

(number line from −5 to 5 with open circle at −2, shaded to the left)

(b) $-3<x$

We can state that "-3 is less than x" or, equivalently, that "x is greater than -3." Be sure you see that $-3<x$ is equivalent to $x>-3$. Although both statements are correct, we usually write the variable first in a simple inequality containing a variable and a numerical value.

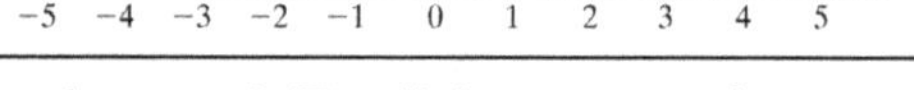

Student Practice

4. State each mathematical relationship in words and then graph it.

(a) $x<3$

(b) $\frac{5}{2}\geq x$

Example

5. Translate each English sentence into an algebraic statement.

(a) The police on the scene said that the car was traveling more than 80 miles per hour. (Use the variable s for speed.)

Since the speed must be greater than 80, we have $s>80$.

(b) The owner of the trucking company said that the payload of a truck must never exceed 4500 pounds. (Use the variable p for payload.)

If the payload of the truck can never exceed 4500 pounds, then the payload must always be less than or equal to 4500 pounds. Thus we write $p\leq 4500$.

Student Practice

6. Translate each English sentence into an algebraic statement.

(a) A student's budget is very tight, so when shopping for a car, her car payment must be less than 125 dollars per month.

(b) The maximum number of people allowed on the boat at one time is 35.

Example	**Student Practice**
7. Solve and graph. $3x+7\geq 13$	**8.** Solve and graph. $6x-4>5$

Subtract 7 from both sides.

$$3x+7\geq 13$$
$$3x+7-7\geq 13-7$$
$$3x\geq 6$$

Divide both sides by 3.

$$3x\geq 6$$
$$\frac{3x}{3}\geq\frac{6}{3}$$
$$x\geq 2$$

−5 −4 −3 −2 −1 0 1 2 3 4 5 x

Example	**Student Practice**
9. Solve and graph. $5-3x>7$	**10.** Solve and graph. $8-3x<9$

Subtract 5 from both sides and simplify.

$$5-5-3x>7-5$$
$$-3x>2$$

Divide by -3. When dividing by a negative number, the inequality is reversed.

$$-3x>2$$
$$\frac{-3x}{-3}<\frac{2}{-3}$$
$$x<-\frac{2}{3}$$

Note the direction of the inequality. The graph is as follows.

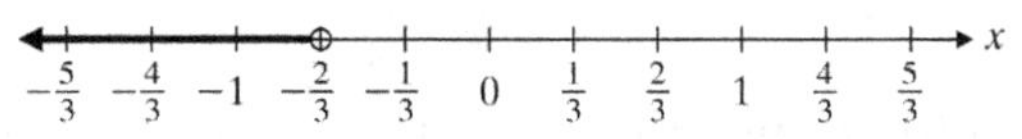

Example	Student Practice
11. Solve and graph. $-\frac{13x}{2} \le \frac{x}{2} - \frac{15}{8}$	**12.** Solve and graph. $-\frac{7x}{3} < \frac{x}{3} + \frac{23}{27}$

Multiply by the LCD, 8, and simplify.

$$8\left(-\frac{13x}{2}\right) \le 8\left(\frac{x}{2}\right) - 8\left(\frac{15}{8}\right)$$
$$-52x \le 4x - 15$$

Subtract both $4x$ from both sides and combine like terms. Then, solve the inequality. Be sure to reverse the direction of the inequality symbol.

$$-52x - 4x \le 4x - 4x - 15$$
$$-56x \le -15$$
$$x \ge \frac{15}{56}$$

The graph is as follows.

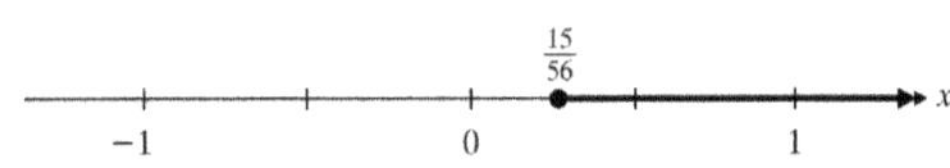

Extra Practice

1. In the statement, replace the question mark with the symbol < or >.

$\frac{1}{3}$? 5

2. Graph. $x < -4$

3. Translate the English sentence into an algebraic statement.

1,000,000 is greater than or equal to the total weight w.

4. Solve and graph. $3x + 5 < 14$

Concept Check

Explain the difference between $12 < x$ and $x < 12$. Would the graphs of these inequalities be the same or different?

MATH COACH

Mastering the skills you need to do well on the test.

Watch the MATH COACH videos in MyMathLab® or on YouTube™ while you work the problems below. These helpful hints will help you avoid making common errors on test problems.

Solving Equations with Both Parentheses and Decimals—

Problem 6 Solve for the variable. $0.8x+0.18-0.4x=0.3(x+0.2)$

> **Helpful Hint:** After removing parentheses, it might be most helpful for you to multiply both sides of the equation by 100 in order to obtain a simpler, equivalent equation without decimals. Check to make sure that you did not make any errors in calculations before solving the equation.

Did you remove the parentheses to get the equation $0.8x+0.18-0.4x=0.3x+0.06$?
Yes ____ No ____

If you answered No, go back and use the distributive property to remove the parentheses. Be careful to place the decimal point in the correct location when multiplying 0.3 and 0.2 together.

Did you multiply each term of the equation by 100 to move the decimal point two places to the right to get the equivalent equation $80x+18-40x=30x+6$?
Yes ____ No ____

If you answered No, stop and carefully complete this step before solving the equation. Remember that you may need to add a 0 to the end of a term in order to move the decimal point two places to the right.

If you answered Problem 6 incorrectly, go back and rework the problem using these suggestions.

Solving Equations with More Than One Set of Parentheses—Problem 12

Solve for the variable. $20-(2x+6)=5(2-x)+2x$

> **Helpful Hint:** Slowly complete the necessary steps to remove each set of parentheses before doing any other steps. Be careful to avoid sign errors.

Did you obtain the equation $20-2x-6=10-5x+2x$ after removing each set of parentheses?
Yes ____ No ____

If you answered No, go back and carefully use the distributive property to remove each set of parentheses. Locate any mistakes you have made and make a note of the type of error discovered.

Did you combine like terms to get the equation $14-2x=10-3x$?
Yes ____ No ____

If you answered No, stop and perform that step correctly.

Now go back and rework the problem using these suggestions.

Solving Equations with Both Fractions and Parentheses—Problem 17

Solve for x. $\frac{2}{3}(x+8)+\frac{3}{5}=\frac{1}{5}(11-6x)$

> **Helpful Hint:** Remove the parentheses first. This is the most likely place to make a mistake. Next, carefully show every step of your work as you multiply each fraction by the LCD. Be sure to check your work.

Did you remove each set of parentheses to obtain the equation $\frac{2}{3}x+\frac{16}{3}+\frac{3}{5}=\frac{11}{5}-\frac{6}{5}x$?

Yes ____ No ____

If you answered No, stop and carefully redo your steps of multiplication, showing every part of your work.

Did you identify the LCD as 15 and then multiply each term by 15 to get $10x+80+9=33-18x$?

Yes ____ No ____

If you answered No, stop and write out your steps slowly.

If you answered Problem 17 incorrectly, go back and rework the problem using these suggestions.

Solving and Graphing Inequalities on a Number Line—Problem 19

Solve and graph the inequality. $2-7(x+1)-5(x+2)<0$

> **Helpful Hint:** Be sure to remove parentheses and combine any like terms on each side of the inequality before solving for the variable. Always verify the following:
> 1) Did you multiply or divide by a negative number? If so, did you reverse the inequality symbol?
> 2) In the graph, is your choice of an open circle or closed circle correct?

Did you remove parentheses to get $2-7x-7-5x-10<0$? Did you combine like terms to obtain the inequality $-15-12x<0$? Next, did you add 15 to both sides of the inequality? Yes ____ No ____

If you answered No to any of these questions, stop now and perform those steps.

Did you remember to reverse the inequality symbol in the last step? Yes ____ No ____

If you answered No, please review the rules for when to reverse the inequality symbol and then go back and perform this step.

Did you use an open circle in your number line graph? Is your arrow pointing to the right? Yes ____ No ____

If you answered No to either question, please review the rules for how to graph an inequality involving the < or > inequality symbols.

Now go back and rework the problem using these suggestions.

Name: ______________________________ Date: ________________
Instructor: ____________________________ Topic: ________________

Module 10 Graphing and Functions
Topic 1 The Rectangular Coordinate System

Vocabulary

graphs • rectangular coordinate system • origin • x-axis • y-axis
ordered pair • coordinates • x-coordinate • y-coordinate • solution

1. The numbers in an ordered pair are often referred to as the ____________ of the point.

2. We can illustrate algebraic relationships with drawings called ____________.

3. The vertical number line above the origin is often called the ____________.

4. The first number in an ordered pair is called the ____________ and it represents the distance from the origin measured along the horizontal axis.

Example	Student Practice
1. Plot the point $(5,2)$ on a rectangular coordinate system. Label this point as A. Since the x-coordinate is 5, we first count 5 units to the right on the x-axis. Then, because the y-coordinate is 2, we count 2 units up from the point where we stopped on the x-axis. This locates the point corresponding to $(5,2)$. We mark this point with a dot and label it A.	**2.** Plot the point $(2,5)$ on the preceding rectangular coordinate system. Label this point as B.

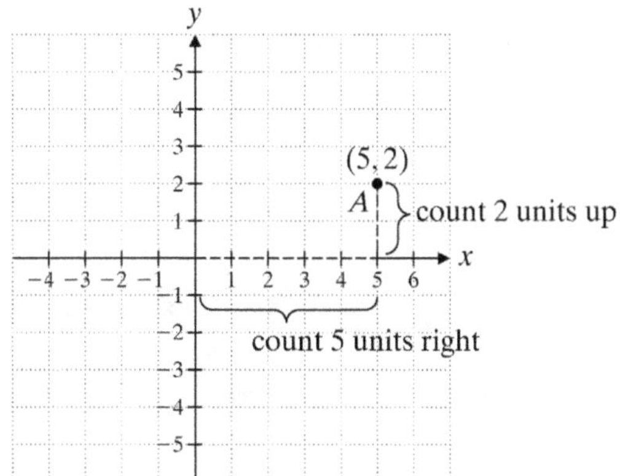

Vocabulary Answers: 1. coordinates 2. graphs 3. y-axis 4. x-coordinate

Example	Student Practice
3. Use the rectangular coordinate system to plot each point. Label the points F and G respectively.	**4.** Use the rectangular coordinate system below to plot each point. Label the points I, J, and K, respectively.

Example 3

(a) $(-5,-3)$

Notice that the x-coordinate, -5, is negative. On the coordinate grid, negative x-values appear to the left of the origin. Thus we will begin by counting 5 squares to the left, starting at the origin. Since the y-coordinate, -3, is negative, we will count 3 units down from the point where we stopped on the x-axis.

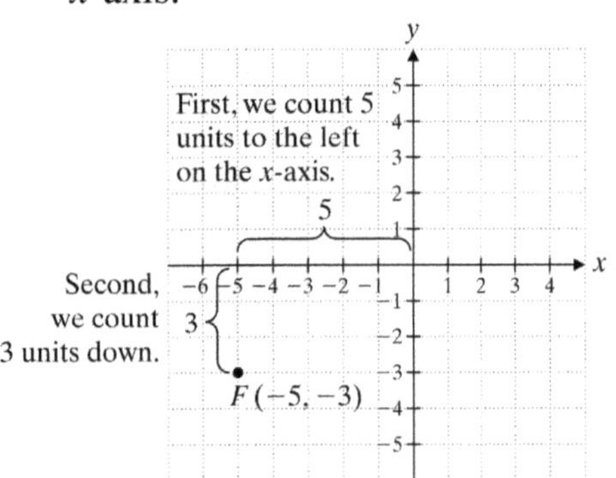

(b) $(2,-6)$

The x-coordinate is positive. Begin by counting 2 squares to the right of the origin. Then count down because the y-coordinate is negative.

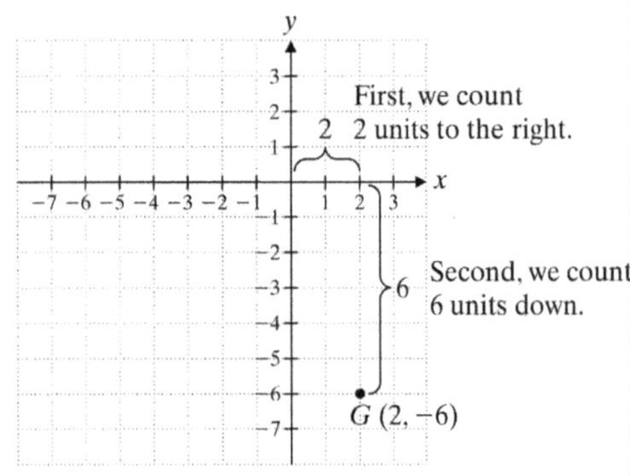

Student Practice 4

(a) $(-1,-4)$

(b) $(-4,3)$

(c) $(3,-5)$

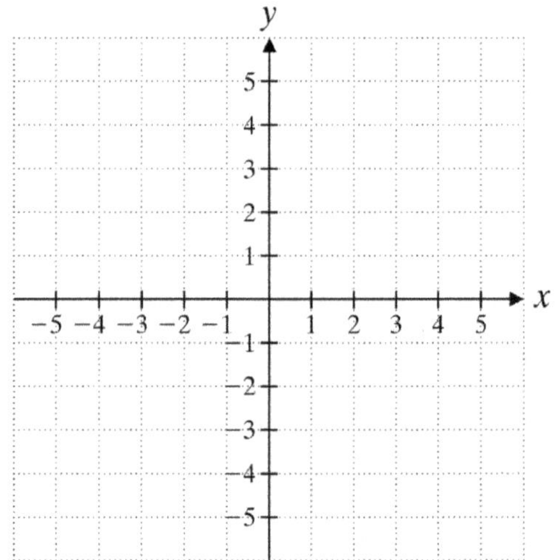

Example	Student Practice
5. What ordered pairs of numbers represent point A and point B on the graph? To find point A, move along the x-axis until you get as close as possible to A, ending at 5. Thus obtaining 5 as the first number of the ordered pair. Then count 4 units upward on a line parallel to the y-axis to reach A. So you obtain 4 as the second number of the ordered pair. Thus point A is $(5,4)$. Use the same approach to find point B: $(-5,-3)$. 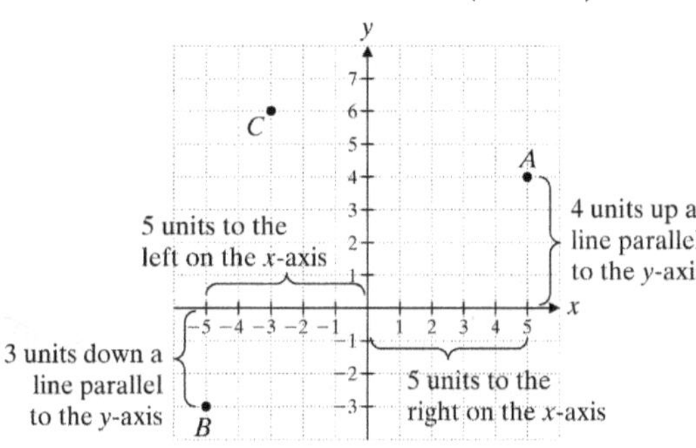	**6.** What ordered pair of numbers represents point C on the graph in Example **5**?
7. Is the ordered pair $(-1,4)$ a solution to the equation $3x+2y=5$? We replace the values for x and y to see if we obtain a true statement. Replace x with -1 and y with 4. $3x+2y=5$ $3(-1)+2(4)\stackrel{?}{=}5$ $-3+8\stackrel{?}{=}5$ $5=5$ The ordered pair $(-1,4)$ is a solution to $3x+2y=5$ because when we replace x with -1 and y with 4, we obtain a true statement.	**8.** Is the ordered pair $(3,-3)$ a solution to the equation $3x+2y=5$?

Example	Student Practice
9. Find the missing coordinate to complete the ordered-pair solution $(0,?)$ to the equation $2x+3y=15$. For the ordered pair $(0,?)$, we know that $x=0$. Replace x with 0 in the equation and solve for y. $2x+3y=15$ $2(0)+3y=15$ $0+3y=15$ $y=5$ Thus we have the ordered pair $(0,5)$.	**10.** Find the missing coordinate to complete the ordered-pair solution $(?,3)$ to the equation $6x+5y=3$.

Extra Practice

1. Plot the point $D:(-1,-3)$.

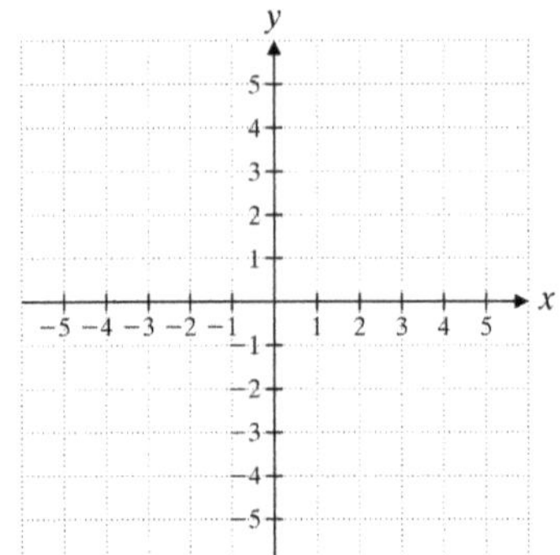

2. Consider the point plotted on the graph below. Give the coordinates for point A.

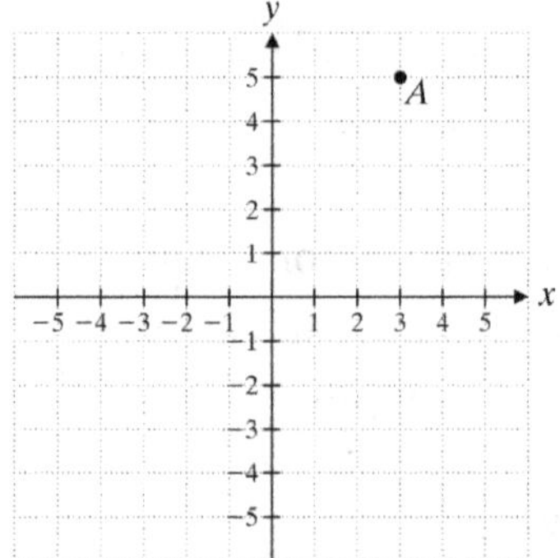

3. Find the missing coordinate to complete the ordered-pair solution to $y=-2x+3$.

(a) $(-1,?)$

(b) $(3,?)$

4. Find the missing coordinate to complete the ordered-pair solution to $4x-2y=12$.

(a) $(-1,?)$

(b) $(4,?)$

Concept Check

Explain how you would find the missing coordinate to complete the ordered-pair solution to the equation $2.5x+3y=12$ if the ordered pair was of the form $(?,-6)$.

Name: ______________________________ Date: ________________
Instructor: ____________________________ Topic :________________

Module 10 Graphing and Functions
Topic 2 Graphing Linear Equations

Vocabulary
linear equation • x-intercept • y-intercept • horizontal • vertical

1. The ____________ of a line is the point where the line crosses the y-axis.

2. The graph of any ____________ in two variables is a straight line.

3. The ____________ of a line is the point where the line crosses the x-axis.

Example	**Student Practice**
1. Find three ordered pairs that satisfy $y=-2x+4$. Then graph the resulting straight line.	**2.** Find three ordered pairs that satisfy $y=2x-3$. Then graph the resulting straight line. Use the given coordinate system.

Let $x=0$, $x=1$, and $x=3$. For each x-value, find the corresponding y-value in the equation. Place each y-value in the table next to its x-value.

x	y
0	4
1	2
3	−2

If we plot these ordered pairs and connect the three points, we get a straight line that is the graph of the equation.

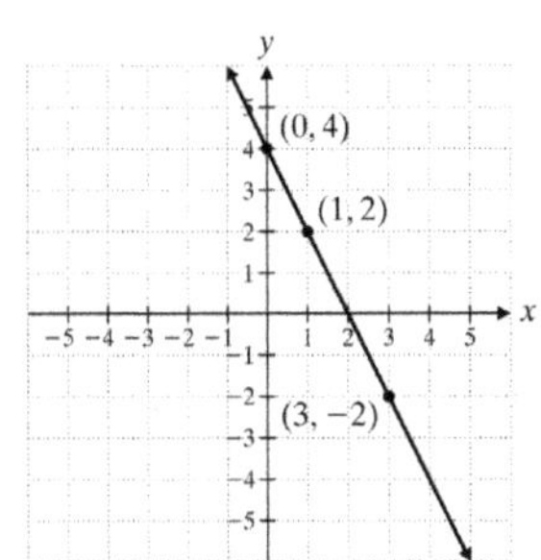

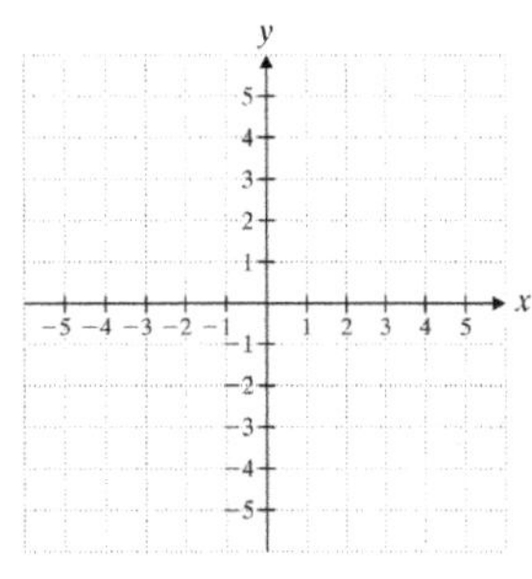

Vocabulary Answers: 1. y-intercept 2. linear equation 3. x-intercept

Example	Student Practice

3. Graph $5x-4y+2=2$.

First, we simplify the equation by subtracting 2 from each side.

$$5x-4y+2=2$$
$$5x-4y+2-2=2-2$$
$$5x-4y=0$$

Since we are free to choose any value of x, $x=0$ is a natural choice. Calculate the value of y when $x=0$.

$$5(0)-4y=0$$
$$-4y=0$$
$$y=0$$

A convenient choice for a replacement of x is a number that is divisible by 4. Let $x=4$ and $x=-4$. Follow the process used above to find the corresponding y-values and place the results in the table of values.

x	y
0	0
4	5
−4	−5

Graph the line.

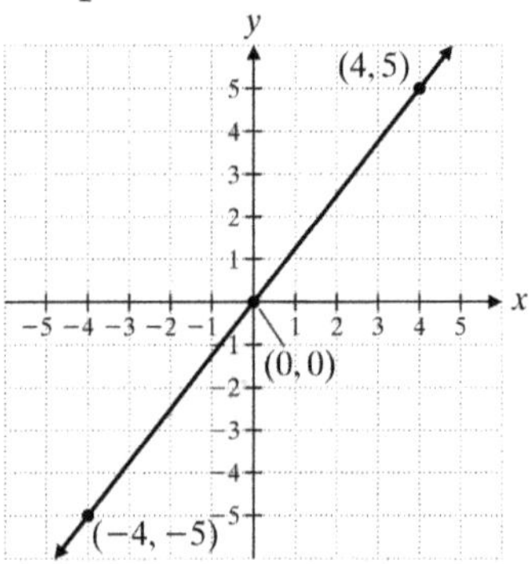

4. Graph $4x-2y+7=7$.

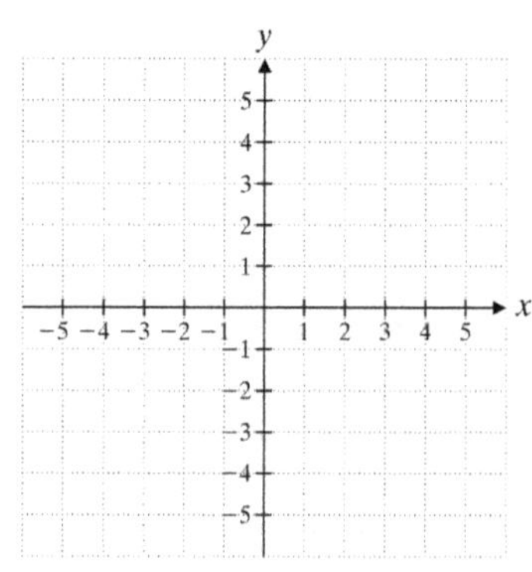

Example	Student Practice

5. Complete **(a)** and **(b)** for the equation $5y-3x=15$.

(a) State the x- and y-intercepts.

Let $y=0$.

$$5(0)-3x=15$$
$$-3x=15$$
$$x=-5$$

$(-5,0)$ is the x-intercept. Now let $x=0$.

$$5y-3(0)=15$$
$$5y=15$$
$$y=3$$

$(0,3)$ is the y-intercept.

(b) Use the intercept method to graph.

Find a third point and then graph. If we let $y=6$, $x=5$. The ordered pair is $(5,6)$.

x	y
−5	0
0	3
5	6

6. Complete **(a)** and **(b)** for the equation $y+2x=-2$.

(a) State the x- and y-intercepts.

(b) Use the intercept method to graph.

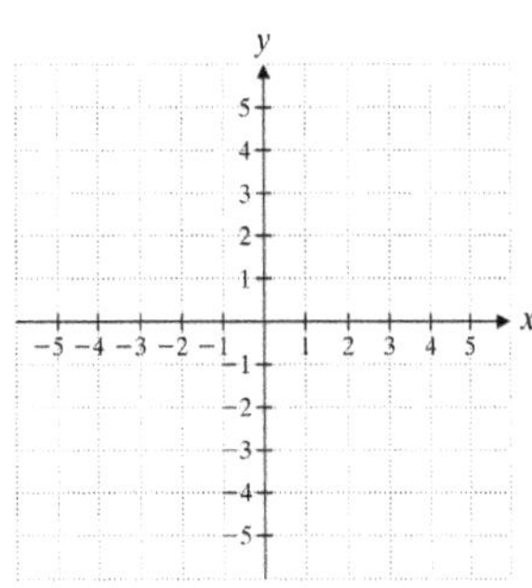

Example	Student Practice
7. Graph $2x+1=11$. Notice that there is only one variable, x, in the equation. Simplifying the equation yields $x=5$. Since the x-coordinate of every point on this line is 5, we can see that the vertical line will be 5 units to the right of the y-axis. 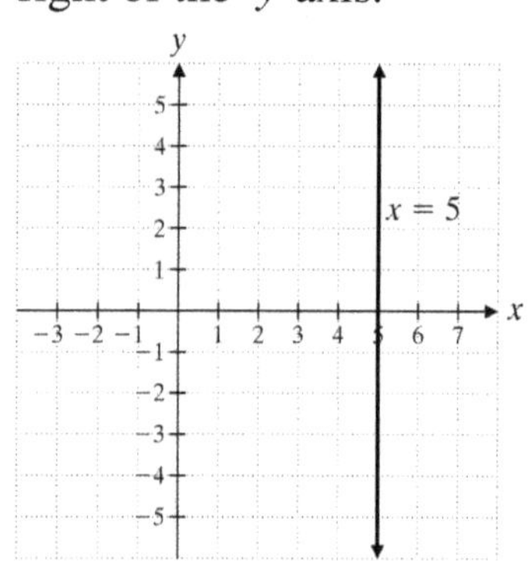	**8.** Graph $y+4=7$. 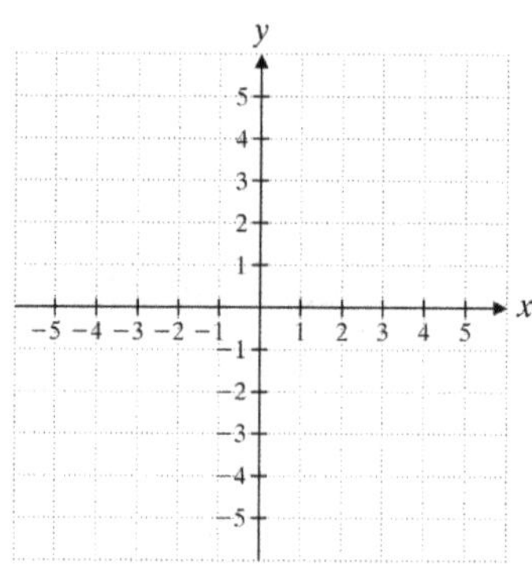

Extra Practice

1. Graph $y=2x-5$ by plotting three points and connecting them.

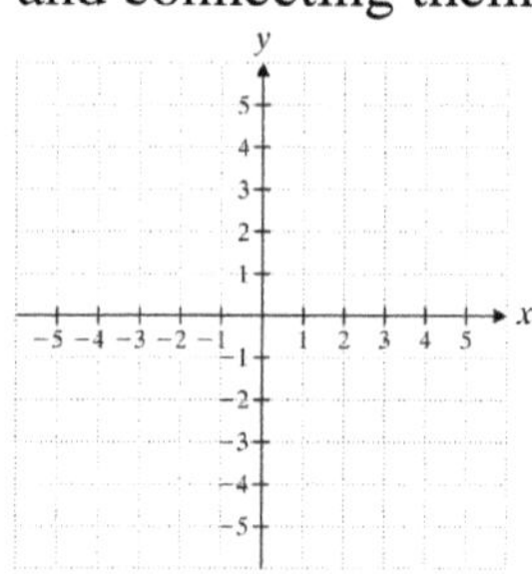

2. Graph $5x+2y=10$ by plotting three points and connecting them.

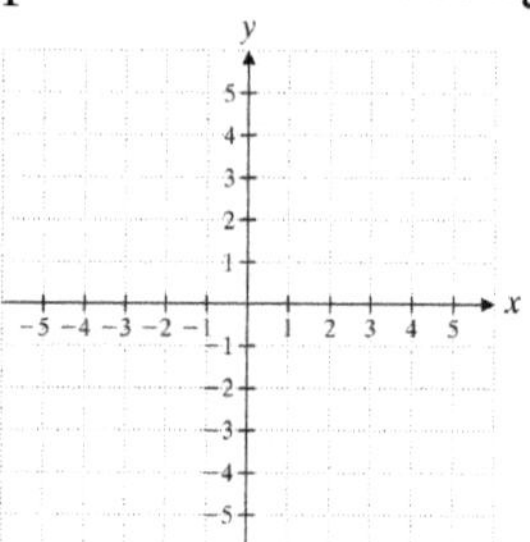

3. Graph $y=5-x$ by plotting intercepts and one other point.

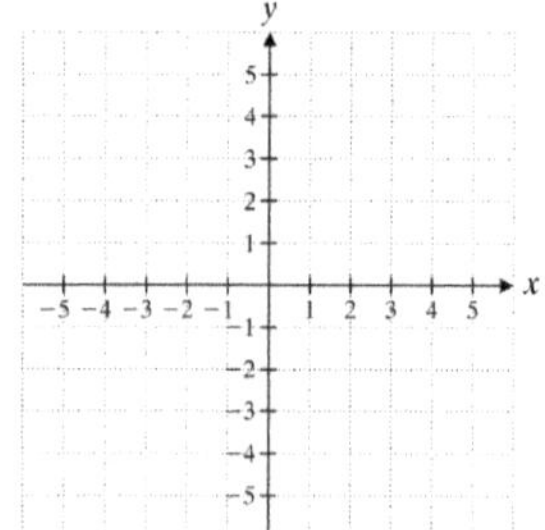

4. Graph $2x-5=y$ by plotting intercepts and one other point.

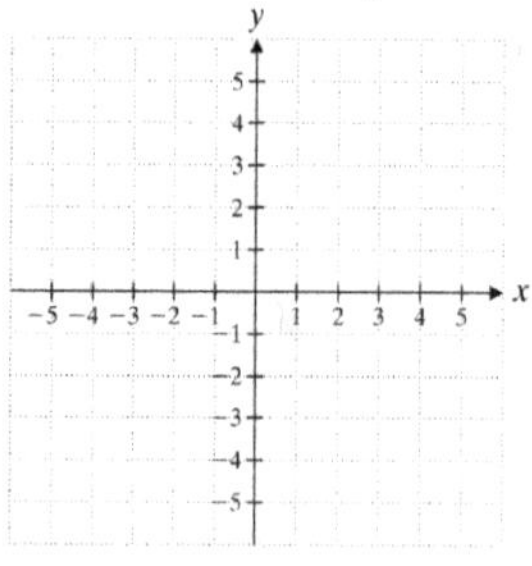

Concept Check

In graphing the equation $3y-7x=0$, what is the most important ordered pair to obtain before drawing a graph of the line? Why is that ordered pair so essential to drawing the graph?

Name: ______________________________ Date: ________________
Instructor: ___________________________ Topic: ________________

Module 10 Graphing and Functions
Topic 3 The Slope of a Line

Vocabulary
slope • positive slope • negative slope • zero slope
undefined slope • slope-intercept form

1. In a coordinate plane, the ____________ of a straight line is defined by the change in y divided by the change in x.

2. A vertical line is said to have ____________.

3. If we know the slope and the y-intercept, we can write the equation of the line in ____________.

Example	Student Practice
1. Find the slope of the line that passes through $(2,0)$ and $(4,2)$. Let $(2,0)$ be the first point (x_1, y_1) and $(4,2)$ be the second point (x_2, y_2). $\text{slope} = m = \dfrac{y_2 - y_1}{x_2 - x_1} = \dfrac{2-0}{4-2} = \dfrac{2}{2} = 1$ Note that the slope of the line will be the same if we let $(4,2)$ be the first point (x_1, y_1) and $(2,0)$ be the second point (x_2, y_2). $m = \dfrac{y_2 - y_1}{x_2 - x_1} = \dfrac{0-2}{2-4} = \dfrac{-2}{-2} = 1$ Thus, it does not matter which point you call (x_1, y_1) and which you call (x_2, y_2).	**2.** Find the slope of the line that passes through $(0,-4)$ and $(-3,-6)$.

Vocabulary Answers: 1. slope 2. undefined slope 3. slope-intercept form

Example	Student Practice
3. Find the slope of the line that passes through the given points. **(a)** $(0,2)$ and $(5,2)$ Calculate the slope. $m=\frac{y_2-y_1}{x_2-x_1}=\frac{2-2}{5-0}=\frac{0}{5}=0$ The slope of a horizontal line is 0. **(b)** $(-4,0)$ and $(-4,-4)$ Calculate the slope. $m=\frac{y_2-y_1}{x_2-x_1}=\frac{-4-0}{-4-(-4)}=\frac{-4}{0}$ Recall that division by 0 is undefined. The slope of a vertical line is undefined.	**4.** Find the slope of the line that passes through the given points. **(a)** $(7,3)$ and $(7,-4)$ **(b)** $(5,3)$ and $(-1,3)$
5. What is the slope and the y-interept of the line $5x+3y=2$? We want to solve for y and get the equation in the form $y=mx+b$. $5x+3y=2$ $3y=-5x+2$ $y=-\frac{5}{3}x+\frac{2}{3}$ $m=-\frac{5}{3}$ and $b=\frac{2}{3}$ The slope is $-\frac{5}{3}$. The y-interept is $\left(0,\frac{2}{3}\right)$.	**6.** What is the slope and the y-interept of the line $9x+3y=12$?

Example	Student Practice
7. Find an equation of the line with slope $\frac{2}{5}$ and y-interept $(0,-3)$.	**8.** Find an equation of the line with slope $\frac{3}{4}$ and y-interept $(0,-7)$.
(a) Write the equation in slope-intercept form, $y=mx+b$. We are given that $m=\frac{2}{5}$ and $b=-3$. Thus we have the following. $y=mx+b$ $y=\frac{2}{5}x+(-3)$ $y=\frac{2}{5}x-3$	**(a)** Write the equation in slope-intercept form, $y=mx+b$.
(b) Write the equation in the form $Ax+By=C$. Clear the equation of fractions so that A, B, and C are integers. $y=\frac{2}{5}x-3$ $5y=5\left(\frac{2x}{5}\right)-5(3)$ $5y=2x-15$ Subtract $2x$ from each side. Then, multiply each term by -1, because the form $Ax+By=C$ is usually written with A as a positive integer. $5y=2x-15$ $-2x+5y=-15$ $2x-5y=15$	**(b)** Write the equation in the form $Ax+By=C$.

Example	Student Practice
9. Graph the equation $y=-\frac{1}{2}x+4$. Begin with the y-intercept. Since $b=4$, plot the point $(0,4)$. The slope, $-\frac{1}{2}$ can be written as $\frac{-1}{2}$. Begin at $(0,4)$ and go down 1 unit and to the right 2 units. This is the point $(2,3)$. Plot the point. Draw the line that connects the two points. This is the graph of the equation $y=-\frac{1}{2}x+4$. 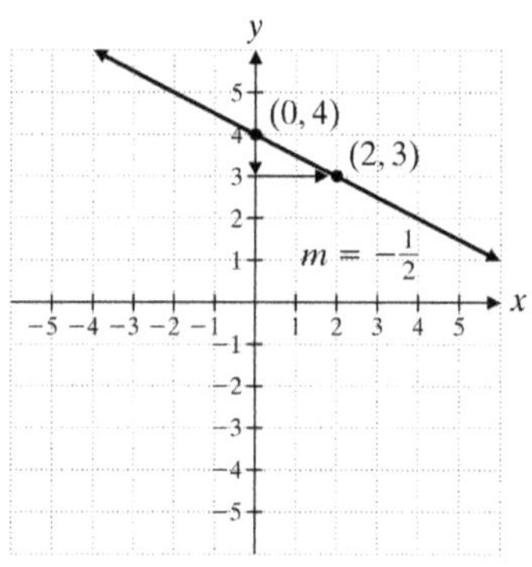	**10.** Graph the equation $y=-\frac{3}{4}x+1$. 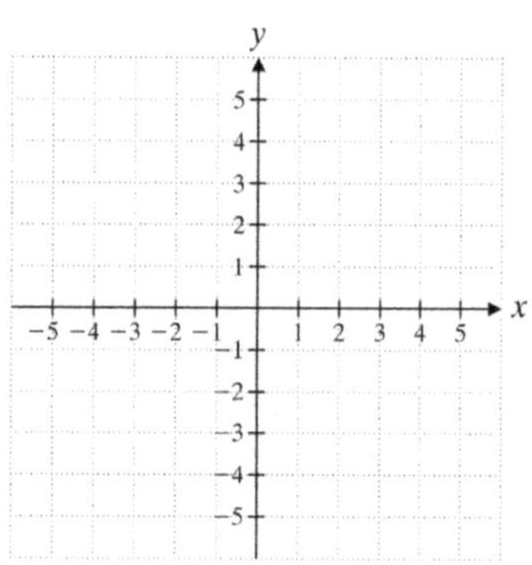

Extra Practice

1. Find the slope of a straight line that passes through the points $(-5,-2)$ and $(1,-4)$.

2. Find the slope and the y-intercept of the line $y=5x$.

3. Write the equation of the line in slope-intercept form given $m=-4$ and the y-intercept is $\left(0,\frac{4}{5}\right)$.

4. Write the equation of the line in slope-intercept form given $m=0$ and the y-intercept is $(0,-2)$.

Concept Check

Consider the formula for slope: $m=\frac{y_2-y_1}{x_2-x_1}$. Explain why we substitute the y coordinates in the numerator.

Name: ______________________________ Date: ________________
Instructor: ____________________________ Topic: ________________

Module 10 Graphing and Functions
Topic 4 Writing the Equation of a Line

Vocabulary
slope • y-intercept • slope-intercept form • vertical units • horizontal units
parallel lines • perpendicular lines

1. Given the slope and a point on the line we can find the ____________.

2. To find the slope of a line given the graph, we count the number of ____________ and horizontal units from one point on the line to another.

3. ____________ have slopes whose product is -1.

4. ____________ have the same slope but different y-intercepts.

Example	Student Practice
1. Find an equation of the line that passes through $(-3,6)$ with slope $-\frac{2}{3}$. We are given the values $m=-\frac{2}{3}$, $x=-3$, and $y=6$. Substitute the given values of x, y, and m into the equation $y=mx+b$. Solve for b. $y=mx+b$ $6=\left(-\frac{2}{3}\right)(-3)+b$ $6=2+b$ $4=b$ Use the values of b and m to write the equation in the form $y=mx+b$. An equation of the line is $y=-\frac{2}{3}x+4$.	**2.** Find an equation of the line that passes through $(2,-5)$ with slope $-\frac{1}{2}$.

Vocabulary Answers: 1. y-intercept 2. vertical units 3. perpendicular lines 4. parallel lines

Example	Student Practice
3. Find an equation of the line that passes through $(2,5)$ and $(6,3)$.	**4.** Find an equation of the line that passes through $(5,4)$ and $(10,1)$.

We first find the slope of the line. Then proceed as in Example **1**.

Substitute $(x_1, y_1) = (2,5)$ and $(x_2, y_2) = (6,3)$ into the formula.

$$m = \frac{y_2 - y_1}{x_2 - x_1}$$

$$m = \frac{y_2 - y_1}{x_2 - x_1} = \frac{3-5}{6-2} = \frac{-2}{4} = -\frac{1}{2}$$

Choose either point, say $(2,5)$, to substitute into $y = mx + b$ as in Example **1**. Then solve for b.

$$y = mx + b$$
$$5 = -\frac{1}{2}(2) + b$$
$$5 = -1 + b$$
$$6 = b$$

Use the values for b and m to write the equation.

An equation of the line is $y = -\frac{1}{2}x + 6$.

Note: We could have substituted the slope and the other point, $(6,3)$, into the slope-intercept form and arrived at the same answer. Try it.

Example	Student Practice

5. What is the equation of the line in the figure below?

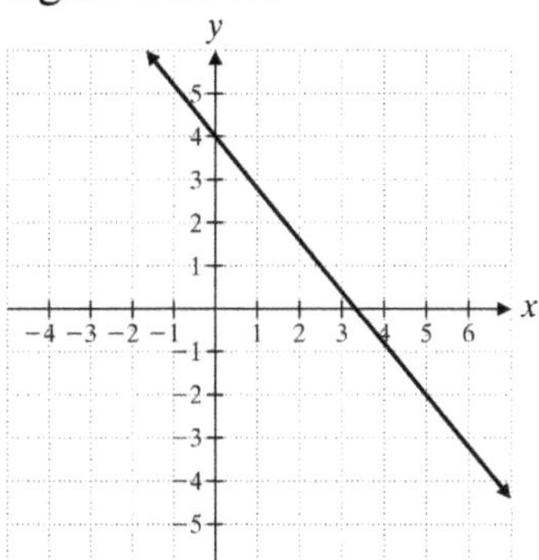

First, look for the y-intercept. The line crosses the y-axis at $(0,4)$. Thus $b=4$.

Second, find the slope.

$$m=\frac{\text{change in } y}{\text{change in } x}$$

Look for another point on the line. We choose $(5,-2)$. Count the number of vertical units from 4 to -2 (rise). Count the number of horizontal units from 0 to 5 (run).

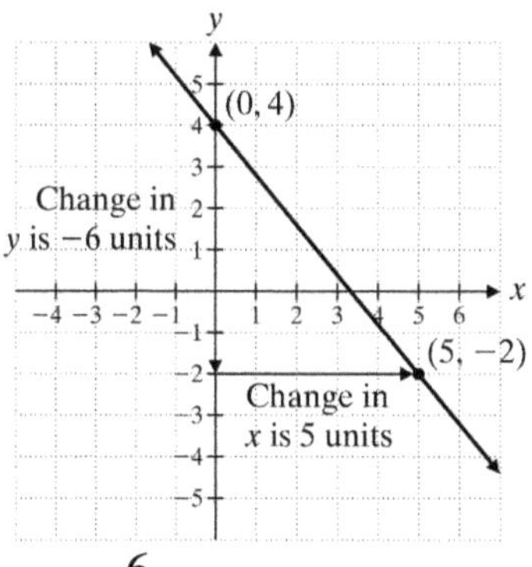

$$m=\frac{-6}{5}$$

Now, using $m=\frac{-6}{5}$ and $b=4$, we can write an equation of the line.

6. What is the equation of the line in the figure below?

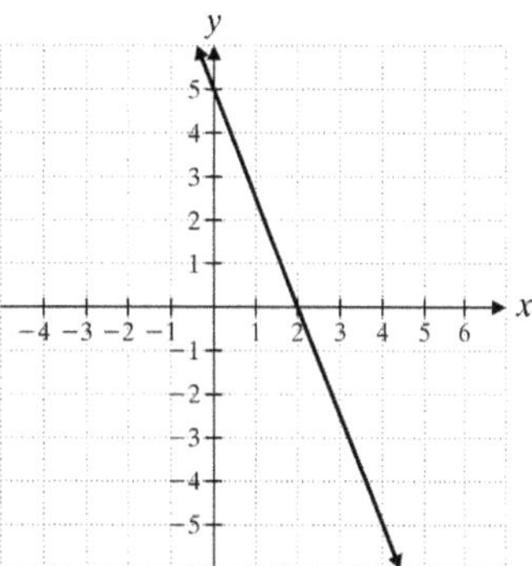

$$y = mx + b$$
$$y = -\frac{6}{5}x + 4$$

Example	Student Practice
7. Line h has a slope of $-\frac{2}{3}$.	**8.** Line h has a slope of $\frac{5}{7}$.
(a) If line f is parallel to line h, what is its slope? Parallel lines have the same slope. Line f has a slope of $-\frac{2}{3}$.	**(a)** If line f is parallel to line h, what is its slope?
(b) If line g is perpendicular to line h, what is its slope? Perpendicular lines have slopes whose product is -1. $m_1m_2 = -1$ $-\frac{2}{3}m_2 = -1$ $m_2 = \frac{3}{2}$ Line g has a slope of $\frac{3}{2}$.	**(b)** If line g is perpendicular to line h, what is its slope?

Extra Practice

1. Find the equation of the line that passes through $(3,1)$ and has slope $-\frac{1}{2}$.
2. Write an equation of the line that passes through $(3,4)$ and $(-1,-16)$.
3. Write an equation of the line that passes through $\left(1,\frac{1}{6}\right)$ and $\left(2,\frac{4}{3}\right)$.
4. Find the equation of a line that passes through $(3,-7)$ and is parallel to $y = -5x + 2$.

Concept Check

How would you find an equation of the line that passes through $(-2,-3)$ and has zero slope?

Name: ______________________________ Date: ________________
Instructor: ___________________________ Topic: ________________

Module 10 Graphing and Functions
Topic 5 Graphing Linear Inequalities

Vocabulary
linear inequality • solution • solid line • dashed line • test point

1. The ____________ of an inequality is the set of all possible ordered pairs that when substituted into the inequality will yield a true statement.

2. If the ____________ is a solution of the inequality, we shade the region on the side of the line that includes the point.

3. We use a ____________ to indicate that the points on the line are included in the solution of the inequality.

Example	Student Practice
1. Graph $5x+3y>15$.	**2.** Graph $4x+3y>12$.

Begin by graphing the line $5x+3y=15$. Since there is no equals sign in the inequality, draw a dashed line to indicate that the line is not part of the solution set. The easiest test point to test is $(0,0)$. Substitute $(0,0)$ for (x,y).

$$5x+3y>15$$
$$5(0)+3(0)>15$$
$$0>15 \quad \text{false}$$

$(0,0)$ is not a solution. Shade the region on the side of the line that does not include $(0,0)$.

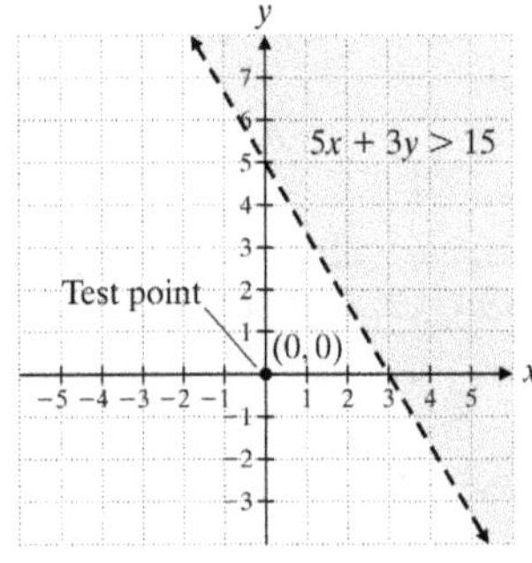

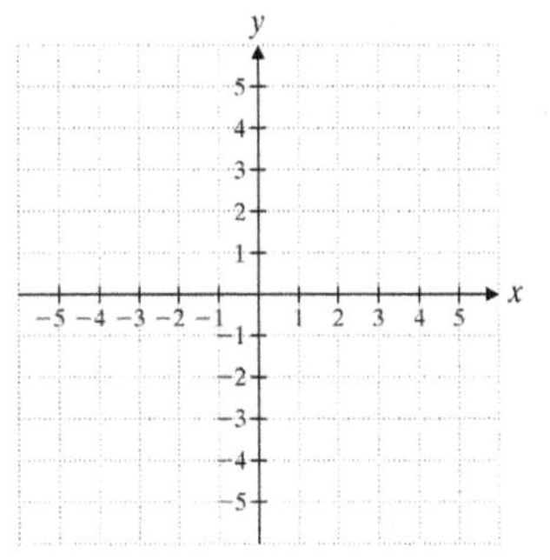

Vocabulary Answers: 1. solution 2. test point 3. solid line

Example	Student Practice
3. Graph $2y \le -3x$.	**4.** Graph $4y \le -5x$.

First, graph $2y = -3x$. Since $\le$ is used, the line will be a solid line.

We see that the line passes through $(0,0)$. We choose another test point. We will choose $(-3,-3)$.

$$2y \le -3x$$
$$2(-3) \le -3(-3)$$
$$-6 \le 9 \quad \text{true}$$

Since $(-3,-3)$ is a solid solution to the inequality, shade the region that includes $(-3,-3)$, that is the region below the line.

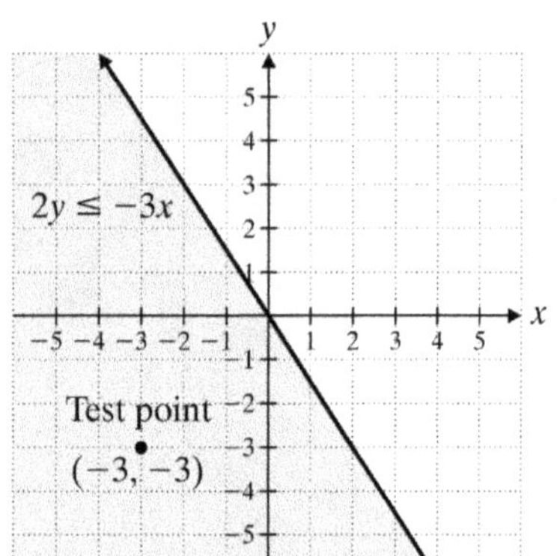

Example	Student Practice
5. Graph $x < -2$.	**6.** Graph $x > -3$.

First, graph $x = -2$. Since < is used, the line will be dashed.

Second, test $(0,0)$ in the inequality.

$$x < -2$$
$$0 < -2 \quad \text{false}$$

Since $(0,0)$ is not a solution to the inequality, shade the region that does not include $(0,0)$, that is the region to the left of the line $x = -2$.

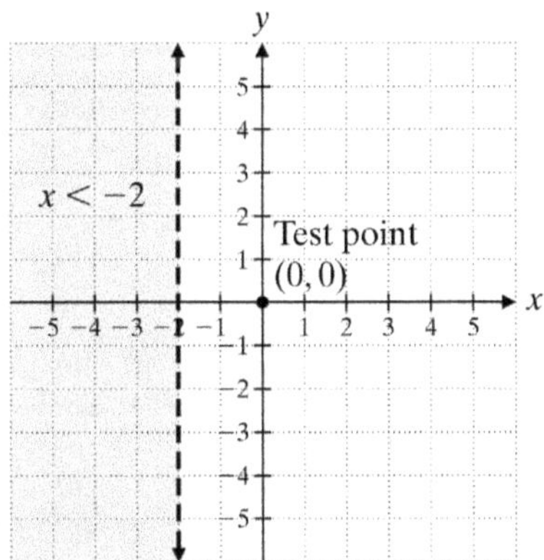

Observe that every point in the shaded region has an x-value that is less than −2.

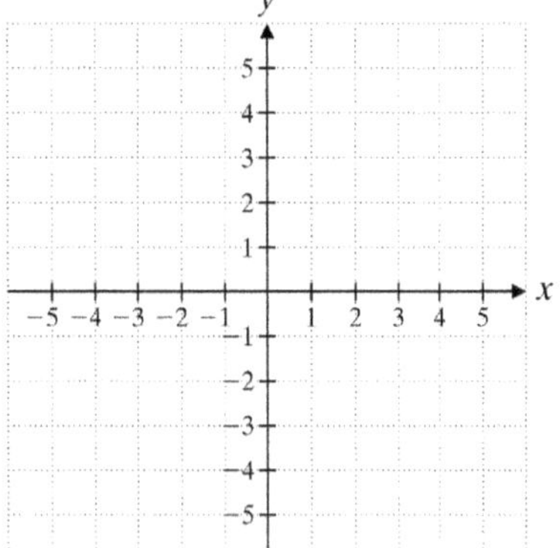

Extra Practice

1. Graph $y < 2x + 1$.

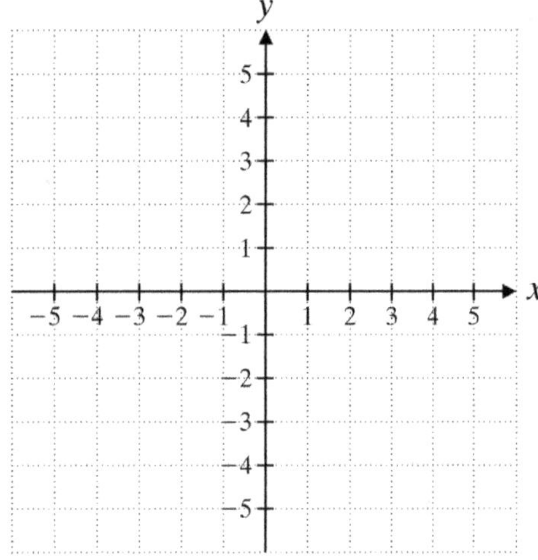

2. Graph $3x - 5y - 10 \geq 0$.

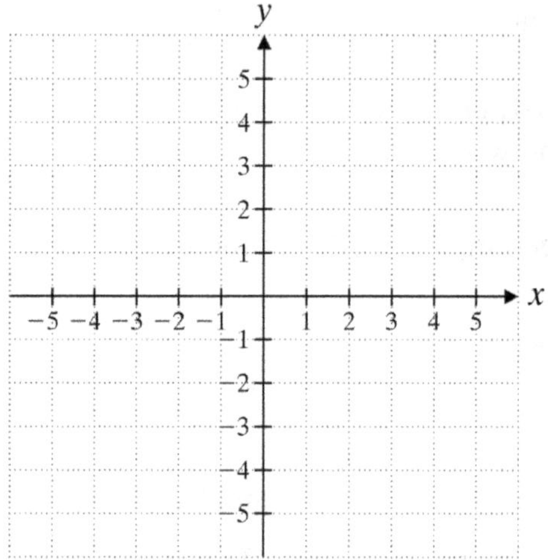

3. Graph $y \leq 4$.

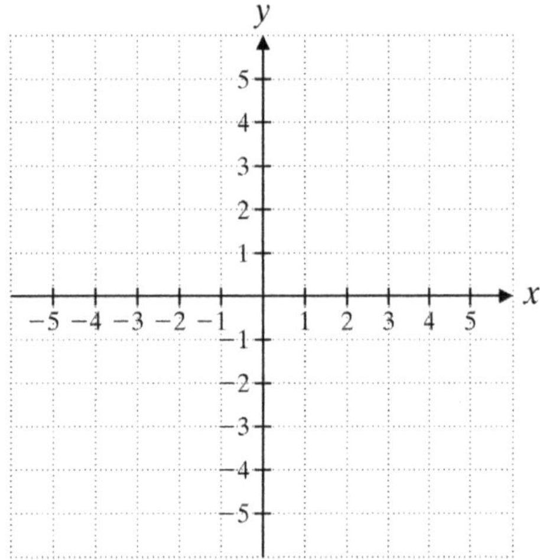

4. Graph $3x + 6y - 9 < 0$.

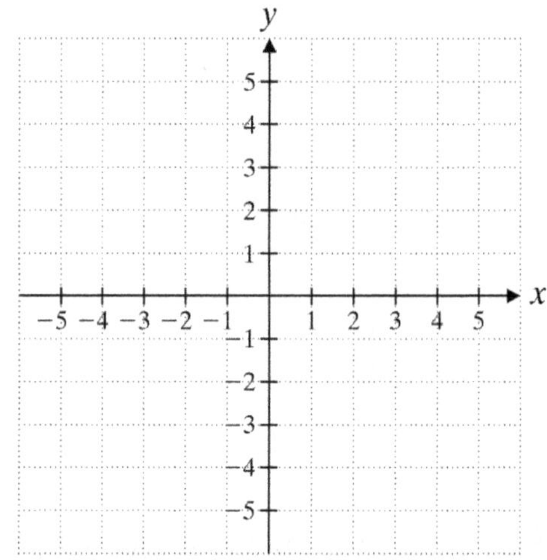

Concept Check

Explain how you would determine if you should shade the region above the line or below the line if you were to graph the inequality $y > -3x + 4$ using $(0,0)$ as a test point.

Name: ______________________________ Date: ________________
Instructor: ____________________________ Topic: ________________

Module 10 Graphing and Functions
Topic 6 Functions

Vocabulary
independent variable • dependent variable • relation • domain • range
function • absolute zero • vertical line test • function notation

1. A(n) ____________ is any set of ordered pairs.

2. A(n) ____________ is a relation in which no two different ordered pairs have the same first coordinate.

3. The ____________ is used to determine whether a relation is a function.

4. All the first coordinates in all of the ordered pairs of the relation make up the ____________ of the relation.

Example	Student Practice
1. State the domain and range of the relation. $\{(5,7),(9,11),(10,7),(12,14)\}$ The domain consists of all the first coordinates in the ordered pairs. The first coordinates are 5, 9, 10, and 12. The range consists of all the second coordinates in the ordered pairs. The second coordinates are 7, 11, 7, and 14. We usually list the values of a domain or range from smallest to largest. The domain is $\{5,9,10,12\}$. The range is $\{7,11,14\}$. Note that we list 7 only once.	**2.** State the domain and range of the relation. $\{(-3,6),(7,1),(-2,1),(4,6)\}$

Vocabulary Answers: 1. relation 2. function 3. vertical line test 4. domain

Example	Student Practice

3. Determine whether the relation is a function.

(a) $\{(3,9),(4,16),(5,9),(6,36)\}$

No two ordered pairs have the same first coordinate. Thus this set of ordered pairs defines a function.

(b) $\{(7,8),(9,10),(12,13),(7,14)\}$

Two different ordered pairs, $(7,8)$ and $(7,14)$ have the same first coordinate. Thus this relation is not a function.

4. Determine whether the relation is a function.

(a) $\{(9,3),(16,4),(9,5),(36,6)\}$

(b) $\{(-3,17),(2,1),(4,-3),(7,17)\}$

5. Graph $y = x^2$.

Begin by constructing a table of values. We select values for x and then determine by the equation the corresponding values of y. We then plot the ordered pairs and connect the points with a smooth curve.

x	$y = x^2$	y
-2	$y = (-2)^2 = 4$	4
-1	$y = (-1)^2 = 1$	1
0	$y = (0)^2 = 0$	0
1	$y = (1)^2 = 1$	1
2	$y = (2)^2 = 4$	4

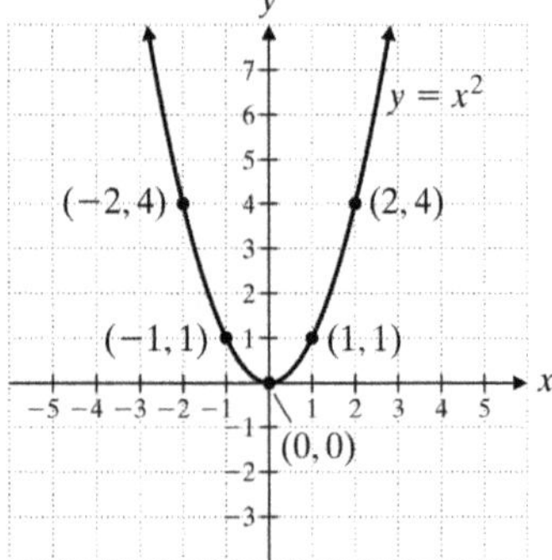

6. Graph $x = y^2$.

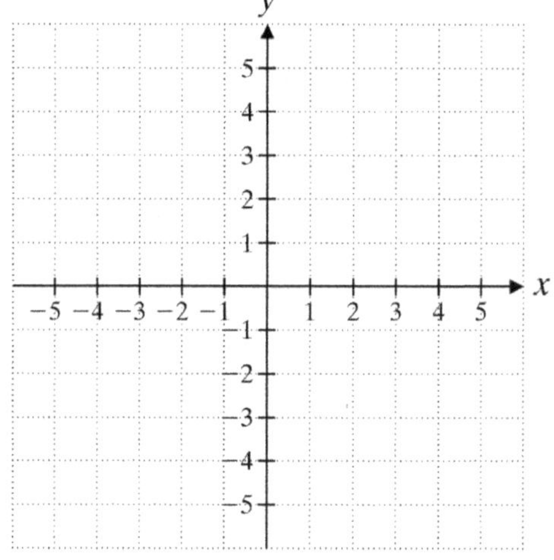

Example	Student Practice
7. Determine whether each of the following is the graph of a function.	**8.** Determine whether each of the following is the graph of a function.

7. (a)

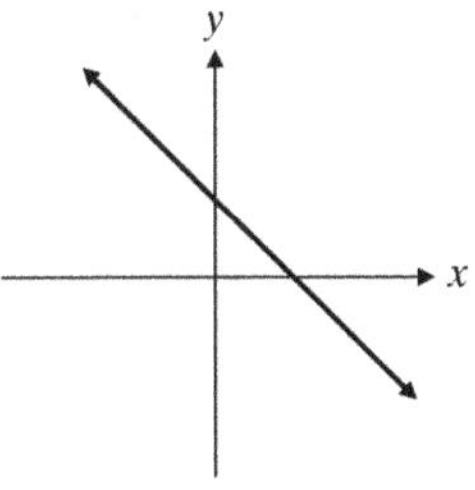

The graph of a straight line is a function. Any vertical line will cross this straight line in only one location.

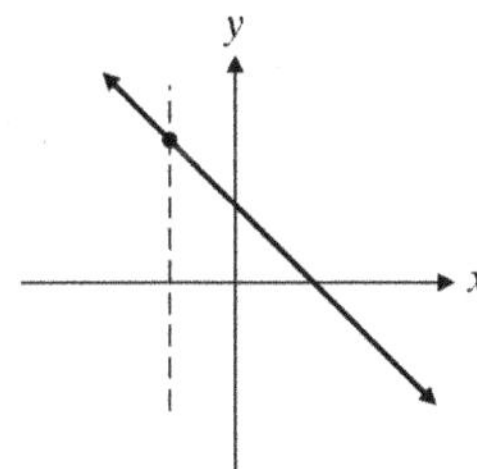

7. (b)

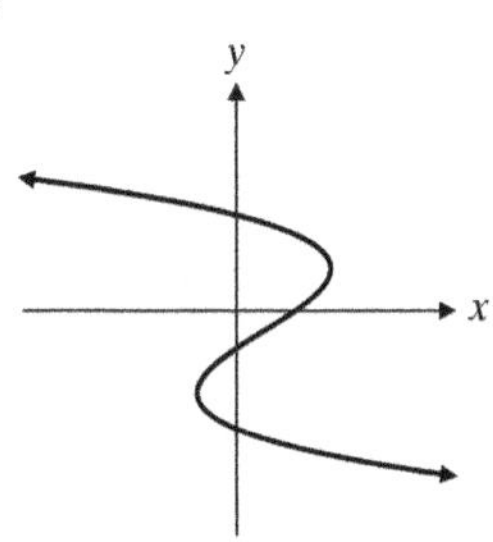

This graph is not the graph of a function. There exists a vertical line that will cross the curve in more than one place.

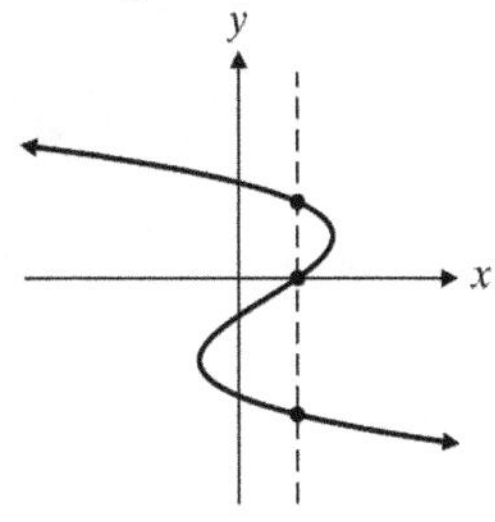

8. (a)

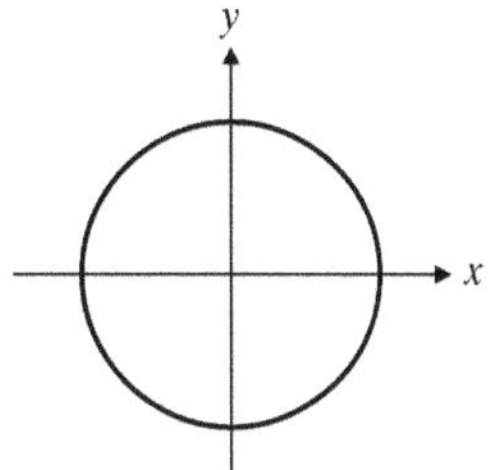

8. (b)

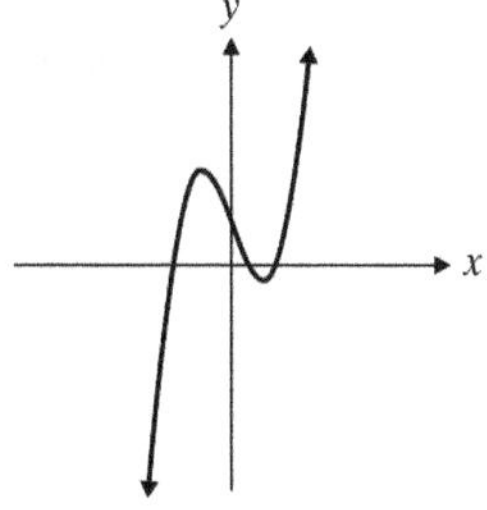

Example	Student Practice
9. If $f(x) = 3x^2 - 4x + 5$, find each of the following.	**10.** If $f(x) = 4x^2 - 2x + 7$, find each of the following.
(a) $f(-2)$ $f(-2) = 3(-2)^2 - 4(-2) + 5$ $= 3(4) + -4(-2) + 5$ $= 12 + 8 + 5 = 25$	**(a)** $f(-3)$
(b) $f(4)$ $f(4) = 3(4)^2 - 4(4) + 5$ $= 3(16) + -4(4) + 5$ $= 48 - 16 + 5 = 37$	**(b)** $f(2)$
(c) $f(0)$ $f(0) = 3(0)^2 - 4(0) + 5 = 5$	**(c)** $f(0)$

Extra Practice

1. Find the domain and range of the relation. Determine whether the relation is a function.

   $\{(2.5,3),(3.5,0),(5.5,-2),(8.5,-6)\}$

2. Given $f(x) = 2x^2 - 3x + 1$, find the indicated values.

   (a) $f(0)$

   (b) $f(-3)$

   (c) $f(3)$

3. Determine whether the relation is a function.

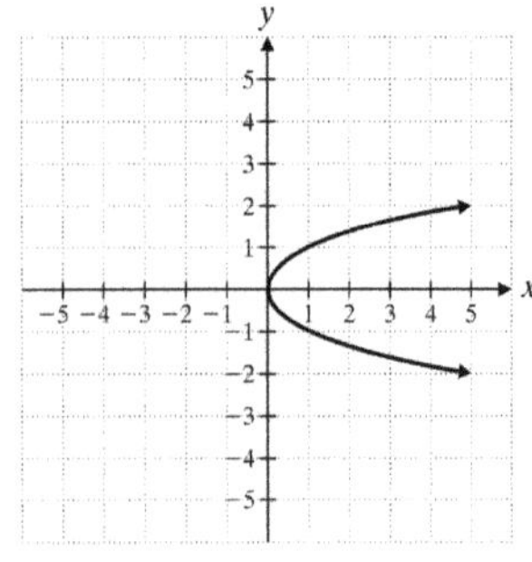

4. Graph $y = -3x^2$.

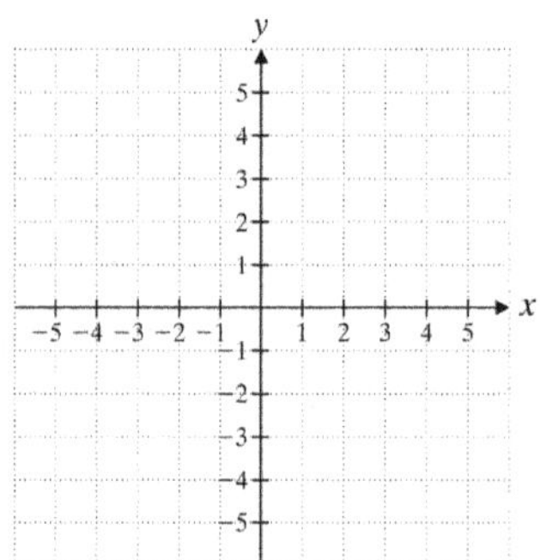

Concept Check

In the relation $\{(3,4),(5,6),(3,8),(2,9)\}$, why is there a different number of elements in the domain than the range?

MATH COACH

Mastering the skills you need to do well on the test.

Watch the MATH COACH videos in MyMathLab® or on YouTube™ while you work the problems below. These helpful hints will help you avoid making common errors on test problems.

Graphing a Linear Equation by Plotting Three Ordered Pairs—Problem 4 Graph $y = \frac{2}{3}x - 4$.

> **Helpful Hint:** Find three ordered pairs that are solutions to the equation. Plot those 3 points. Then draw a line through the points. When the equation is solved for y and there are fractional coefficients on x, it is sometimes a good idea to choose values for x that result in integer values for y. This will make graphing easier.

Choosing values for x that result in integer values for y means choosing 0 or multiples of 3 since 3 is the denominator of the fractional coefficient on x. When we multiply by these values, the result becomes an integer.

Did you choose values for x that result in values for y that are not fractions? Yes ____ No ____

If you answered No, try using $x = 0$, $x = 3$, and $x = 6$. Solve the equation for y in each case to find the y-coordinate. Remember that it will make the graphing process easier if you choose values for x that will clear the fraction from the equation.

Did you plot the three points and connect them with a line?
Yes ____ No ____

If you answered No, go back and complete this step.

If you feel more comfortable using the slope-intercept method to solve this problem, simply identify the y-intercept from the equation in $y = mx + b$ form, and then use the slope m to find two other points.

If you answered Problem 4 incorrectly, go back and rework the problem using these suggestions.

Write the Equation of a Line Given Two Points—Problem 10
Find an equation for the line passing through $(5, -4)$ and $(-3, 8)$.

> **Helpful Hint:** When given two points (x_1, y_1) and (x_2, y_2), we can find the slope m using the slope formula $m = \frac{y_2 - y_1}{x_2 - x_1}$. Then you can substitute m into the equation $y = mx + b$ along with the coordinates of one of the points to find b, the y-intercept.

When you substituted the points into the slope formula to find m, did you obtain either $m = \frac{8-(-4)}{-3-5}$ or $m = \frac{-4-8}{5-(-3)}$?

Yes ____ No ____

If you answered No, check your work to make sure you substituted the points correctly. Be careful to avoid any sign errors.

(continued on next page)

Did you use $m = -\frac{3}{2}$ and either of the points given when you substituted the values into the equation $y = mx + b$ to find the value of b? Yes ____ No ____

If you answered No, stop and make a careful substitution for $m = -\frac{3}{2}$ and either $x = 5$ and $y = -4$ or $x = -3$ and $y = 8$. See if you can solve the resulting equation for b.

Now go back and rework the problem using these suggestions.

Graphing Linear Inequalities in Two Variables—Problem 12
Graph the region described by $-3x - 2y > 10$.

> **Helpful Hint:** First graph the equation $-3x - 2y = 10$. Determine if the line should be solid or dashed. Then pick a test point to see if it satisfies the inequality $-3x - 2y > 10$. If the test point satisfies the inequality, shade the side of the line on which the point lies. If the test point does not satisfy the inequality, shade the opposite side of the line.

Examine your work. Does the line $-3x - 2y = 10$ pass through the point $(0,-5)$? Yes ____ No ____

If you answered No, substitute $x = 0$ into the equation and solve for y. Check the calculations for each of the points you plotted to find the graph of this equation.

Did you draw a solid line? Yes ____ No ____

If you answered Yes, look at the inequality symbol. Remember that we only use a solid line with the symbols $\le$ and $\ge$. A dashed line is used for $<$ and $>$.

Did you shade the area above the dashed line? Yes ____ No ____

If you answered Yes, stop now and use $(0,0)$ as a test point and substitute it into the inequality $-3x - 2y > 10$. Then use the Helpful Hint to determine which side to shade.

If you answered Problem 12 incorrectly, go back and rework the problem using these suggestions.

Using Function Notation to Evaluate a Function—Problem 16(a) and 16(b)
For $f(x) = -x^2 - 2x - 3$: **(a)** find $f(0)$. **(b)** find $f(-2)$.

> **Helpful Hint:** Replace x with the number indicated. It is a good idea to place parentheses around the value to avoid any sign errors. Then use the order of operations to evaluate the function in each case.

(a) Did you replace x with 0 and write

$$f(0) = -(0)^2 - 2(0) - 3?$$

Yes ____ No ____

If you answered No, take time to go over your steps one more time, remembering that 0 times any number is 0.

(b) Did you replace x with -2 and write

$$f(0) = -(0)^2 - 2(0) - 3?$$

Yes ____ No ____

If you answered No, go over your steps again, remembering to place parentheses around -2. Note that $(-2)^2 = 4$ and therefore $-(-2)^2 = -4$.

Now go back and rework the problem again using these suggestions.

Name: ______________________________ Date: ________________
Instructor: ___________________________ Topic: ______________

Module 11 Systems of Linear Equations and Inequalities
Topic 1 Systems of Linear Equations in Two Variables

Vocabulary
systems of two linear equations in two variables • solution to a system • inconsistent
dependent • substitution method • addition method • independent
consistent system • no solution • identity

1. If a system of equations has no solution, it is said to be ___________.

2. In the ___________, we choose one equation and solve for one variable. Then we substitute this expression into the other equation.

3. A system with infinitely many solutions is said to be a(n) ___________.

Example	Student Practice
1. Determine whether $(3,-2)$ is a solution to the following system. $x+3y=-3$ $4x+3y=6$ Substitute $(3,-2)$ into the first equation to see whether the ordered pair is a solution to the first equation. $(3)+3(-2)\stackrel{?}{=}-3$ $3-6\stackrel{?}{=}-3$ $-3=-3$ Likewise, determine whether $(3,-2)$ is a solution to the second equation. $4(3)+3(-2)\stackrel{?}{=}6$ $6=6$ Since $(3,-2)$ is a solution to each equation, it is a solution to the system.	**2.** Determine whether $(3,2)$ is a solution to the following system. $2x+y=8$ $3x+5y=19$

Vocabulary Answers: 1. inconsistent 2. substitution method 3. dependent

Example	Student Practice
3. Solve this system of equations by graphing.	**4.** Solve this system of equations by graphing.

3. Solve this system of equations by graphing.

$$2x+3y=12$$
$$x-y=1$$

Graph each line. Then, find the point of intersection.

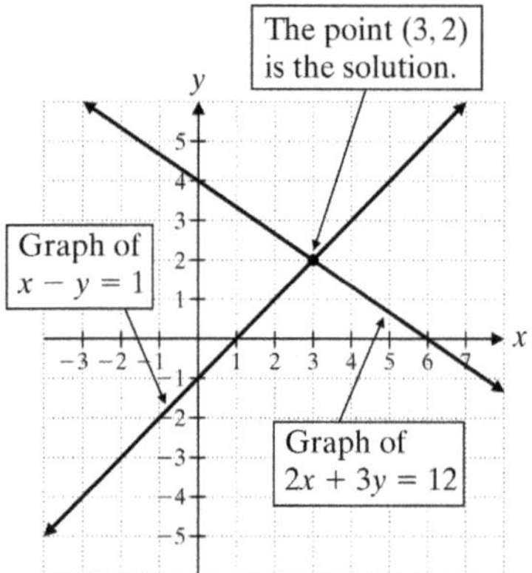

The solution is $(3,2)$. The graphing method does not always lead to an accurate result because it involves a visual estimation of the point of intersection. Verify the solution by substituting $(3,2)$ into each equation.

$$2(3)+3(2)\overset{?}{=}12 \qquad 3-2\overset{?}{=}1$$
$$12=12 \qquad 1=1$$

Thus, we have verified that the solution to the system is $(3,2)$.

4. Solve this system of equations by graphing.

$$-x+y=-6$$
$$3x+5y=2$$

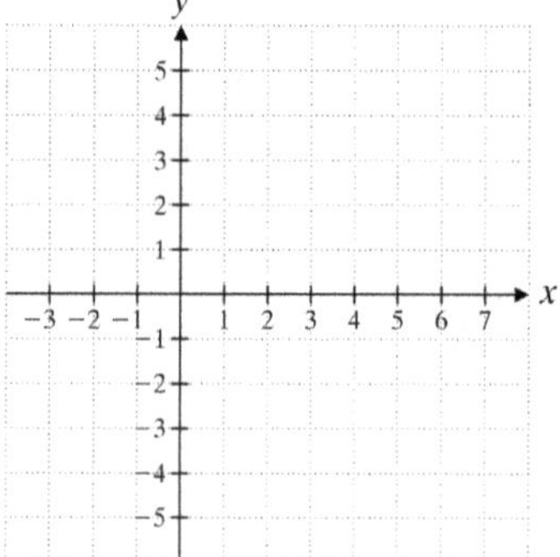

Example	Student Practice
5. Find the solution to the following system of equations. Use the substitution method.	**6.** Find the solution to the following system of equations. Use the substitution method.

Example

5. Find the solution to the following system of equations. Use the substitution method.

$$x+3y=-7 \quad (1)$$
$$4x+3y=-1 \quad (2)$$

Solve for x in equation (1) because x has a coefficient of 1, $x=-7-3y$ (3). Now, substitute this expression for x in equation (2) and solve for y.

$$\begin{aligned} 4x+3y&=-1 \\ 4(-7-3y)+3y&=-1 \\ -28-12y+3y&=-1 \\ -28-9y&=-1 \\ -9y&=27 \\ y&=-3 \end{aligned}$$

Substitute $y=-3$ into equation (1) to find x.

$$\begin{aligned} x+3(-3)&=-7 \\ x-9&=-7 \\ x&=2 \end{aligned}$$

The solution is $(2,-3)$. Verify the solution in both of the original equations.

$$2+3(-3)\stackrel{?}{=}-7 \qquad 4(2)+3(-3)\stackrel{?}{=}-1$$
$$2-9\stackrel{?}{=}-7 \qquad 8-9\stackrel{?}{=}-1$$
$$-7=-7 \qquad -1=-1$$

The solution checks.

Student Practice

6. Find the solution to the following system of equations. Use the substitution method.

$$x-2y=4 \quad (1)$$
$$2x-5y=5 \quad (2)$$

Example	Student Practice
7. Solve the following system by the addition method.	**8.** Solve the following system by the addition method.

$$\frac{x}{4}+\frac{y}{6}=-\frac{2}{3} \quad (1)$$

$$\frac{x}{5}+\frac{y}{2}=\frac{1}{5} \quad (2)$$

Clear equation (1) of fractions by multiplying each term by 12. Clear equation (2) of fraction by multiplying each term by 10. Now we have an equivalent system that does not contain fractions.

$$3x+2y=-8 \quad (3)$$

$$2x+5y=2 \quad (4)$$

To eliminate the variable x, we multiply equation (3) by 2 and (4) by 3. We now have the following equivalent system. Add the equations, then solve for y.

$$\begin{aligned} 6x+4y&=-16 \\ -6x-15y&=-6 \\ \hline -11y&=-22 \\ y&=2 \end{aligned}$$

Substitute $y=2$ into equation (3) and solve for x.

$$\begin{aligned} 3x+2(2)&=-8 \\ 3x+4&=-8 \\ 3x&=-12 \\ x&=-4 \end{aligned}$$

The solution to the system is $(-4,2)$. The check is left to the student.

8. (Student Practice)

$$\frac{x}{6}+\frac{y}{9}=\frac{11}{9} \quad (1)$$

$$\frac{x}{3}+\frac{y}{7}=\frac{43}{21} \quad (2)$$

Example	Student Practice
9. If possible, solve the system. $2x+8y=16 \quad (1)$ $4x+16y=-8 \quad (2)$ To eliminate y, we'll multiply equation (1) by -2. $-2(2x)+(-2)(8y)=(-2)(16)$ $-4x-16y=-32 \quad (3)$ We now have the following equivalent system. $-4x-16y=-32 \quad (3)$ $4x+16y=-8 \quad (2)$ When we add equations (3) and (2), we get $0=-40$, which is false. Thus, we conclude that this system is inconsistent, and there is no solution.	**10.** If possible, solve the system. $x+4y=5 \quad (1)$ $-3x-12y=15 \quad (2)$
11. If possible, solve the system. $0.5x-0.2y=1.3 \quad (1)$ $-1.0x+0.4y=-2.6 \quad (2)$ Clear the decimals by multiplying the equations by 10. $5x-2y=13 \quad (3)$ $-10x+4y=-26 \quad (4)$ Eliminate y by multiplying each term of equation (3) by 2. $10x-4y=26 \quad (5)$ $-10x+4y=-26 \quad (4)$ $0=0$ This resulting statement is an identity. The equations are dependent and there is an infinite number of solutions.	**12.** If possible, solve the system. $1.4x+0.6y=-7.4 \quad (1)$ $-0.7x-0.3y=3.7 \quad (2)$

Example	Student Practice
13. Select a method and solve each system of equations.	**14.** Select a method and solve each system of equations.
(a) $\begin{aligned} x+y&=3080 \\ 2x+3y&=8740 \end{aligned}$	**(a)** $\begin{aligned} x+y&=2235 \\ -4x+3y&=2155 \end{aligned}$
Since there are x- and y-variables that have coefficients of 1, select the substitution method. The solution to the system is $(500,2580)$. The check is left to the student.	
(b) $\begin{aligned} 5x-2y&=19 \\ -3x+7y&=35 \end{aligned}$	**(b)** $\begin{aligned} 4x+7y&=2 \\ -5x-8y&=-4 \end{aligned}$
Since none of the x- or y-variables have coefficients of 1 or -1, select the addition method. The solution to the system is $(7,8)$. The check is left to the student.	

Extra Practice

If possible, solve the system. If there is not a unique solution to a system, state a reason.

1. $\begin{aligned} 3x+y&=-1 \\ 4x+2y&=0 \end{aligned}$

2. $\begin{aligned} y&=3x+2 \\ -15x+5y&=0 \end{aligned}$

3. $\begin{aligned} 6x+3y&=12 \\ 6x-3y&=-12 \end{aligned}$

4. $\begin{aligned} 10y-2x&=8 \\ y&=\frac{1}{5}x+\frac{4}{5} \end{aligned}$

Concept Check

Explain what happens when you go through the steps to solve the following system of equations. Why does this happen?

$$\begin{aligned} 6x-4y&=8 \\ -9x+6y&=-12 \end{aligned}$$

Name: ________________________________ Date: ________________

Instructor: ________________________________ Topic: ________________

Module 11 Systems of Linear Equations and Inequalities
Topic 2 Systems of Linear Equations in Three Variables

Vocabulary

systems of three linear equations in three variables • solution
ordered triple • substitution method • addition method

1. The solution to a system of three linear equations in three unknowns is called a(n) ____________.

2. To solve a system of three linear equations in three variables, first use the ____________ to eliminate any variable from any pair of equations.

Example	Student Practice
1. Determine whether $(2,-5,1)$ is a solution to the following system.	**2.** Determine whether $(2,8,-7)$ is a solution to the following system.

Example

1. Determine whether $(2,-5,1)$ is a solution to the following system.

$$3x+y+2z=3$$
$$4x+2y-z=-3$$
$$x+y+5z=2$$

Substitute $x=2,\ y=-5,$ and $z=1$ into the first equation to see whether the ordered triple is a solution to the first equation.

$$3(2)+(-5)+2(1)\stackrel{?}{=}3$$
$$6-5+2\stackrel{?}{=}3$$
$$3=3$$

Likewise, determine whether $(2,-5,1)$ is a solution to the second and third equations.

$$4(2)+4(-5)-1\stackrel{?}{=}-3 \qquad 2+(-5)+5(1)\stackrel{?}{=}2$$
$$8-10-1\stackrel{?}{=}-3 \qquad 2-5+5\stackrel{?}{=}2$$
$$-3=-3 \qquad 2=2$$

Since $(2,-5,1)$ is a solution to each equation, it is a solution to the system.

Student Practice

2. Determine whether $(2,8,-7)$ is a solution to the following system.

$$4x-y+3z=-21$$
$$x-y-2z=8$$
$$14x-3y+z=-3$$

Vocabulary Answers: 1. ordered triple 2. addition method

Example	Student Practice
3. Find the solution to (that is, solve) the following system of equations. $-2x+5y+z=8 \quad (1)$ $-x+2y+3z=13 \quad (2)$ $x+3y-z=5 \quad (3)$ Add equations (1) and (3) to eliminate z. $-2x+5y+z=8 \quad (1)$ $x+3y-z=5 \quad (3)$ $-x+8y=13 \quad (4)$ Now, choose a different pair of the original equations and eliminate the same variable. Multiply equation (3) by 3 and call it equation (6). Add the result to equation (2). $-x+2y+3z=13 \quad (2)$ $3x+9y-3z=15 \quad (6)$ $2x+11y=28 \quad (5)$ Solve the resulting system of two linear equations in two unknowns. $-x+8y=13 \quad (4)$ $2x+11y=28 \quad (5)$ Thus, $x=3$ and $y=2$. Substitute these values into one of the original equations to find z. $-2(3)+5(2)+z=8$ $-6+10+z=8$ $z=4$ The solution to the system is $(3,2,4)$. The check is left to the student.	**4.** Find the solution to (that is, solve) the following system of equations. $3x+2y+9z=10 \quad (1)$ $4x-2y-z=2 \quad (2)$ $4x+5y+12z=32 \quad (3)$

Example	**Student Practice**
5. Solve the system.	**6.** Solve the system.

Example

5. Solve the system.

$$\begin{aligned} 4x+3y+3z &= 4 \quad (1) \\ 3x \quad +2z &= 2 \quad (2) \\ 2x-5y \quad &= -4 \quad (3) \end{aligned}$$

Use equations (2) and (1) to obtain an equation that contains only x and y. Multiply equation (1) by 2 and equation (2) by -3 to obtain the following system.

$$\begin{aligned} 8x+6y+6z &= 8 \quad (4) \\ -9x \quad -6z &= -6 \quad (5) \\ \hline -x+6y \quad &= 2 \quad (6) \end{aligned}$$

Notice that equation (3) already has no z-term. Solve the system formed by equations (3) and (6).

$$\begin{aligned} 2x-5y &= -4 \quad (3) \\ -x+6y &= 2 \quad (6) \end{aligned}$$

Thus, $x=3$ and $y=2$. Substitute these values into one of the original equations containing z. Use equation (2) since it only has two variables.

$$\begin{aligned} 3x+2z &= 2 \\ 3(-2)+2z &= 2 \\ 2z &= 8 \\ z &= 4 \end{aligned}$$

The solution to the system is $(-2,0,4)$. The check is left to the student.

Student Practice

6. Solve the system.

$$\begin{aligned} 5x+2y+3z &= -16 \quad (1) \\ 3x+7y \quad &= 11 \quad (2) \\ 9x \quad -2z &= 1 \quad (3) \end{aligned}$$

Extra Practice

1. Solve the system.

$$\begin{aligned} x+4y-\ z&=-15\\ 2x-\ y-2z&=-12\\ 3x-\ y+\ z&=\ 1 \end{aligned}$$

2. Solve the system.

$$\begin{aligned} 0.2x+0.4y+0.6z&=\ 0\\ 0.1x+\ y-0.5z&=-1.6\\ 0.2x-0.5y+0.1z&=\ 0.4 \end{aligned}$$

3. Solve the system.

$$\begin{aligned} x-y\quad&=\ 0\\ y-z&=\ 5\\ x+y+z&=13 \end{aligned}$$

4. Solve the system, if possible.

$$\begin{aligned} x+\ y-\ z&=4\\ 2x-5y+\ z&=1\\ 3x+3y-3z&=0 \end{aligned}$$

Concept Check

Explain how you would eliminate the variable z and obtain two equations with only the variables x and y in the following system.

$$\begin{aligned} 2x+4y-2z&=-22\\ 4x+3y+5z&=-10\\ 5x-2y+3z&=\ 13 \end{aligned}$$

Name: ______________________________ Date: ________________
Instructor: ___________________________ Topic: ________________

Module 11 Systems of Linear Equations and Inequalities
Topic 3 Applications of Systems of Linear Equations

Vocabulary
understand the problem • $D = RT$ • check • write a system of equations

1. When solving applied problems using equations, the first step is to ___________.

2. The final step in the problem solving process is to ___________ the answer.

Example	Student Practice
1. For the paleontology lecture on campus, advance tickets cost \$5 and tickets at the door cost \$6. The ticket sales this year came to \$4540. The department chairman wants to raise prices next year to \$7 for advance tickets and \$9 for tickets at the door. He said that if exactly the same number of people attend next year, the ticket sales at these new prices will total \$6560. If he is correct, how many tickets were sold in advance this year? How many tickets were sold at the door? Let $x =$ the number of tickets bought in advance and $y =$ the number of tickets bought at the door. The total sales for advance tickets will be $5x$ and for door tickets, $6y$. Thus, we have $5x + 6y = 4540$. The equation for next year's sales is $7x + 9y = 6560$. $5x + 6y = 4540$ $7x + 9y = 6560$ Solve the system to find that 500 advance tickets were sold and 340 door tickets were sold. The check is left to the student.	**2.** For the concert on campus, advance tickets cost \$15 and tickets at the door cost \$20. The ticket sales this year came to \$6250. The activities coordinator wants to raise prices next year to \$17 for advance tickets and \$25 for tickets at the door. She said that if exactly the same number of people attend next year, the ticket sales at these new prices will total \$7375. If she is correct, how many tickets were sold in advance this year? How many tickets were sold at the door?

Vocabulary Answers: 1. understand the problem 2. check

Example	Student Practice

3. An airplane travels between two cities that are 1500 miles apart. The trip against the wind takes 3 hours. The return trip with the wind takes $2\frac{1}{2}$ hours. What is the speed of the plane in still air (in other words, how fast would the plane travel if there were no wind)? What is the speed of the wind?

Let $x =$ the speed of the plane in still air and let $y =$ the speed of the wind. The wind speed opposes the plane's speed in still air, so we must subtract, $x - y$. The wind speed is added to the planes speed in still air, and we add, $x + y$. Using the formula $(\text{rate})(\text{time}) = \text{distance}$, we have the equations $(x - y)(3) = 1500$ and $(x + y)(2.5) = 1500$. Remove parentheses to get the following system.

$$3x - 3y = 1500 \quad (1)$$
$$2.5x + 2.5y = 1500 \quad (2)$$

Solve the system using the addition method.

$$15x - 15y = 7500$$
$$15x + 15y = 9000$$
$$30x = 16{,}500$$
$$x = 550$$

Substitute this result into equation (1) to find y.

$$3(550) - 3y = 1500$$
$$y = 50$$

The speed of the plane in still air is 550 miles per hour and the speed of the wind is 50 miles per hour.

4. A boat travels upstream 40 miles. The trip against the current takes 4 hours. The return trip with the current takes 2 hours. What is the speed of the boat in still water (in other words, how fast would the boat travel if there was no current)? What is the speed of the current?

Example	Student Practice
5. A trucking firm has three sizes of trucks. The biggest truck holds 10 tons of gravel, the next size holds 6 tons, and the smallest holds 4 tons. The firm has a contract to provide 15 trucks to haul 104 tons of gravel. To reduce fuel costs the firm's manager wants to use two more of the fuel-efficient 10-ton trucks than the 6-ton trucks. How many trucks of each type should she use?	**6.** A trucking firm has three sizes of trucks. The biggest truck holds 12 tons of gravel, the next size holds 8 tons, and the smallest holds 6 tons. The firm has a contract to provide 16 trucks to haul 154 tons of gravel. To reduce fuel costs the firm's manager wants to use three more of the fuel-efficient 12-ton trucks than the 8-ton trucks. How many trucks of each type should he use?

Let $x=$ the number of 10-ton trucks used, $y=$ the number of 6-ton trucks used, and $z=$ the number of 4-ton trucks used. Fifteen trucks will be used. So, $x+y+z=15$. There are 104 tons of cargo to be carried and the different trucks carry 10, 6, and 4 tons, respectively. So, $10x+6y+4z=104$. Two more 10-ton trucks than 6-ton trucks are used. So, $x-y=2$. The system is as follows.

$$\begin{aligned} x+y+z &= 15 \quad (1)\\ 10x+6y+4z &= 104 \quad (2)\\ x-y &= 2 \quad (3) \end{aligned}$$

Eliminate z using equations (1) and (2). Use the result and equation (3) to find x.

$$\begin{aligned} 3x+y &= 22\\ x-y &= 2\\ \hline 4x &= 24\\ x &= 6 \end{aligned}$$

Use the result to find y and then z using equations (3) and (1).

$$6-y=2 \qquad 6+4+z=15$$
$$y=4 \quad \text{and} \quad z=5$$

The manager needs six 10-ton trucks, four 6-ton trucks, and five 4-ton trucks.

Extra Practice

1. The sum of two numbers is 102. If three times the smaller number is subtracted from twice the larger number, the result is 49. Find the two numbers.

2. The Revel family farm has 500 acres of land. It costs \$60 to plant an acre of soybeans and \$36 to plant an acre of corn. If the Revels want to spend a total of \$22,440 on planting, how many acres of each crop should they plant?

3. Devon bought 15 items at the office supply store. She spent a total of \$26.75. The binders cost \$2.40, the pens cost \$1.85, and the erasers cost \$0.60. Devon bought 4 more pens than erasers. How many of each item did she buy?

4. A total of 250 people attended a movie. The tickets cost \$11 for adults, \$8 for students, and \$7 for senior citizens. The ticket sales totaled \$2318. The manager found that if they had raised the prices to \$14 for adults, \$10 for students, and \$8 for senior citizens, they would have made \$2892. How many tickets of each type were sold?

Concept Check

A plane flew 1200 miles with a tail wind in 2.5 hours. The return trip against the wind took 3 hours. This situation is represented by the following system. Explain how you would set up two equations using the given information if the plane flew 1500 miles instead of 1200 miles.

$$\begin{aligned} 2.5x + 2.5y &= 1200 \\ 3x - 3y &= 1200 \end{aligned}$$

Name: ______________________________ Date: ________________

Instructor: ____________________________ Topic: ________________

Module 11 Systems of Linear Equations and Inequalities
Topic 4 Systems of Linear Inequalities

Vocabulary

system of linear inequalities in two variables • vertex • intersection

1. The solution to a system of inequalities is the ____________ of the solution sets of the individual inequalities of the system.

2. We call two linear inequalities in two variables a(n) ____________.

3. In the solution to a system of linear inequalities, a point where the boundary lines intersect is called a(n) ____________ of the solution.

Example	Student Practice
1. Graph the solution to the system.	**2.** Graph the solution to the system.

Example

1. Graph the solution to the system.

$$y \le -3x+2$$
$$-2x+y \ge -1$$

The graph of $y \le -3x+2$ is the region on and below the line $y=-3x+2$. The graph of $-2x+y \ge -1$ is the region on and above the line $-2x+y=-1$. The two solutions are graphed on one rectangular coordinate system below. The darker shaded region is the intersection of the two graphs. Thus, the solution to the system of two inequalities is the darker shaded region and its boundary lines.

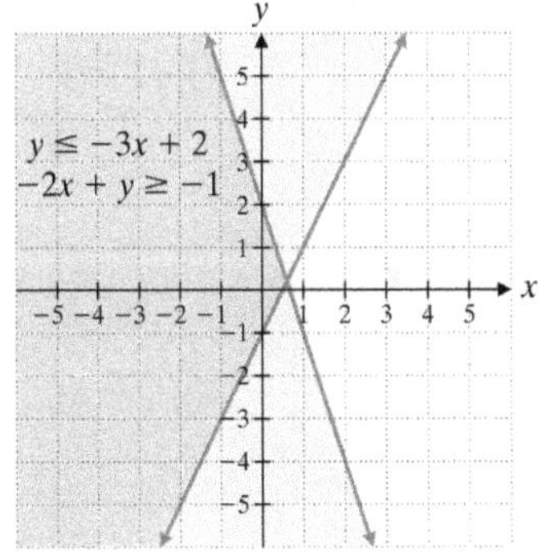

Student Practice

2. Graph the solution to the system.

$$y \le -x+4$$
$$-5x+y \ge -1$$

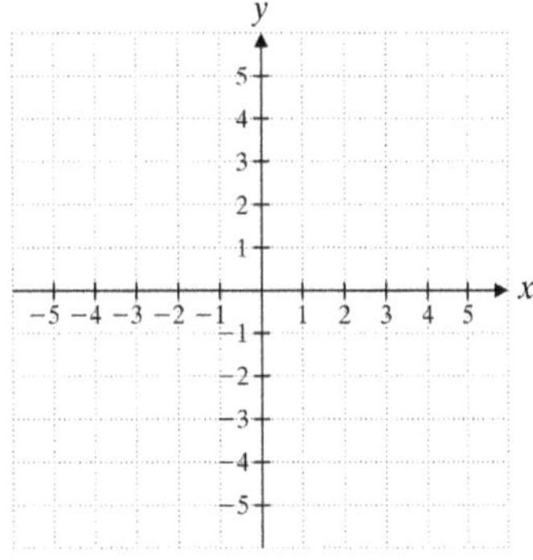

Vocabulary Answers: 1. intersection 2. system of linear inequalities in two variables 3. vertex

Example	Student Practice
3. Graph the solution to the system.	**4.** Graph the solution to the system.

3. Graph the solution to the system.

$y < 4$

$y > \frac{3}{2}x - 2$

The graph of $y < 4$ is the region below the line $y = 4$. It does not include the line since the inequality symbol is $<$. Thus, we use a dashed line to indicate that the boundary line is not part of the answer. The graph of $y > \frac{3}{2}x - 2$ is the region above the line $y = \frac{3}{2}x - 2$. Again, we use a dashed line to indicate that the boundary line is not part of the answer. The final solution is the darker shaded region. The solution does not include the dashed boundary lines.

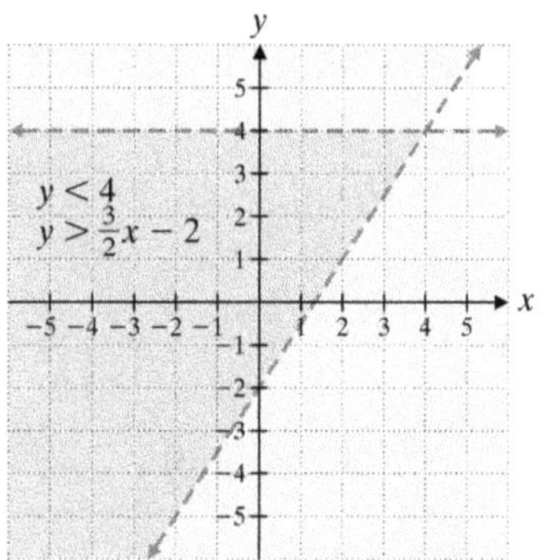

Student Practice

4. Graph the solution to the system.

$y > -3$

$y < -\frac{2}{3}x - 3$

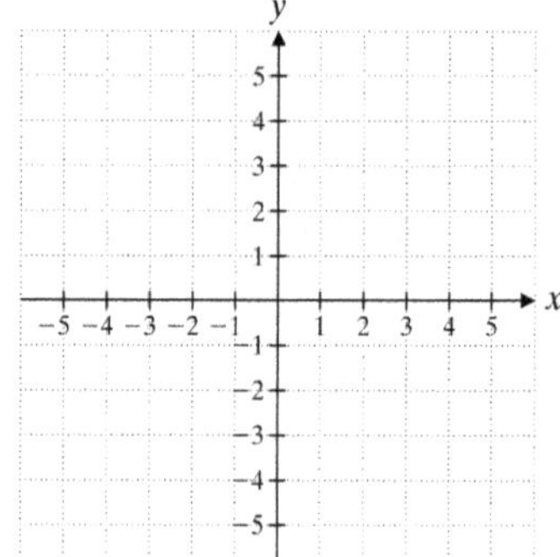

Example	**Student Practice**
5. Graph the solution of the following system of inequalities. Find the coordinates of any points where boundary lines intersect.	**6.** Graph the solution of the following system of inequalities. Find the coordinates of any points where boundary lines intersect.

5. Graph the solution of the following system of inequalities. Find the coordinates of any points where boundary lines intersect.

$$x+y\le 5$$
$$x+2y\le 8$$
$$x\ge 0$$
$$y\ge 0$$

The graph of $x+y\le 5$ is the region on and below the line $x+y=5$. The graph of $x+2y\le 8$ is the region on and below the line $x+2y=8$. We solve the system containing the equations $x+y=5$ and $x+2y=8$ to find that their point of intersection is $(2,3)$. The graph of $x\ge 0$ is the y-axis and all the region to the right of the y-axis. The graph of $y\ge 0$ is the x-axis and all the region above the x-axis. Thus, the solution to the system is the shaded region and its boundary lines. There are four points where the boundary lines intersect. These points are called the vertices of the solution. Thus, the vertices of the solution are $(0,0)$, $(0,4)$, $(2,3)$, and $(5,0)$.

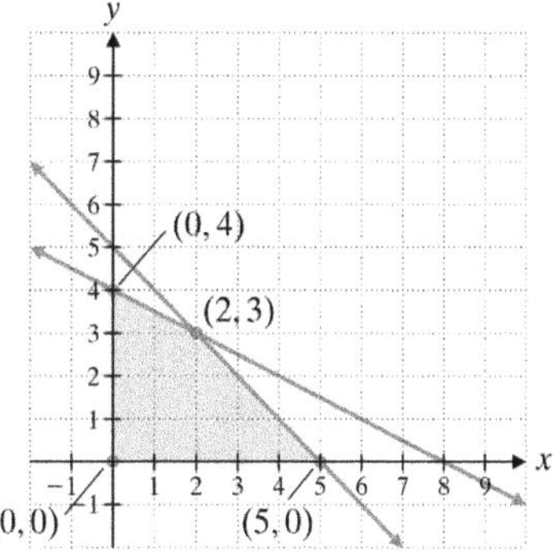

6. Graph the solution of the following system of inequalities. Find the coordinates of any points where boundary lines intersect.

$$x+y\le 5$$
$$x+3y\le 9$$
$$x\ge 0$$
$$y\ge 0$$

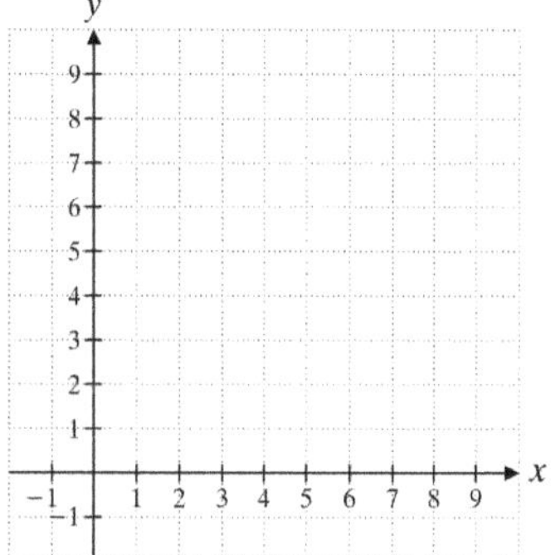

Extra Practice

1. Graph the solution for the following system.

$$4x + 2y \geq -4$$
$$y > x - 3$$

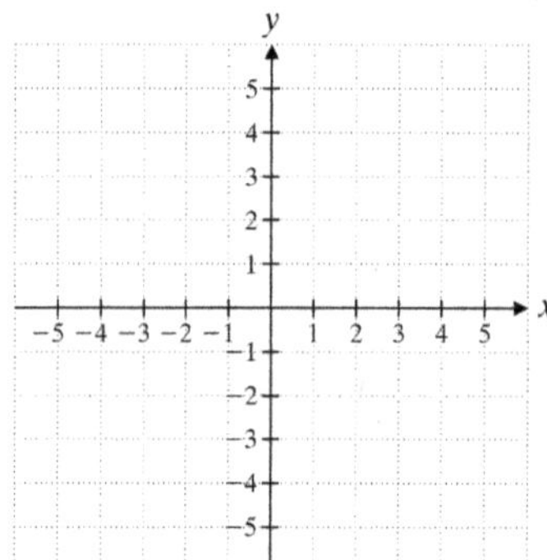

2. Graph the solution for the following system.

$$x < -3$$
$$y \geq -5$$

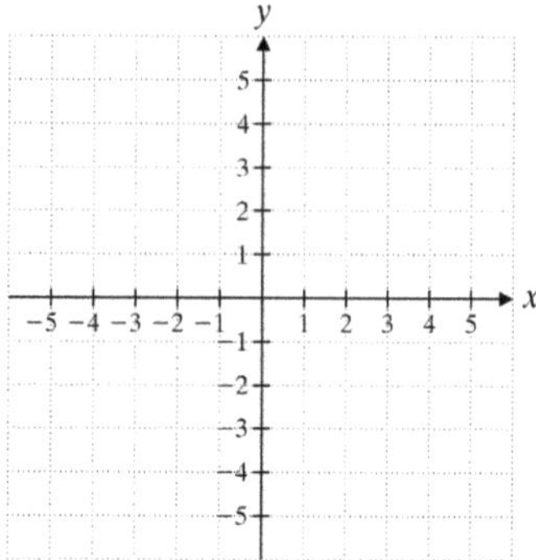

3. Graph the solution for the following system.

$$y + x \leq 4$$
$$y + x > 1$$

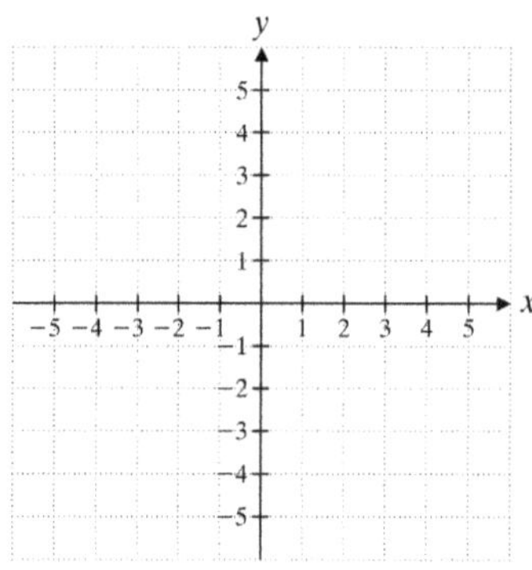

4. Graph the solution to the following system of inequalities. Find the vertices of the solution.

$$y > -3x - 5$$
$$y < 4$$
$$-2x + y > 0$$

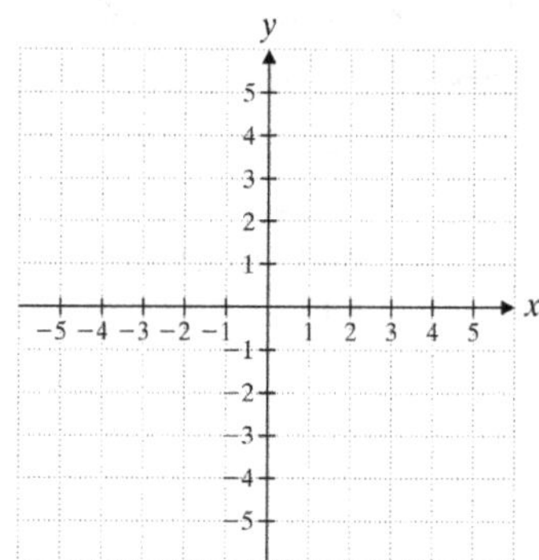

Concept Check

Explain how you would graph the region described by the following.

$y > x + 2$

$x < 3$

MATH COACH

Mastering the skills you need to do well on the test.

Watch the MATH COACH videos in MyMathLab® or on YouTube™ while you work the problems below. These helpful hints will help you avoid making common errors on test problems.

Solving a System of Linear Equations with Fractions—Problem 5

Solve the system. $\frac{1}{3}x+\frac{5}{6}y = 2$

$\frac{3}{5}x - y = -\frac{7}{5}$

> **Helpful Hint:** First clear the fractions by multiplying the first equation by its LCD and multiplying the second equation by its LCD. Then choose the appropriate method to solve this system.

Did you multiply the first equation by the LCD, 6? Did you obtain the equation $2x+5y=12$? Yes ____ No ____

If you answered No, consider why 6 is the LCD and perform the multiplication again. Remember to multiply each term of the equation by the LCD.

Did you multiply the second equation by 5? Did you obtain the equation $3x-5y=-7$? Yes ____ No ____

If you answered No, go back and perform those multiplications again. Be careful as you carry out calculations.

Did you use the addition (elimination) method to complete the solution?
Yes ____ No ____

If you answered No, consider why this method might make the solution easiest to find.

If you answered Problem 5 incorrectly, go back and rework the problem using these suggestions.

Solving a System of Three Linear Equations in Three Variables—Problem 7 Solve the system.

(1) $3x+5y-2z=-5$
(2) $2x+3y-z=-2$
(3) $2x+4y+6z=18$

> **Helpful Hint:** Try to eliminate one of the variables from the original first and second equations. Then eliminate this same variable from the original second and third equations. The result should be a system of two equations in two variables.

Did you choose the variable z to eliminate?
Yes ____ No ____

If you answered No, consider why this might be the easiest variable to eliminate out of the three.

Did you multiply equation (2) by -2 and add the result to equation (1) to obtain the equation $-x-y=-1$?
Yes ____ No ____

If you answered No, stop and perform those operations using only original equations (2) and (1).

Next did you multiply the original equation (2) by 6 and add the result to equation (3) to obtain the equation $14x+22y=6$? Yes ____ No ____

If you answered No, stop and perform those operations using only original equations (2) and (3).

Make sure that after performing these steps, your result is a system of two linear equations in two variables. Once you solve this system in two variables, remember to go back to one of the original equations and substitute in the resulting values for x and y to find the value of z.

Now go back and rework the problem using these suggestions.

Solving an Application Using a System of Three Equations in Three Variables—Problem 11

The math club is selling items with the college logo to raise money. Sam bought 4 pens, a mug, and a T-shirt for \$20.00. Alicia bought 2 pens and 2 mugs for \$11.00. Ramon bought 6 pens, a mug, and 2 T-shirts for \$33.00. What was the price of each pen, mug, and T-shirt?

> **Helpful Hint:** Read through the word problem carefully. Specify what each variable represents. Then construct appropriate equations for Sam, Alicia, and Ramon. Solve the resulting system of three linear equations in three variables.

Did you realize that this problem requires a system of three equations in three variables?
Yes ____ No ____

Did you let x = the price of the pens, y = the price of the mugs, and z = the price of the T-shirts?
Yes ____ No ____

If you answered No to either question, go back and read the problem carefully again. Consider that you have three items with unknown prices. Consider that information is provided for three people—Sam, Alicia, and Ramon.

Did you write Sam's equation as $4x+y+z=20$?

Yes ____ No ____

If you answered No, reread the second sentence in the word problem. Apply the information using the variables listed above. Now translate this information into a linear equation in three variables.

Did you write Alicia's equation as $2x+2y=11$? Yes ____ No ____

If you answered No, reread the third sentence in the word problem. Notice that we do not use the variable z because Alicia did not buy any T-shirts.

Now see if you can write a third equation for Ramon that totals \$33. Then solve the resulting system.

If you answered Problem 11 incorrectly, go back and rework the problem using these suggestions.

Graphing a System of Linear Inequalities—Problem 14

Solve the system of linear inequalities by graphing. $3x+y>8$
$x-2y>5$

> **Helpful Hint:** Remember to use dashed lines when the inequality symbols are $>$ or $<$. Take one inequality at a time and graph the border line. Then shade above or below each border line based on test points. The intersection of the two shaded regions is the solution to the system.

First graph the border line $3x+y=8$. Did you see that this line passes through $(3,-1)$ and $(1,5)$? Did you see that the line should be dashed?
Yes ____ No ____

If you answered No to any of these questions, stop and examine the first inequality again. Notice the inequality symbol used. Then carefully substitute $x=3$ into the equation and solve for y. Now you have two ordered pairs to use when graphing the bordered line.

Now you must decide to shade above or below the first border line. Did you substitute $(0,0)$ into $3x+y>8$ and decide to shade above the line?
Yes ____ No ____

If you answered No, remember that $3(0)+0$ is not greater than 8. So we do not shade on the side of the line that contains $(0,0)$. We must shade above the line.

Follow this procedure for the second border line. Remember that your solution is the intersection of the two shaded regions.

Now go back and rework the problem using these suggestions.

Name: ______________________ Date: ______________

Instructor: ______________________ Topic: ______________

Module 12 Exponents and Polynomials
Topic 1 The Rules of Exponents

Vocabulary

exponent • base • exponential expression • the product rule
numerical coefficient • the quotient rule

1. In the exponential expression x^a, x is called the ____________.

2. In the exponential expression x^a, a is called the ____________.

3. Simplifying the exponential expression $\frac{x^a}{x^b}$ requires using the ____________.

4. A(n) ____________ is a number that is multiplied by a variable, such as the 4 in $4x^2$.

Example	**Student Practice**
1. Multiply.	**2.** Multiply.
(a) $x^3 \cdot x^6$	**(a)** $z^4 \cdot z$
The expressions have the same base so we can add the exponents.	
$x^3 \cdot x^6 = x^{3+6} = x^9$	
(b) $x \cdot x^5$	**(b)** $2^2 \cdot 2^6$
Every variable that does not have a written exponent is understood to have an exponent of 1. Thus, $x = x^1$.	
$x \cdot x^5 = x^{1+5} = x^6$	

Vocabulary Answers: 1. base 2. exponent 3. quotient rule 4. numerical coefficient

Example	Student Practice
3. Multiply. $(5ab)\left(-\frac{1}{3}a\right)(9b^2)$ Multiply the numerical coefficients and group like bases. $(5ab)\left(-\frac{1}{3}a\right)(9b^2)$ $=(5)\left(-\frac{1}{3}\right)(9)(a\cdot a)(b\cdot b^2)$ Use the rule for multiplying expression with exponents. Add the exponents. $(5)\left(-\frac{1}{3}\right)(9)(a\cdot a)(b\cdot b^2)=-15a^2b^3$	**4.** Multiply. $(-4x^3)(xy^2)(2x^2y)$
5. Divide. $\frac{2^{16}}{2^{11}}$ The expressions have the same base so we can subtract the exponents. $\frac{2^{16}}{2^{11}}=2^{16-11}=2^5$	**6.** Divide. $\frac{x^9}{x^5}$
7. Divide. $\frac{12^{17}}{12^{20}}$ The expressions have the same base so we can subtract the exponents. Notice that the larger exponent is in the denominator. $\frac{12^{17}}{12^{20}}=\frac{1}{12^{20-17}}=\frac{1}{12^3}$	**8.** Divide. $\frac{n^6}{n^{10}}$

Example	Student Practice
9. Simplify. $\dfrac{4x^0y^2}{8^0y^5z^3}$ Any number (except 0) to the 0 power equals 1. $\dfrac{4x^0y^2}{8^0y^5z^3}=\dfrac{4(1)y^2}{(1)y^5z^3}$ $=\dfrac{4y^2}{y^5z^3}$ $=\dfrac{4}{y^3z^3}$	**10.** Simplify. $\dfrac{(2y^4)(3x^2y)}{12x^0y^{10}}$
11. Simplify. **(a)** $(x^3)^5$ We are raising a power to a power, so multiply exponents. $(x^3)^5=x^{3\cdot5}=x^{15}$ **(b)** $(-1)^8$ Since n is even, a positive number results. $(-1^8)=+1$	**12.** Simplify. **(a)** $(y^8)^4$ **(b)** $(-1)^{13}$
13. Simplify. $\left(\dfrac{x}{y}\right)^5$ The fraction within the parentheses is raised to a power. Raise both the numerator and the denominator to that power. $\left(\dfrac{x}{y}\right)^5=\dfrac{x^5}{y^5}$	**14.** Simplify. $\left(\dfrac{y}{z^2}\right)^3$

Example	Student Practice
15. Simplify. $\left(\frac{-3x^2z^0}{y^3}\right)^4$ Simplify inside the parentheses first. Apply the rules for raising a power to a power and simplify. $\left(\frac{-3x^2z^0}{y^3}\right)^4=\left(\frac{-3x^2}{y^3}\right)^4$ $=\frac{(-3)^4x^8}{y^{12}}=\frac{81x^8}{y^{12}}$	**16.** Simplify. $\left(\frac{4x^5y}{-16x^0y^3}\right)^3$

Extra Practice

1. Multiply. Leave your answer in exponent form. $(12x^4y^2)(2x^2y^3)$

2. Simplify. Leave your answer in exponent form. Assume that all variables in any denominator are nonzero. $\frac{9a^3}{a^3}$

3. Simplify. Leave your answer in exponent form. Assume that all variables in any denominator are nonzero. $\left(\frac{b}{b^3}\right)^2$

4. Simplify. Leave your answer in exponent form. Assume that all variables in any denominator are nonzero. $\left(\frac{9x^2y^0}{14x^5}\right)^2$

Concept Check

Explain the steps you would need to follow to simplify the expression. $\frac{(4x^3)^2}{(2x^4)^3}$

Name: ______________________________ Date: ______________
Instructor: ____________________________ Topic: ______________

Module 12 Exponents and Polynomials
Topic 2 Negative Exponents and Scientific Notation

Vocabulary
exponent • base • negative exponent • scientific notation • significant digits

1. All digits in a number, excluding the zeros that allow the decimal point to be properly located, are considered ____________.

2. In the number 1.23×10^5, "10" is referred to as the ____________ .

3. A useful way to express very large or very small numbers is to use ____________ .

4. When using scientific notation to express a number less than zero, it is necessary to use a ____________ on the base of ten.

Example	Student Practice
1. Evaluate. 3^{-2} First write the expression with a positive exponent. Then evaluate. $3^{-2}=\frac{1}{3^2}=\frac{1}{9}$	**2.** Evaluate. 7^{-4}
3. Simplify. Write the expression with no negative exponents. $\left(3x^{-4}y^2\right)^{-3}$ Use the power to power rule. $\left(3x^{-4}y^2\right)^{-3}=3^{-3}x^{12}y^{-6}$ Now write the expression with positive exponents only and simplify. $3^{-3}x^{12}y^{-6}=\frac{x^{12}}{3^3y^6}=\frac{x^{12}}{27y^6}$	**4.** Simplify. Write the expression with no negative exponents. $\left(2xy^{-2}z^3\right)^{-4}$

Vocabulary Answers: 1. significant digits 2. base 3. scientific notation 4. negative exponent

Example	Student Practice
5. Write in scientific notation. 157,000,000 Move the decimal point 8 places to the left and multiply by 100,000,000. $157{,}000{,}000 = 1.\underbrace{57000000}_{\text{8 places}} \times 1\underbrace{00000000}_{\text{8 zeros}}$ $= 1.57 \times 10^{8}$	**6.** Write in scientific notation. 2564
7. Write in scientific notation. **(a)** 0.061 Move the decimal point 2 places to the right and multiply by 10^{-2}. $0.061 = 6.1 \times 10^{-2}$ **(b)** 0.000052 $0.000052 = 5.2 \times 10^{-5}$	**8.** Write in scientific notation. **(a)** 0.0013 **(b)** 0.000001
9. Write in decimal notation. **(a)** 1.568×10^{2} The exponent 2 tells us to move the decimal point 2 places to the right. $1.568 \times 10^{2} = 156.8$ **(b)** 7.432×10^{-3} The exponent -3 tells us to move the decimal point 3 places to the left. $7.432 \times 10^{-3} = 7.432 \times \frac{1}{1000}$ $= 0.007432$	**10.** Write in decimal notation. **(a)** 5.28×10^{-4} **(b)** 3.2214×10^{6}

Example	Student Practice
11. The approximate distance from Earth to the star Polaris is 208 parsecs (pc). A parsec is a distance of approximately 3.09×10^{13} km. How long would it take a space probe traveling at 40,000 km/hr to reach the star? Round to three significant digits.	**12.** Use the information in example **15** to answer the following. How long would it take the space probe to reach a star that is 500 parsecs from Earth? Round to three significant digits.

Understand the problem. We need to change the distance from parsecs to kilometers using the given relationship.

$$208 \text{ pc} = \frac{(208 \text{ pc})(3.09\times10^{13} \text{ km})}{1 \text{ pc}}$$

$$= 642.72\times10^{13}$$

Write an equation using the distance formula, $\text{distance} = \text{rate}\times\text{time}$, or $d = r\times t$.

Substitute known values and change all given values to scientific notation.

$$6.4272\times10^{15} \text{ km} = \frac{4\times10^{4} \text{ km}}{1 \text{ hr}}\times t$$

Multiply both sides by the reciprocal of $\frac{4\times10^{4} \text{ km}}{1 \text{ hr}}$ and simplify.

$$6.4272\times10^{15} \text{ km}\times\frac{1 \text{ hr}}{4\times10^{4} \text{ km}} = t$$

$$\frac{(6.4272\times10^{15} \cancel{\text{km}})(1 \text{ hr})}{4\times10^{4} \cancel{\text{km}}} = t$$

$$1.6068\times10^{11} \text{ hr} = t$$

Round to three significant figures.

$1.6068\times10^{11} \text{ hr} \approx 1.61\times10^{11} \text{hr}$

The check is left to the student.

Extra Practice

1. Simplify. Express your answer with positive exponents. Assume that all variables are nonzero. 7^{-3}

2. Simplify. Express your answer with positive exponents. Assume that all variables are nonzero. $\left(\frac{3a^3b^{-2}}{c^3}\right)^{-4}$

3. Write in decimal notation. 6.34×10^3

4. Evaluate by using scientific notation and the laws of exponents. Leave your answer in scientific notation. $\frac{0.0046}{0.023}$

Concept Check

Explain how you would simplify a problem like the following so that your answer has only positive exponents. $\left(4x^{-3}y^4\right)^{-3}$

Name: ________________________________ Date: ________________
Instructor: ____________________________ Topic: ________________

Module 12 Exponents and Polynomials
Topic 3 Fundamental Polynomial Operations

Vocabulary
polynomial • multivariable polynomial • degree of a term • degree of a polynomial
monomial • binomial • trinomial • decreasing order • evaluate

1. A(n) ____________ is a polynomial with two terms.

2. In a term, the sum of the exponents of all of the variables is called the ____________.

3. When a polynomial is written in ____________, the value of each exponent decreases as we move from left to right.

4. The highest degree of all of the terms in a polynomial is called the ____________.

Example	Student Practice
1. State the degree of the polynomial, and whether it is a monomial, a binomial, or a trinomial.	**2.** State the degree of the polynomial, and whether it is a monomial, a binomial, or a trinomial.
(a) $5xy+3x^3$ This polynomial is of degree 3. It has two terms, so it is a binomial.	**(a)** $7y^3z^4$
(b) $-7a^5b^2$ The sum of the exponents is $5+2=7$. Therefore, this polynomial is of degree 7. It has one term, so it is a monomial.	**(b)** x^2+5x-4
(c) $8x^4-9x-15$ This polynomial is of degree 4. It has three terms, so it is a trinomial.	**(c)** $2xy^6+7x^2$

Vocabulary Answers: 1. binomial 2. degree of a term 3. decreasing order 4. degree of a polynomial

Example	Student Practice
3. Add. $\left(5x^2-6x-12\right)+\left(-3x^2-9x+5\right)$ Group like terms. $\left(5x^2-6x-12\right)+\left(-3x^2-9x+5\right)$ $=\left[5x^2+\left(-3x^2\right)\right]+\left[-6x+(-9x)\right]+\left[-12+5\right]$ Add like terms. $\left[(5-3)x^2\right]+\left[(-6-9)x\right]+\left[-12+5\right]$ $=2x^2+(-15x)+(-7)$ $=2x^2-15x-7$	**4.** Add. $\left(-x^2+x-5\right)+\left(5x^2-9x+10\right)$
5. Add. $\left(\frac{1}{2}x^2-6x+\frac{1}{3}\right)+\left(\frac{1}{5}x^2-2x-\frac{1}{2}\right)$ The numerical coefficients of polynomials may be any real number. Thus, polynomials may have numerical coefficients that are decimals or fractions. To add, first group like terms. $\left(\frac{1}{2}x^2-6x+\frac{1}{3}\right)+\left(\frac{1}{5}x^2-2x-\frac{1}{2}\right)$ $=\left[\frac{1}{2}x^2+\frac{1}{5}x^2\right]+\left[-6x+(-2x)\right]+\left[\frac{1}{3}+\left(-\frac{1}{2}\right)\right]$ Add like terms. $\left[\left(\frac{1}{2}+\frac{1}{5}\right)x^2\right]+\left[(-6-2)x\right]+\left[\frac{1}{3}+\left(-\frac{1}{2}\right)\right]$ $=\left[\left(\frac{5}{10}+\frac{2}{10}\right)x^2\right]+(-8x)+\left[\frac{2}{6}-\frac{3}{6}\right]$ $=\frac{7}{10}x^2-8x-\frac{1}{6}$	**6.** Add. $\left(\frac{3}{4}x^2-2x+\frac{1}{8}\right)+\left(x^2+\frac{2}{5}x+\frac{1}{2}\right)$

Example	Student Practice
7. Subtract. $(7x^2-6x+3)-(5x^2-8x-12)$ We change the sign of each term in the second polynomial and then add. $(7x^2-6x+3)-(5x^2-8x-12)$ $=(7x^2-6x+3)+(-5x^2+8x+12)$ $=(7-5)x^2+(-6+8)x+(3+12)$ $=2x^2+2x+15$	**8.** Subtract. $(-5x^2-3x+1)-(-2x^2+7x-4)$
9. Automobiles sold in the United States have become more fuel efficient over the years due to regulations from Congress. The number of miles per gallon obtained by the average automobile in the United States can be described by the polynomial $0.3x+12.9$, where x is the number of years since 1970. (*Source:* U.S. Federal Highway Administration.) Use this polynomial to estimate the number of miles per gallon obtained by the average automobile in 1972. The year 1972 is two years later than 1970, so $x=2$. Thus, the number of miles per gallon obtained by the average automobile in 1972 can be estimated by evaluating $0.3x+12.9$ when $x=2$. $0.3(2)+12.9=0.6+12.9$ $=13.5$ We estimate that the average car in 1972 obtained 13.5 miles per gallon.	**10.** Use the information in example **9** to answer the following. Estimate the number of miles per gallon that will be obtained by the average automobile in the United States in 2015.

Extra Practice

1. State the degree of the polynomial and whether it is a monomial, a binomial, or a trinomial. $23x^6 - 14x^3 + 5$

2. Add. $(6.4x-3)+(4.4x-11)$

3. Subtract.
$(2r^4 - 3r^2 + 14) - (-3r^4 - 2r^2 + 6)$

4. Buses sold in the U.S. have become more fuel efficient over the years. The number of miles per gallon obtained by a certain type of bus can be described by the polynomial $0.37x + 5.31$, where x is the number of years since 1970. Use this polynomial to estimate the number of miles per gallon obtained by this type of bus in 1980.

Concept Check

Explain how you would determine the degree of the following polynomial and how you would decide if it is a monomial, a binomial, or a trinomial. $2xy^2 - 5x^3y^4$

Name: ______________________________ Date: ________________
Instructor: ____________________________ Topic: ________________

Module 12 Exponents and Polynomials
Topic 4 Multiplying Polynomials

Vocabulary

polynomial • monomial • binomial • distributive property • FOIL

1. A ____________ is a polynomial with only one term.

2. The process of using the distributive property to multiply two binomials is often referred to as ____________.

3. The ____________ states that for all real numbers *a*, *b*, and *c*, $a(b+c) = ab + ac$.

Example	**Student Practice**
1. Multiply. $3x^2(5x-2)$ Use the distributive property and multiply each term by $3x^2$. $3x^2(5x-2) = 3x^2(5x) + 3x^2(-2)$ Simplify. $3x^2(5x) + 3x^2(-2)$ $= (3\cdot5)(x^2\cdot x) + (3)(-2)x^2$ $= 15x^3 - 6x^2$	**2.** Multiply. $5y^2(-y+4)$
3. Multiply. $(x^2-2x+6)(-2xy)$ Use the distributive property and multiply each term by $-2xy$. $(x^2-2x+6)(-2xy)$ $= -2x^3y + 4x^2y - 12xy$	**4.** Multiply. $(x^2-7x-10)(2x^2y)$

Vocabulary Answers: 1. monomial 2. FOIL 3. distributive property

Example	Student Practice
5. Multiply. $(2x-1)(3x+2)$ Multiply the First terms, $2x$ and $3x$. $(2x)(3x)=6x^2$ Multiply the Outer terms, $2x$ and 2. $(2x)(2)=4x$ Multiply the Inner terms, -1 and $3x$. $(-1)(3x)=-3x$ Multiply the Last terms, -1 and 2. $(-1)(2)=-2$ Add the results and combine like terms. First + Outer + Inner + Last $=6x^2 + 4x - 3x - 2$ $=6x^2+x-2$	**6.** Multiply. $(3x-1)(5x+3)$
7. Multiply. $(3x+2y)(5x-3z)$ Multiply the First terms, $3x$ and $5x$. $(3x)(5x)=15x^2$ Multiply the Outer terms, $3x$ and $-3z$. $(3x)(-3z)=-9xz$ Multiply the Inner terms, $2y$ and $5x$. $(2y)(5x)=10xy$ Multiply the Last terms, $2y$ and $-3z$. $(2y)(-3z)=-6yz$ Add the results and combine like terms. $(3x+2y)(5x-3z)$ $=15x^2-9xz+10xy-6yz$	**8.** Multiply. $(7x-3y)(x+2z)$

Example	Student Practice
9. Multiply. $(7x-2y)^2$ When we square a binomial, it is the same as multiplying the binomial by itself. $(7x-2y)(7x-2y)$ Multiply the First terms, $7x$ and $7x$. $(7x)(7x)=49x^2$ Multiply the Outer terms, $7x$ and $-2y$. $(7x)(-2y)=-14xy$ Multiply the Inner terms, $-2y$ and $7x$. $(-2y)(7x)=-14xy$ Multiply the Last terms, $-2y$ and $-2y$. $(-2y)(-2y)=4y^2$ Add the results and combine like terms. $(7x-2y)^2=49x^2-14xy-14xy+4y^2$ $=49x^2-28xy+4y^2$	**10.** Multiply. $(4x+3y)^2$
11. Multiply. $(3x^2+4y^3)(2x^2+5y^3)$ Use the FOIL method and the rules for multiplying expressions with exponents. $(3x^2+4y^3)(2x^2+5y^3)$ $=6x^4+15x^2y^3+8x^2y^3+20y^6$ $=6x^4+23x^2y^3+20y^6$	**12.** Multiply. $(10x^2+5y^4)(2x^2-3y^4)$

Example	Student Practice
13. The width of a living room is $(x+4)$ feet. The length of the room is $(3x+5)$ feet. What is the area of the room in square feet?	**14.** The width of a brick patio is $(2x+1)$ feet. The length of the patio is $(4x-2)$ feet. What is the area of the patio in square feet?

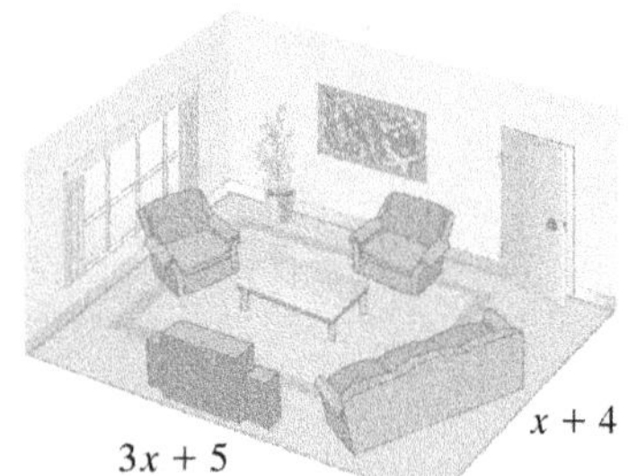

Use the area formula and solve.

$A = (\text{length})(\text{width})$

$A = (3x+5)(x+4)$

$A = 3x^2 + 12x + 5x + 20$

$A = 3x^2 + 17x + 20$

There are $\left(3x^2 + 17x + 20\right)$ square feet in the room.

Extra Practice

1. Multiply. $5x\left(-2x^3 + 3x\right)$

2. Multiply. $(x-3)(x-11)$

3. Multiply. $(3x-7y)(5x+8y)$

4. Multiply. $\left(2x^2 - 3y^3\right)\left(4x^2 + 5y^3\right)$

Concept Check

Explain how you would multiply $(7x-3)^2$.

Name: ______________________________ Date: ________________
Instructor: ___________________________ Topic: ________________

Module 12 Exponents and Polynomials
Topic 5 Multiplication: Special Cases

Vocabulary

polynomial • term • binomial • FOIL • square of a sum
square of a difference • vertical multiplication • horizontal multiplication

1. One method of multiplying polynomials that uses a method similar to that used in arithmetic for multiplying whole numbers is called ___________.

2. A binomial of the form $(a-b)^2$ is referred to as the ___________.

3. One way to multiply polynomials with more than two terms is to use ___________ horizontal multiplication.

Example	Student Practice
1. Multiply. $(7x+2)(7x-2)$ Use the rule for multiplying a sum and a difference. $(a+b)(a-b)=a^2-b^2$ Here, $a=7x$ and $b=2$. $(7x+2)(7x-2)=(7x)^2-(2)^2$ $=49x^2-4$ The check is left to the student.	**2.** Multiply. $(8y+4)(8y-4)$
3. Multiply. $(5x-8y)(5x+8y)$ Use the rule for multiplying a sum and a difference. Here, $a=5x$ and $b=8y$. $(5x-8y)(5x+8y)=(5x)^2-(8y)^2$ $=25x^2-64y^2$	**4.** Multiply. $(2x-9y)(2x+9y)$

Vocabulary Answers: 1. vertical multiplication 2. square of a difference 3. FOIL

Example	Student Practice
5. Multiply. $(5y-2)^2$	**6.** Multiply. $(7x+y)^2$

Use a rule for a binomial squared.

$(a+b)^2 = a^2 + 2ab + b^2$

$(a-b)^2 = a^2 - 2ab + b^2$

Here, $a = 5y$ and $b = 2$.

$$\begin{aligned}&(5y-2)^2\\&=(5y)^2-(2)(5y)(2)+(2)^2\\&=25y^2-20y+4\end{aligned}$$

Example	Student Practice
7. Multiply vertically. $(3x^3+2x^2+x)(x^2-2x-4)$	**8.** Multiply vertically. $(x^2+3x-6)(2x^2-4x-5)$

Place one polynomial over the other. Find the partial products and line them up under the original polynomials. Note that the answers for each partial product are placed so that like terms are underneath each other.

$$\begin{array}{r}3x^3+2x^2+x\\\underline{x^2-2x-4}\\-12x^3-8x^2-4x\\-6x^4-4x^3-2x^2\qquad\\3x^5-2x^4-x^3\qquad\qquad\end{array}$$

Find the sum of the three partial products.

$$\begin{array}{r}3x^3+2x^2+x\\\underline{x^2-2x-4}\\-12x^3-8x^2-4x\\-6x^4-4x^3-2x^2\qquad\\\underline{3x^5-2x^4-x^3\qquad\qquad}\\3x^5-4x^4-15x^3-10x^2-4x\end{array}$$

Example	Student Practice
9. Multiply horizontally. $(x^2+3x+5)(x^2-2x-6)$ Use the distributive property repeatedly. $(x^2+3x+5)(x^2-2x-6)$ $=x^2(x^2-2x-6)+3x(x^2-2x-6)$ $+5(x^2-2x-6)$ $=x^4-2x^3-6x^2+3x^3-6x^2-18x$ $+5x^2-10x-30$ $=x^4+x^3-7x^2-28x-30$	**10.** Multiply horizontally. $(x^2-8x+10)(x^2+x+4)$
11. Multiply. $(2x-3)(x+2)(x+1)$ Multiply the first pair of binomials. Note that it does not matter which two binomials are multiplied first. $(2x-3)(x+2)=2x^2+4x-3x-6$ $=2x^2+x-6$ Replace the first two factors with their resulting product. $(2x^2+x-6)(x+1)$ Multiply again and combine like terms. $(2x^2+x-6)(x+1)$ $=(2x^2+x-6)x+(2x^2+x-6)1$ $=2x^3+x^2-6x+2x^2+x-6$ $=2x^3+3x^2-5x-6$	**12.** Multiply. $(x-2)(x+3)(2x+4)$

Extra Practice

1. Multiply. Use the special formula that applies. $(x+10)(x-10)$

2. Multiply. Use the special formula that applies. $(5a+3b)(5a-3b)$

3. Multiply. $(8x^2-2x+3)(3x+1)$

4. Multiply. $(x+2)(x-1)(x-5)$

Concept Check

Using the formula $(a+b)^2=a^2+2ab+b^2$, explain how to multiply $(6x-9y)^2$.

Name: ______________________________ Date: ________________
Instructor: ____________________________ Topic: ________________

Module 12 Exponents and Polynomials
Topic 6 Dividing Polynomials

Vocabulary
polynomial • subtraction • long division • descending order • binomial • monomial

1. To divide a polynomial by a ___________, divide each term of the numerator by the denominator, then write the sum of the results.

2. ___________ is a process used to divide polynomials when the divisor has two or more terms.

3. When dividing a polynomial by a binomial, you must first place the terms of each in ___________, inserting a 0 for any missing terms.

4. When performing long division with polynomials, take great care on the___________ step when negative numbers are involved.

Example	Student Practice
1. Divide. $\dfrac{8y^6-8y^4+24y^2}{8y^2}$	**2.** Divide. $\dfrac{21x^5+7x^3-28x^2}{7x^2}$

Divide each term of the polynomial by the monomial.

$$\frac{8y^6-8y^4+24y^2}{8y^2}=\frac{8y^6}{8y^2}-\frac{8y^4}{8y^2}+\frac{24y^2}{8y^2}$$

Use the property $\dfrac{x^a}{x^b}=x^{a-b}$ to divide each term.

$$\frac{8y^6}{8y^2}-\frac{8y^4}{8y^2}+\frac{24y^2}{8y^2}=y^4-y^2+3$$

Vocabulary Answers: 1. monomial 2. descending order 3. long division 4. subtraction

Example	Student Practice
3. Divide. $\left(x^3+5x^2+11x+4\right)\div(x+2)$	**4.** Divide. $\left(2x^3+4x^2+3x+6\right)\div(x+1)$

Notice the terms are already in descending order with no missing terms. Divide the first term of the polynomial by the first term of the binomial.

$$\begin{array}{r} x^2 \\ x+2\overline{\big)x^3+5x^2+11x+4} \end{array}$$

Multiply x^2 by $x+2$ and subtract the result from the first two terms.

$$\begin{array}{r l} & x^2+3x \\ x+2 & \overline{\big)x^3+5x^2+11x+4} \\ & \underline{x^3+2x^2} \\ & \quad 3x^2+11x \end{array}$$

Continue this process until the degree of the remainder is less than the degree of the divisor.

$$\begin{array}{r l} & x^2+3x+5 \\ x+2 & \overline{\big)x^3+5x^2+11x+4} \\ & \underline{x^3+2x^2} \\ & \quad 3x^2+11x \\ & \quad \underline{3x^2+6x} \\ & \qquad\quad 5x+4 \\ & \qquad\quad \underline{5x+10} \\ & \qquad\qquad -6 \end{array}$$

The remainder is written as the numerator of a fraction that has the binomial divisor as its denominator.

$$x^2+3x+5+\frac{-6}{x+2}$$

The check is left to the student.

Example	Student Practice

5. Divide. $(5x^3 - 24x^2 + 9) \div (5x + 1)$

Insert $0x$ into the polynomial to represent the missing x-term.

$(5x^3 - 24x^2 + 0x + 9) \div (5x + 1)$

Divide the first term of the polynomial, $5x^3$, by the first term of the binomial, $5x$.

$$\begin{array}{r}
x^2 \\
5x+1\overline{)5x^3 - 24x^2 + 0x + 9} \\
\underline{5x^3 + x^2} \\
-25x^2
\end{array}$$

Divide $-25x^2$ by $5x$ and then divide $5x$ by $5x$. Be cautious with the negative signs.

$$\begin{array}{r}
x^2 - 5x + 1 \\
5x+1\overline{)5x^3 - 24x^2 + 0x + 9} \\
\underline{5x^3 + x^2} \\
-25x^2 + 0x \\
\underline{-25x^2 - 5x} \\
5x + 9 \\
\underline{5x + 1} \\
8
\end{array}$$

Write the remainder as the numerator of a fraction that has the binomial divisor as its denominator.

The answer is $x^2 - 5x + 1 + \dfrac{8}{5x+1}$.

The check is left to the student.

6. Divide. $(8x^3 - 8x + 5) \div (2x + 1)$

Example	Student Practice
7. Divide and check. $(12x^3-11x^2+8x-4)\div(3x-2)$	**8.** Divide and check. $(8x^3-4x^2+6)\div(2x-3)$

$$\begin{array}{r} 4x^2-x+2 \\ 3x-2\overline{)12x^3-11x^2+8x-4} \\ \underline{12x^3-8x^2} \\ -3x^2+8x \\ \underline{-3x^2+2x} \\ 6x-4 \\ \underline{6x-4} \\ 0 \end{array}$$

Check the answer.

$$(3x-2)(4x^2-x+2)=12x^3-11x^2+8x-4$$

Extra Practice

1. Divide. $\dfrac{12a^7-4a^5+8a^3-2a^2}{2a^2}$

2. Divide. $(12y^4-18y^3+27y^2)\div 3y^2$

3. Divide and check. $\dfrac{3x^3-5x^2+7x-5}{x-1}$

Divide and check. $\dfrac{y^3-2y-4}{y-1}$

Concept Check

Explain how you would check if $x^2+2x+8+\dfrac{13}{x-2}$ is the correct answer to the problem $(x^3+4x-3)\div(x-2)$. Perform the check. Does the answer check?

MATH COACH

Mastering the skills you need to do well on the test.

Watch the MATH COACH videos in MyMathLab® or on YouTube™ while you work the problems below. These helpful hints will help you avoid making common errors on test problems.

Raising Monomials to a Power—Problem 8 Simplify $\dfrac{(3x^2)^3}{(6x)^2}$.

> **Helpful Hint:** Do the problem in three stages. First, use the power to a power rule to raise the numerator to the third power. Second, raise the denominator to the second power. Then divide the monomials using the rules of exponents. Be sure to simplify any fractions.

Did you use the power to a power rule to raise both 3^1 and x^2 to the third power in the numerator and both 6^1 and x^1 to the second power in the denominator?
Yes ____ No ____

If you answered No, stop and review the power to a power rule before completing these steps again.

Did you remember to simplify the fraction $\dfrac{27}{36}$?

Yes ____ No ____

Finally, did you remember to use the quotient rule to subtract the exponents in the x terms?
Yes ____ No ____

If you answered No to either of these questions, go back and examine your work carefully before completing these steps again.

If you answered Problem 8 incorrectly, go back and rework the problem using these suggestions.

Simplifying Monomials Involving Negative Exponents—Problem 11

Simplify and write with only positive exponents. $\dfrac{3x^{-3}y^2}{x^{-4}y^{-5}}$

> **Helpful Hint:** First, use the definition of a negative exponent to rewrite the expression using only positive exponents. Then use the rules for exponents to simplify the resulting expression.

Did you remove the negative exponents by rewriting the expression as $\dfrac{3x^4y^2y^5}{x^3}$? Yes ____ No ____

If you answered No, review the definition of negative exponents in Section 4.2 and complete this step again.

Did you use the quotient rule to simplify the x terms and the product rule to simplify the y terms? Yes ____ No ____

If you answered No, review the rules for exponents in Sections 4.1 and 4.2 and simplify the expression again.

Now go back and rework the problem using these suggestions.

Multiplying Three Binomials—Problem 20 Multiply $(3x+2)(2x+1)(x-3)$.

Helpful Hint: A good approach is to start by multiplying the first two binomials. Then multiply that result by the third binomial. Be careful to avoid sign errors when multiplying, and be careful to write down the correct exponent for each term.

Did you use the FOIL method to multiply the first two binomials and obtain $6x^2+7x+2$? Yes ____ No ____

If you answered No, stop and complete this step.

Did you multiply the result above by $(x-3)$?
Yes ____ No ____

Did you multiply each term of $6x^2+7x+2$ by x?
Yes ____ No ____

Did you multiply each term of $6x^2+7x+2$ by -3?
Yes ____ No ____

If you answered No to any of these questions, go back and examine each step of the multiplication carefully.

Be sure to write the correct exponent each time that you multiply. Be careful to avoid sign errors when multiplying by −3. Then combine like terms before writing your final answer.

If you answered Problem 20 incorrectly, go back and rework the problem using these suggestions.

Dividing a Polynomial by a Binomial—Problem 27 Divide $\left(2x^3-6x-36\right)\div(x-3)$.

Helpful Hint: Review the procedure for dividing a polynomial by a binomial in Section 4.6. Make sure you understand each step. Be sure you understand where the expression $0x^2$ came from in the dividend. Be careful with subtraction. Write out the subtraction steps to avoid sign errors.

Did you write the division problem in the form
$x-3\overline{)2x^3+0x^2-6x-36}$? Yes ____ No ____

If you answered No, remember that every power must be represented. We must use $0x^2$ as a placeholder so that we can perform our division.

When you carried out the first step of division, did you obtain $2x^2$ as the first part of your answer?
Yes ____ No ____

When you multiplied $x-3$ by $2x^2$, and then subtracted, did you get the result $6x^2$? Yes ____ No ____

If you answered No to these questions, stop and examine your first division step carefully. Make sure that you subtracted carefully too. Remember to write out the subtraction steps: $0x^2-\left(-6x^2\right)=6x^2$.

Next, did you bring down $-6x$ from the dividend to obtain $6x^2-6x$?
Yes ____ No ____

If you answered No, go back and look at the dividend again and see how to obtain this result.

Now go back and rework the problem using these suggestions.

Name: ______________________________ Date: ________________
Instructor: ___________________________ Topic: ________________

Module 13 Factoring
Topic 1 Removing a Common Factor

Vocabulary

factor • to factor • common factor • greatest common factor

1. When two or more numbers, variables, or algebraic expressions are multiplied, each is called a ___________.

2. A ___________ is a factor that both terms have in common.

3. When you are asked ___________ a number or an algebraic expression, you are being asked, "What factors, when multiplied, will give that number or expression?"

4. When we factor, we begin by looking for the ___________.

Example	Student Practice
1. Factor. **(a)** $3x-6y$ Begin by looking for a common factor, a factor that both terms have in common. Then rewrite the expression as a product. $3x-6y=3(x-2y)$ This is true because $3(x-2y)=3x-6y$.	**2.** Factor. **(a)** $5y+15z$
(b) $9x+2xy$ $9x+2xy=x(9+2y)$ This is true because $x(9+2y)=9x+2xy$. Notice that factoring is using the distributive property in reverse.	**(b)** $12y-5yz$

Vocabulary Answers: 1. factor 2. common factor 3. to factor 4. greatest common factor

Example	Student Practice
3. Factor $24xy+12x^2+36x^3$. Remember to remove the greatest common factor. We start by finding the greatest common factor of 24, 12, and 36. You may want to factor each number, or you may notice that 12 is a common factor. 12 is the greatest numerical common factor. Notice also that x is a factor of each term. Thus, $12x$ is the greatest common factor. $24xy+12x^2+36x^3=12x\left(2y+x+3x^2\right)$	**4.** Factor $44y^3+55y^2-11xy$. Remember to remove the greatest common factor.
5. Factor. **(a)** $12x^2-18y^2$ Note that the largest integer that is common to both terms is 6 (not 3 or 2). $12x^2-18y^2=6\left(2x^2-3y^2\right)$ **(b)** $x^2y^2+3xy^2+y^3$ Although y is common to all of the terms, we factor out y^2 since 2 is the largest exponent of y that is common to all terms. We do not factor out x, since x is not common to all of the terms. $x^2y^2+3xy^2+y^3=y^2\left(x^2+3x+y\right)$	**6.** Factor. **(a)** $21m^2-28n^2$ **(b)** $m^3n^2+9m^2n^2+3m^4$

Example	Student Practice
7. Factor. $8x^3y+16x^2y^2-24x^3y^3$ We see that 8 is the largest integer that will divide evenly into the three numerical coefficients. We can factor x^2 out of each term. We can also factor y out of each term. $8x^3y+16x^2y^2-24x^3y^3$ $=8x^2y\left(x+2y-3xy^2\right)$ Check. $8x^2y\left(x+2y-3xy^2\right)$ $=8x^3y+16x^2y^2-24x^3y^3$	**8.** Factor. $27xy^3-36x^2y^2-9x^3y^3$
9. Factor. $9a^3b^2+9a^2b^2$ We observe that both terms contain a common factor of 9. We can also factor a^2 and b^2 out of each term. Thus, the greatest common factor is $9a^2b^2$. $9a^3b^2+9a^2b^2=9a^2b^2(a+1)$	**10.** Factor. $11m^4n^2+11m^3n^2$
11. Factor. $7x^2(2x-3y)-(2x-3y)$ The common factor of the terms is $(2x-3y)$. What happens when we factor out $(2x-3y)$? What are we left with in the second term? Recall that $(2x-3y)=1(2x-3y)$. $7x^2(2x-3y)-(2x-3y)$ $=7x^2(2x-3y)-\mathbf{1}(2x-3y)$ $=(2x-3y)\left(7x^2-1\right)$	**12.** Factor. $5y^2(3x+4y)-(3x+4y)$

Example	Student Practice
13. A computer programmer is writing a program to find the total area of 4 circles. She uses the formula $A = \pi r^2$. The radii of the circles are a, b, c, and d, respectively. She wants the final answer to be in factored form with the value of π occurring only once, in order to minimize the rounding error. Write the total area of the 4 circles with a formula that has π occurring only once. For each circle, $A = \pi r^2$, where $r = a$, b, c, or d. We add the area of each of the 4 circles. The total area is $\pi a^2 + \pi b^2 + \pi c^2 + \pi d^2$. In factored form the total area = $\pi\left(a^2 + b^2 + c^2 + d^2\right)$.	**14.** Find the total area of 3 circles using the formula $A = \pi r^2$. The radii of the circles are m, $2n$, and $3z$, respectively. Write the total area of the 3 circles with a formula that has π occurring only once.

Extra Practice

1. Remove the largest possible common factor. Check your answer by multiplication. $3x^3 + 12x^2 - 21x$

2. Remove the largest possible common factor. Check your answer by multiplication. $60x^2y + 18xy - 24x$

3. Remove the largest possible common factor. Check your answer by multiplication. $8a(x+3y) - b(x+3y)$

4. Remove the largest possible common factor. Check your answer by multiplication. $5a(4x-3) - (4x-3)$

Concept Check

Explain how you would remove the greatest common factor from the following polynomial.
$36a^3b^2 - 72a^2b^3$

Name: ______________________________ Date: ________________
Instructor: ____________________________ Topic: ________________

Module 13 Factoring
Topic 2 Factoring by Grouping

Vocabulary
factoring by grouping • common factor • commutative property • FOIL

1. Sometimes you will need to factor out a negative ___________ from the second two terms to obtain two terms that contain the same parenthetical expression.

2. A procedure used to factor a four-term polynomial is called ___________.

3. Rearrange the order using the ___________ of addition.

4. To check, we multiply the two binomials using the ___________ procedure.

Example	**Student Practice**
1. Factor. $x(x-3)+2(x-3)$ Observe each term: $\underbrace{x(x-3)}_{\text{first term}}+\underbrace{2(x-3)}_{\text{second term}}$ The common factor of the first and second terms is the quantity $(x-3)$. $x(x-3)+2(x-3)=(x-3)(x+2)$	**2.** Factor. $z(2z+5)-4(2z+5)$
3. Factor. $2x^2+3x+6x+9$ Factor out a common factor of x from the first two terms. Factor out a common factor of 3 from the second two terms. $2x^2+3x+6x+9=x(2x+3)+3(2x+3)$ The expression in parentheses is now a common factor of the terms. $x(2x+3)+3(2x+3)=(2x+3)(x+3)$	**4.** Factor. $12x^2+4x+15x+5$

Vocabulary Answers: 1. common factor 2. factoring by grouping 3. commutative property 4. FOIL

Example	Student Practice
5. Factor. $4x+8y+ax+2ay$ Factor out a common factor of 4 from the first two terms. Factor out a common factor of a from the second two terms. $4x+8y+ax+2ay$ $=4(x+2y)+a(x+2y)$ The common factor of the terms is the expression in parentheses, $x+2y$. $4(x+2y)+a(x+2y)=(x+2y)(4+a)$	**6.** Factor. $7y+21z+xy+3xz$
7. Factor. $bx+4y+4b+xy$ Rearrange the terms so that the first two terms have a common factor. $bx+4y+4b+xy=bx+4b+xy+4y$ Factor out the common factor b from the first two terms and the common factor y from the second two terms. $bx+4b+xy+4y=b(x+4)+y(x+4)$ $=(x+4)(b+y)$	**8.** Factor. $nz+6y+6z+ny$
9. Factor. $2x^2+5x-4x-10$ Factor out the common factor x from the first two terms and the common factor -2 from the second two terms. $2x^2+5x-4x-10$ $=x(2x+5)-4x-10$ $=x(2x+5)-2(2x+5)$ $=(2x+5)(x-2)$	**10.** Factor. $2x^2+3x-6x-9$

Example	Student Practice
11. Factor. $2ax-a-2bx+b$ Factor out the common factor a from the first two terms and the common factor $-b$ from the second two terms. $2ax-a-2bx+b=a(2x-1)-b(2x-1)$ Since the two resulting terms contain the same parenthetical expression, we can complete the factoring. $a(2x-1)-b(2x-1)=(2x-1)(a-b)$	**12.** Factor. $5nz-n-5mz+m$
13. Factor and check your answer. $8ad+21bc-6bd-28ac$ Use the commutative property of addition to rearrange the order so the first two terms have a common factor. $8ad+21bc-6bd-28ac$ $=8ad-6bd-28ac+21bc$ Factor out the common factor $2d$ from the first two terms and the common factor $-7c$ from the last two terms. $=2d(4a-3b)-7c(4a-3b)$ Factor out the common factor $(4a-3b)$. $=(4a-3b)(2d-7c)$ To check, we multiply the two binomials using the FOIL procedure. $(4a-3b)(2d-7c)$ $=8ad-28ac-6bd+21bc$ $=8ad+21bc-6bd-28ac$	**14.** Factor and check your answer. $9wz+35xy-15xz-21wy$

Extra Practice

1. Factor by grouping. Check you answer.
$x(x+1)-2(x+1)$

2. Factor by grouping. Check you answer.
$x^2-2x+5x-10$

3. Factor by grouping. Check you answer.
$6x^2+15x-4x-10$

4. Factor by grouping. Check you answer.
$12ax+15ay+8xb+10by$

Concept Check

Explain how you would factor the following polynomial.

$10ax+b^2+2bx+5ab$

Name: ______________________________ Date: ______________
Instructor: ___________________________ Topic: _____________

Module 13 Factoring
Topic 3 Factoring of the Form $x^2 + bx + c$

Vocabulary
first terms • outer and inner terms • second terms • last terms

1. When multiplying an expression of the form $(x+m)(x+n)$, the product of the __________ in the factors produces the first term of the polynomial.

2. When multiplying an expression of the form $(x+m)(x+n)$, the sum of the __________ in the factors gives the coefficient of the middle term of the polynomial.

3. When multiplying an expression of the form $(x+m)(x+n)$, the sum of the products of the __________ in the factors produces the middle term of the polynomial.

4. When multiplying an expression of the form $(x+m)(x+n)$, the product of the __________ of the factors gives the last term of the polynomial.

Example	Student Practice
1. Factor. $x^2+7x+12$ The answer will be of the form $(x+m)(x+n)$. We want to find the two numbers, m and n, that you can multiply to get 12 and add to get 7. The numbers are 3 and 4. $x^2+7x+12=(x+3)(x+4)$	**2.** Factor. $x^2+9x+18$
3. Factor. $x^2+12x+20$ We want two numbers that have a product of 20 and a sum of 12. The numbers are 10 and 2. $x^2+12x+20=(x+10)(x+2)$	**4.** Factor. $x^2+14x+33$

Vocabulary Answers: 1. first terms 2. second terms 3. outer and inner terms 4. last terms

Example	Student Practice
5. Factor. $x^2-8x+15$ We want two numbers that have a product of $+15$ and a sum of -8. They must be negative numbers since the sign of the middle term is negative and the sign of the last term is positive. The sum $-5+(-3)$ is -8 and the product $(-5)(-3)$ is $+15$. $x^2-8x+15=(x-5)(x-3)$ Multiply using FOIL to check.	**6.** Factor. $x^2-11x+28$
7. Factor. $x^2-3x-10$ We want two numbers whose product is -10 and whose sum is -3. The two numbers are -5 and $+2$. $x^2-3x-10=(x-5)(x+2)$	**8.** Factor. $x^2-2x-24$
9. Factor and check your answer. $y^2+10y-24$ The two numbers whose product is -24 and whose sum is $+10$ are $+12$ and -2. $y^2+10y-24=(y+12)(y-2)$ It is very easy to make a sign error in these problems. Make sure that you mentally check your answer by FOIL to obtain the original expression. Check. $(y+12)(y-2)=y^2-2y+12y-24$ $=y^2+10y-24$	**10.** Factor and check your answer. $z^2+12z-28$

Example	Student Practice
11. Factor. $y^4 - 2y^2 - 35$ Notice that $y^4 = (y^2)(y^2)$. This will be the first term in each set of parentheses. $(y^2 \quad)(y^2 \quad)$ The last term of the polynomial is negative. Thus, the signs of m and n will be different. $(y^2 + \quad)(y^2 - \quad)$ Now think of factors of 35 whose difference is 2. $(y^2 + 5)(y^2 - 7)$ Multiply using FOIL to check.	**12.** Factor. $x^4 - 3x^2 - 10$
13. Factor. $3x^2 + 9x - 162$ First factor out the common factor 3 from each term of the polynomial. $3x^2 + 9x - 162 = 3(x^2 + 3x - 54)$ Then factor the remaining polynomial. $3(x^2 + 3x - 54) = 3(x-6)(x+9)$ The final answer is $3(x-6)(x+9)$. Be sure to include the 3. Check. $3(x-6)(x+9) = 3(x^2 + 3x - 54)$ $= 3x^2 + 9x - 162$	**14.** Factor. $4x^2 - 16x - 84$

Example	Student Practice
15. Find a polynomial in factored form for the shaded area in the figure.	**16.** Find a polynomial in factored form for the shaded area in the figure.

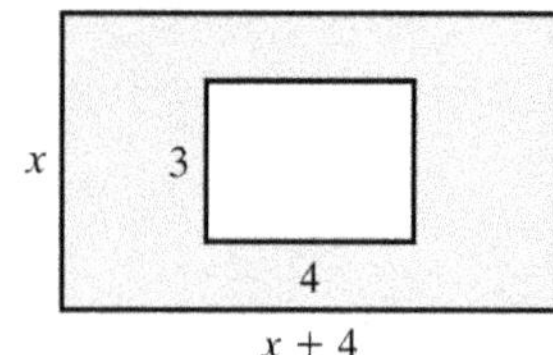

To obtain the shaded area, we find the area of the larger rectangle and subtract from it the area of the smaller rectangle. Thus, we have the following:

$$\text{shaded area} = x(x+4)-(4)(3)$$
$$= x^2+4x-12$$

Now we factor this polynomial to obtain the shaded area $=(x+6)(x-2)$.

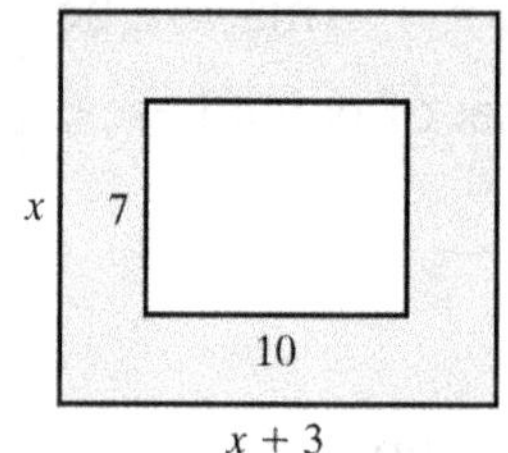

Extra Practice

1. Factor. x^2+6x+8

2. Factor. $a^2-7a+12$

3. Factor. x^2-x-12

4. Factor. $2x^2+18x+28$

Concept Check

Explain how you would completely factor $4x^2-4x-120$.

Name: ______________________________ Date: ________________
Instructor: ____________________________ Topic: ________________

Module 13 Factoring
Topic 4 Factoring Trinomials of the Form $ax^2 + bx + c$

Vocabulary
trial-and-error method • grouping method • multiplying • greatest common factor

1. The ____________ requires listing the possible factoring combinations and computing the middle terms by the FOIL method.

2. Always check by ____________ the factors to see if the original trinomial is obtained.

3. First factor out the ____________ if any, and then factor the trinomial.

4. The ____________ for factoring trinomials of the form ax^2+bx+c requires writing the polynomial with four terms and then factoring the polynomial by grouping.

Example	Student Practice
1. Factor. $2x^2+5x+3$ In order for the coefficient of the x^2-term of the polynomial to be 2, the coefficients of the x-terms in the factors must be 2 and 1. Thus, $2x^2+5x+3=(2x\quad)(x\quad)$. In order for the last term of the polynomial to be 3, the constants in the factors must be 3 and 1. Since all signs in the polynomial are positive, each factor in parentheses will be positive. We have two possibilities. We check them by multiplying. $(2x+1)(x+3)=2x^2+7x+3$ $(2x+3)(x+1)=2x^2+5x+3$ The correct answer is $(2x+3)(x+1)$.	**2.** Factor. $2x^2+13x+20$

Vocabulary Answers: 1. trial-and-error method 2. multiplying 3. greatest common factor 4. grouping method

Example	Student Practice
3. Factor. $4x^2 - 13x + 3$	**4.** Factor. $3x^2 - 4x + 1$

The different factorizations of 4 are $(2)(2)$ and $(1)(4)$. The factorization of 3 is $(1)(3)$. Let us list the possible factoring combinations and compute the middle term by the FOIL method. Note that the signs of the constants in both factors will be negative.

Possible Factors	Middle Term
$(2x-3)(2x-1)$	$-8x$
$(4x-3)(x-1)$	$-7x$
$(4x-1)(x-3)$	$-13x$

The correct middle term is $-13x$. The correct answer is $(4x-1)(x-3)$.

Example	Student Practice
5. Factor. $3x^2 - 2x - 8$	**6.** Factor. $4x^2 - 5x - 21$

The factorization of 3 is $(1)(3)$. The factorizations of 8 are $(8)(1)$ and $(4)(2)$. Let us list only one-half of the possibilities.

Possible Factors	Middle Term
$(x+8)(3x-1)$	$+23x$
$(x+1)(3x-8)$	$-5x$
$(x+4)(3x-2)$	$+10x$
$(x+2)(3x-4)$	$+2x$

The middle term $+2x$ is only incorrect because the sign is wrong. So we just reverse the signs of the constraints.

The correct answer is $(x-2)(3x+4)$.

Example	Student Practice
7. Factor by grouping. $2x^2+5x+3$ First find the grouping number. The grouping number is $(2)(3)=6$. The factors of 6 are $6\cdot 1$ and $3\cdot 2$. We choose the numbers 3 and 2 because their product is 6 and their sum is 5. Next, write $5x$ as the sum $2x+3x$. Then, factor by grouping. $2x^2+5x+3=2x^2+2x+3x+3$ $=2x(x+1)+3(x+1)$ $=(x+1)(2x+3)$ Multiply to check. $(x+1)(2x+3)=2x^2+3x+2x+3$ $=2x^2+5x+3$	**8.** Factor by grouping. $2x^2+5x+3$
9. Factor by grouping. $3x^2-2x-8$ First find the grouping number. The grouping number is $(3)(-8)=-24$. Now we want two numbers whose product is -24 and whose sum is -2. They are -6 and 4. So we write $-2x$ as the sum $-6x+4x$. Then, we factor by grouping. $3x^2-6x+4x-8=3x(x-2)+4(x-2)$ $=(x-2)(3x+4)$	**10.** Factor by grouping. $5x^2-13x-6$

Example	Student Practice
11. Factor. $9x^2+3x-30$ We first factor out the common factor 3 from each term of the trinomial. $9x^2+3x-30=3(3x^2+1x-10)$ We then factor the trinomial by the grouping method or by the trial-and-error method. $3(3x^2+1x-10)=3(3x-5)(x+2)$	**12.** Factor. $8x^2+22x-6$
13. Factor. $32x^2-40x+12$ We first factor out the greatest common factor 4 from each term of the trinomial. $32x^2-40x+12=4(8x^2-10x+3)$ We then factor the trinomial by the grouping method or by the trial-and-error method. $4(8x^2-10x+3)=4(2x-1)(4x-3)$	**14.** Factor. $30x^2-65x+30$

Extra Practice

1. Factor by the trial-and-error method. Check your answer by using FOIL. $4x^2-12x+5$

2. Factor by the grouping method. Check your answer by using FOIL. $7x^2-4x-3$

3. Factor by any method. $5x^2+43x-18$

4. Factor by first factoring out the greatest common factor. $15y^2+51y-36$

Concept Check

Explain how you would factor $10x^3+18x^2y-4xy^2$.

Name: ______________________________ Date: ________________
Instructor: ____________________________ Topic: ________________

Module 13 Factoring
Topic 5 Special Cases of Factoring

Vocabulary
difference of two squares • negative • perfect-square trinomials • greatest common factor

1. The difference of two squares formula only works if the last term is ___________.

2. A ___________ is a trinomial where the first and last terms are perfect squares and the middle term is twice the product of the values whose squares are the first and last terms.

3. The ___________ can be factored into the sum and difference of those values that were squared.

4. For some polynomials, we first need to factor out the ___________.

Example	**Student Practice**
1. Factor. $9x^2-1$ We see that the polynomial is in the form of the difference of two squares. $9x^2$ is a square and 1 is a square. $9x^2=(3x)^2$ and $1=(1)^2$. So using the formula we can write the following. $9x^2-1=(3x+1)(3x-1)$	**2.** Factor. $36x^2-1$
3. Factor. $25x^2-16$ $25x^2=(5x)^2$ and $16=(4)^2$. Again we use the formula for the difference of two squares. $25x^2-16=(5x+4)(5x-4)$	**4.** Factor. $64x^2-9$

Vocabulary Answers: 1. negative 2. perfect-square trinomials 3. difference of two squares 4. greatest common factor

Example	Student Practice
5. Factor. $81x^4-1$ Because $81x^4=(9x^2)^2$ and $1=(1)^2$, we see that $81x^4-1=(9x^2+1)(9x^2-1)$. The factoring is not complete. We can factor $9x^2-1$, because $9x^2-1=(3x+1)(3x-1)$. $81x^4-1=(9x^2+1)(3x+1)(3x-1)$	**6.** Factor. $16x^4-1$
7. Factor. x^2+6x+9 This is a perfect-square trinomial. The first and last terms are perfect squares because $x^2=(x)^2$ and $9=(3)^2$. The middle term, $6x$, is twice the product of 3 and x. Since x^2+6x+9 is a perfect-square trinomial we can use the formula $a^2+2ab+b^2=(a+b)^2$ with $a=x$ and $b=3$. So we have $x^2+6x+9=(x+3)^2$.	**8.** Factor. $x^2+10x+25$

Example	Student Practice
9. Factor. **(a)** $49x^2+42xy+9y^2$ This is a perfect square trinomial, because $49x^2=(7x)^2$, $9y^2=(3y)^2$, and $42xy=2(7x\cdot 3y)$. $49x^2+42xy+9y^2=(7x+3y)^2$ **(b)** $36x^4-12x^2+1$ This is a perfect square trinomial, because $36x^4=(6x^2)^2$, $1=(1)^2$, and $12x^2=2(6x^2\cdot 1)$. $36x^4-12x^2+1=(6x^2-1)^2$	**10.** Factor. **(a)** $64x^2+80xy+25y^2$ **(b)** $49x^4-14x^2+1$
11. Factor. $49x^2+35x+4$ This is not a perfect-square trinomial! Although the first and last terms are perfect squares since $(7x)^2=49x^2$ and $(2)^2=4$, the middle term, $35x$, is not double the product of 2 and $7x$. $35x\neq 28x$! So we must factor by trial and error or by grouping to obtain $49x^2+35x+4=(7x+4)(7x+1)$.	**12.** Factor. $36x^2+25x+4$

Example	Student Practice
13. Factor. $12x^2-48$ We see that the greatest common factor is 12. First we factor out 12. Then we use the difference-of-two-squares formula, $a^2-b^2=(a+b)(a-b)$. $12x^2-48=12(x^2-4)$ $=12(x+2)(x-2)$	**14.** Factor. $3x^2-75$
15. Factor. $24x^2-72x+54$ First we factor out the greatest common factor, 6. Then we use the perfect-square-trinomial formula, $a^2-2ab+b^2=(a-b)^2$. $24x^2-72x+54=6(4x^2-12x+9)$ $=6(2x-3)^2$	**16.** Factor. $48x^2-72x+27$

Extra Practice

1. Factor using the difference-of-two-squares formula. $25a^2-36b^2$

2. Factor by using the perfect-square trinomial formula. $36a^2+60ab+25b^2$

3. Factor by using the difference-of-two-squares. x^4-100

4. Factor by using the perfect-square trinomial formula. $9x^4-30x^2+25$

Concept Check

Explain how to factor the polynomial $24x^2+120x+150$.

Name: ______________________________ Date: ________________
Instructor: ____________________________ Topic: ________________

Module 13 Factoring
Topic 6 A Brief Review of Factoring

Vocabulary
perfect-square trinomial • difference of two squares • factor by grouping • prime
trinomial of the form x^2+bx+c • trinomial of the form ax^2+bx+c • common factor

1. Some polynomials have a ____________ consisting of a number, a variable, or both.

2. When we ____________, we rearrange the order if the first two terms do not have a common factor.

3. If we cannot factor a polynomial by elementary methods, we will identify it as a ____________ polynomial.

4. In a ____________ there are three terms and the first and last terms are perfect squares.

Example	Student Practice
1. Factor.	**2.** Factor.
(a) $25x^3-10x^2+x$	**(a)** $9y^3+6y^2+y$
Factor out the common factor x. The other factor is a perfect-square trinomial.	
$25x^3-10x^2+x=x\left(25x^2-10x+1\right)$ $=x(5x-1)^2$	
(b) $20x^2y^2-45y^2$	**(b)** $-4x^3+28x^2+32x$
Factor out the common factor $5y^2$. The other factor is a difference of squares.	
$20x^2y^2-45y^2=5y^2\left(4x^2-9\right)$ $=5y^2(2x+3)(2x-3)$	

Vocabulary Answers: 1. common factor 2. factor by grouping 3. prime 4. perfect-square trinomial

Example	Student Practice
3. Factor. $ax^2 - 9a + 2x^2 - 18$ We factor by grouping since there are four terms. Factor out the common factor a from the first two terms and 2 from the second two terms. $ax^2 - 9a + 2x^2 - 18$ $= a(x^2 - 9) + 2(x^2 - 9)$ Factor out the common factor $(x^2 - 9)$. $a(x^2 - 9) + 2(x^2 - 9) = (a+2)(x^2 - 9)$ Factor $x^2 - 9$ using the difference-of-two squares formula. $(a+2)(x^2 - 9) = (a+2)(x-3)(x+3)$	**4.** Factor. $bx^2 - 16b - 3x^2 + 48$
5. Factor, if possible. **(a)** $x^2 + 6x + 12$ The factors of 12 are $(1)(12)$ or $(2)(6)$ or $(3)(4)$. None of these pairs add up to 6, the coefficient of the middle term. Thus, the problem cannot be factored by the methods in this chapter. **(b)** $25x^2 + 4$ We have a formula to factor the difference of two squares. There is no way to factor the sum of two squares. That is, $a^2 + b^2$ cannot be factored.	**6.** Factor, if possible. **(a)** $x^2 + 5x + 17$ **(b)** $x^2 - x + 21$

Extra Practice

1. Factor, if possible. Be sure to factor completely. Factor out the greatest common factor first, if one exists.
   $x^2 + 9$

2. Factor, if possible. Be sure to factor completely. Factor out the greatest common factor first, if one exists.
   $63x - 7x^3$

3. Factor, if possible. Be sure to factor completely. Factor out the greatest common factor first, if one exists.
   $-x^3 + 12x^2 + 45x$

4. Factor, if possible. Be sure to factor completely. Factor out the greatest common factor first, if one exists.
   $x^2 + 2x + 3x + 6$

Concept Check

Explain how to completely factor $2x^2 + 6xw - 5x - 15w$.

Name: ______________________________ Date: ________________

Instructor: ____________________________ Topic: _______________

Module 13 Factoring
Topic 7 Solving Quadratic Equations by Factoring

Vocabulary

quadratic equation • standard form • real roots • zero factor property

1. Many quadratic equations have two real number solutions, also called ____________.

2. The ____________ states that if $a \cdot b = 0$, then $a = 0$ or $b = 0$.

3. A ____________ is a polynomial equation in one variable that contains a variable term of degree 2 and no terms of higher degree.

Example	Student Practice
1. Solve the equation to find the two roots. $2x^2 + 13x - 7 = 0$	**2.** Solve the equation to find the two roots. $3x^2 + 8x - 3 = 0$

Example 1 (continued):

The equation is in standard form. Factor. Then set each factor equal to 0 and solve the equations to find the two roots.

$$2x^2 + 13x - 7 = 0$$
$$(2x - 1)(x + 7) = 0$$
$$2x - 1 = 0 \quad x + 7 = 0$$
$$x = \frac{1}{2} \qquad x = -7$$

Check. If $x = \frac{1}{2}$ and if $x = -7$ then we have the following.

$$2\left(\frac{1}{2}\right)^2 + 13\left(\frac{1}{2}\right) - 7 = \frac{1}{2} + \frac{13}{2} - \frac{14}{2} = 0$$
$$2(-7)^2 + 13(-7) - 7 = 98 - 91 - 7 = 0$$

Thus $\frac{1}{2}$ and -7 are both roots of the equation $2x^2 + 13x - 7 = 0$.

Vocabulary Answers: 1. real roots 2. zero factor property 3. quadratic equation

Example	Student Practice
3. Solve the equation to find the two roots. $7x^2-3x=0$ The equation is in standard form. Here $c=0$. Factor out the common factor. Then set each factor equal to 0 by the zero factor property. Solve the equations to find the two roots. $7x^2-3x=0$ $x(7x-3)=0$ $x=0 \quad 7x-3=0$ $7x=3$ $x=\frac{3}{7}$ The two roots are 0 and $\frac{3}{7}$. Check. Verify that 0 and $\frac{3}{7}$ are the roots of $7x^2-3x=0$.	**4.** Solve the equation to find the two roots. $5x^2-4x=0$
5. Solve. $x^2=12-x$ The equation is not in standard form. Add x and -12 to both sides of the equation so that the left side is equal to zero. Then factor and set each factor equal to 0. Solve the equations for x. $x^2=12-x$ $x^2+x-12=0$ $(x-3)(x+4)=0$ $x-3=0 \quad x+4=0$ $x=3 \quad x=-4$ The check is left to the student.	**6.** Solve. $x^2=63-2x$

Example	Student Practice
7. Carlos lives in Mexico City. He has a rectangular brick walkway in front of his house. The length of the walkway is 3 meters longer than twice the width. The area of the walkway is 44 square meters. Find the length and width of the rectangular walkway.	**8.** The length of a rectangle is 31 inches shorter than triple the width. The rectangle has an area of 60 square inches. Find the length and width of the rectangle.

Let $w =$ the width in meters.

Then $2w+3 =$ the length in meters.

Next, write an equation.

$$\text{area} = (\text{width})(\text{length})$$
$$44 = w(2w+3)$$

Now, solve and state the answer.

$$44 = w(2w+3)$$
$$44 = 2w^2 + 3w$$
$$0 = 2w^2 + 3w - 44$$
$$0 = (2w+11)(w-4)$$
$$2w+11=0 \qquad w-4=0$$
$$w = -5\frac{1}{2} \qquad w = 4$$

Since it would not make sense to have a rectangle with a negative number as a width, $-5\frac{1}{2}$ is not a valid solution.

Since $w = 4$, the width of the walkway is 4 meters. The length is $2w+3$, so we have $2(4)+3=8+3=11$. Thus, the length of the walkway is 11 meters.

The check is left to the student.

Example	Student Practice
9. A tennis ball is thrown upward with an initial velocity of 8 meters/second. Suppose the initial height above the ground is 4 meters. At what time t will the ball hit the ground?	**10.** A baseball is thrown upward with an initial velocity of 9 meters/second. Suppose the initial height above the ground is 18 meters. At what time t will the ball hit the ground?

In this case $S=0$ since the ball will hit the ground. The initial upward velocity is $v=8$ meters/second. The initial height is 4 meters, so $h=4$.

$$S=-5t^2+vt+h$$
$$0=-5t^2+8t+4$$
$$5t^2-8t-4=0$$
$$(5t+2)(t-2)=0$$
$$5t+2=0 \qquad t-2=0$$
$$t=-\frac{2}{5} \qquad t=2$$

We want a positive time for t in seconds; thus we do not use $t=-\frac{2}{5}$. Therefore, the ball will strike the ground 2 seconds after it is thrown.

The check is left to the student.

Extra Practice

1. Solve for the roots of the quadratic equation. Check your answer.
   $2x^2-5x-3=0$

2. Solve for the roots of the quadratic equation. Check your answer.
   $3x^2=27$

3. Solve for the roots of the quadratic equation. Check your answer.
   $x^2-35=2x$

4. Solve for the roots of the quadratic equation. Check your answer.
   $x^2-7x=-12$

Concept Check

Explain how you would solve the following problem: A rectangle has an area of 65 square feet. The length of the rectangle is 3 feet longer than double the width. Find the length and the width of the rectangle.

MATH COACH

Mastering the skills you need to do well on the test.

Watch the MATH COACH videos in MyMathLab® or on YouTube™ while you work the problems below. These helpful hints will help you avoid making common errors on test problems.

Factoring the Difference of Two Squares—Problem 2

Factor completely. $16x^2 - 81$

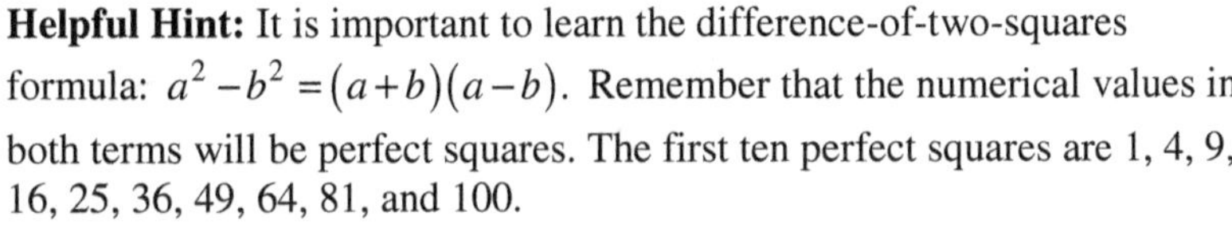

Helpful Hint: It is important to learn the difference-of-two-squares formula: $a^2 - b^2 = (a+b)(a-b)$. Remember that the numerical values in both terms will be perfect squares. The first ten perfect squares are 1, 4, 9, 16, 25, 36, 49, 64, 81, and 100.

Did you remember that $(4x)^2 = 16x^2$? Yes ____ No ____

Did you remember that $9^2 = 81$? Yes ____ No ____

If you answered No to these questions, stop and review the list of the first ten perfect squares. Consider that $4^2 = 16$ and $x \cdot x = x^2$.

Do you see how $16x^2 - 81$ can be factored using the formula $a^2 - b^2$? Yes ____ No ____

If you answered No, stop and review the difference-of-two-squares formula again. Make sure that one set of parentheses contains a + sign and the other set of parentheses contains a − sign.

If you answered Problem 2 incorrectly, go back and rework the problem using these suggestions.

Factoring a Perfect-Square Trinomial—Problem 4

Factor completely. $9a^2 - 30a + 25$

Helpful Hint: Remember the perfect-square-trinomial formula: $a^2 - 2ab + b^2 = (a-b)^2$. You must verify two things to determine if you can use this formula:
(1) The numerical values in the first term and the last term must be perfect squares.
(2) The middle term must equal "twice the product of the values whose squares are the first and last terms."

Did you remember that $(3a)^2 = 9a^2$? Yes ____ No ____

Did you remember that $5^2 = 25$? Yes ____ No ____

If you answered No to these questions, stop and review the first ten perfect squares. Consider that $3^2 = 9$ and $a \cdot a = a^2$.

Do you see how $9a^2 - 30a + 25$ can be factored using the formula $(a-b)^2$? Yes ____ No ____

If you answered No, check to see if the middle term, $30a$, equals twice the product of $3a$ and 5.

Now go back and rework the problem using these suggestions.

Factoring a Polynomial with Four Terms by Grouping—Problem 6
Factor completely. $10xy+15by-8x-12b$

Helpful Hint: Look for common factors first. We can find the greatest common factors of the first two terms and factor. Then we can find the greatest common factor of the second two terms and factor. Make sure that you obtain the same binomial factor for each step. Be careful with $+/-$ signs.

Did you identify $5y$ as the greatest common factor of the first two terms: $10xy+15by$?
Yes ____ No ____

If you answered No, remember that to factor completely, we must remove all common factors from both terms. Stop now and complete this step.

Did you identify 4 as the greatest common factor of the second two terms: $-8x-12b$?
Yes ____ No ____

If you answered Yes to these questions, then you obtained $(2x+3b)$ in the first term and $(-2x-3b)$ in the second term. These are not the same binomial factor. Stop and consider how to get the same binomial factor of $(2x+3b)$.

In your final answer, is the binomial factor of $(2x+3b)$ listed once?
Yes ____ No ____

If you answered Problem 6 incorrectly, go back and rework the problem using these suggestions.

Factoring a Polynomial with a Common Factor—Problem 19 Factor completely. $3x^2-3x-90$

Helpful Hint: Look for the greatest common factor of all three terms as your first step. Don't forget to include this common factor as part of your answer. Always check your final product to make sure that it matches the original polynomial. Do this by multiplying.

Did you obtain $(3x-18)(x+5)$ or $(3x-15)(x-6)$ as your answer?
Yes ____ No ____

If you answered Yes, then you forgot to factor out the greatest common factor 3 as your first step.

Do you see how to factor x^2-x-30?
Yes ____ No ____

If you answered No, remember that we are looking for two numbers with a product of -30 and a sum of -1.

Be sure to double-check your final answer to be sure there are no common factors and include 3 in your final answer.

Now go back and rework the problem using these suggestions.

Name: ______________________________ Date: ________________
Instructor: ___________________________ Topic: _______________

Module 14 Rational Expressions and Equations
Topic 1 Simplifying Rational Expressions

Vocabulary

rational expression • fractional algebraic expression • basic rule of fractions
simplify the fraction • factors

1. Only ___________ of both the numerator and the denominator can be divided out.

2. A ___________ is a polynomial divided by another polynomial.

3. Dividing out common factors is how to ___________.

4. The ___________ states that for any rational expression $\frac{a}{b}$ and any polynomials a, b, and c, $\frac{ac}{bc}=\frac{a}{b}$.

Example	Student Practice
1. Reduce. $\frac{21}{39}$ Use the rule $\frac{ac}{bc}=\frac{a}{b}$. Let $c=3$. $\frac{21}{39}=\frac{7\cdot \cancel{3}}{13\cdot \cancel{3}}=\frac{7}{13}$	**2.** Reduce. $\frac{56}{77}$
3. Simplify. $\frac{4x+12}{5x+15}$ $\frac{4x+12}{5x+15}=\frac{4(x+3)}{5(x+3)}$ $=\frac{4\cancel{(x+3)}}{5\cancel{(x+3)}}$ $=\frac{4}{5}$	**4.** Simplify. $\frac{6x+8}{15x+20}$

Vocabulary Answers: 1. factors 2. rational expression 3. simplify the fraction 4. basic rule of fractions

Example	Student Practice
5. Simplify. $\frac{x^2+9x+14}{x^2-4}$ $\frac{x^2+9x+14}{x^2-4}=\frac{(x+7)(x+2)}{(x-2)(x+2)}$ $=\frac{(x+7)\cancel{(x+2)}}{(x-2)\cancel{(x+2)}}$ $=\frac{x+7}{x-2}$	**6.** Simplify. $\frac{x^2+3x-28}{x^2-16}$
7. Simplify. $\frac{x^3-9x}{x^3+x^2-6x}$ $\frac{x^3-9x}{x^3+x^2-6x}=\frac{x(x^2-9x)}{x(x^2+x-6)}$ $=\frac{\cancel{x}\cancel{(x+3)}(x-3)}{\cancel{x}\cancel{(x+3)}(x-2)}$ $=\frac{x-3}{x-2}$	**8.** Simplify. $\frac{x^3+8x^2+12x}{x^3-36x}$
9. Simplify. $\frac{5x-15}{6-2x}$ The variable terms in the numerator and in the denominator are opposite in sign. Likewise the numerical terms are opposite in sign. Factor out a negative number from the denominator. $\frac{5x-15}{6-2x}=\frac{5(x-3)}{-2(-3+x)}$ $=\frac{5\cancel{(x-3)}}{-2\cancel{(-3+x)}}$ $=-\frac{5}{2}$	**10.** Simplify. $\frac{8x-24}{42-14x}$

Example	Student Practice
11. Simplify. $\dfrac{2x^2-11x+12}{16-x^2}$	**12.** Simplify. $\dfrac{6x^2-7x-20}{25-4x^2}$

Factor the numerator and the denominator. Observe that $(x-4)$ and $(4-x)$ are opposites.

$$\frac{2x^2-11x+12}{16-x^2}=\frac{(x-4)(2x-3)}{(4-x)(4+x)}$$

Factor -1 out of $(+4-x)$ to obtain $-1(-4+x)$.

$$\frac{(x-4)(2x-3)}{(4-x)(4+x)}=\frac{(x-4)(2x-3)}{-1(-4+x)(4+x)}$$

$$=\frac{\cancel{(x-4)}(2x-3)}{-1\cancel{(-4+x)}(4+x)}$$

$$=\frac{(2x-3)}{-1(4+x)}$$

$$=-\frac{2x-3}{4+x}$$

Example	Student Practice
13. Simplify. $\dfrac{x^2-7xy+12y^2}{2x^2-7xy-4y^2}$	**14.** Simplify. $\dfrac{10x^2-9xy-9y^2}{18x^2-13xy-21y^2}$

$$\frac{x^2-7xy+12y^2}{2x^2-7xy-4y^2}=\frac{(x-4y)(x-3y)}{(2x+y)(x-4y)}$$

$$=\frac{\cancel{(x-4y)}(x-3y)}{(2x+y)\cancel{(x-4y)}}$$

$$=\frac{x-3y}{2x+y}$$

Example	Student Practice
15. Simplify. $\dfrac{6a^2+ab-7b^2}{36a^2-49b^2}$	**16.** Simplify. $\dfrac{25a^2-16b^2}{10a^2-3ab-4b^2}$
$\dfrac{6a^2+ab-7b^2}{36a^2-49b^2}=\dfrac{(6a+7b)(a-b)}{(6a+7b)(6a-7b)}$	
$=\dfrac{\cancel{(6a+7b)}(a-b)}{\cancel{(6a+7b)}(6a-7b)}$	
$=\dfrac{a-b}{6a-7b}$	

Extra Practice

1. Simplify. $\dfrac{3x+12}{x^2+4x}$

2. Simplify. $\dfrac{9x^2-24x+16}{9x^2-16}$

3. Simplify. $\dfrac{81-x^2}{3x^2-21x-54}$

4. Simplify. $\dfrac{25x^2-20xy+4y^2}{15x^2-xy-2y^2}$

Concept Check

Explain why it is important to completely factor both the numerator and the denominator when simplifying $\dfrac{x^2y-y^3}{x^2y+xy^2-2y^3}$.

Name: ______________________________ Date: ________________

Instructor: ____________________________ Topic: ________________

Module 14 Rational Expressions and Equations

Topic 2 Multiplying and Dividing Rational Expressions

Vocabulary

reciprocals • multiply • divide • greatest common factor

1. You should always check for the ___________ as your first step when multiplying rational expressions.

2. Two numbers are ___________ of each other if their product is 1.

3. To ___________ two rational expressions, multiply the numerators and multiply the denominators.

4. To ___________ two rational expressions, invert the second fraction and multiply it by the first fraction.

Example	**Student Practice**
1. Multiply. $\frac{x^2-x-12}{x^2-16}\cdot\frac{2x^2+7x-4}{x^2-4x-21}$ Factoring is always the first step. $\frac{x^2-x-12}{x^2-16}\cdot\frac{2x^2+7x-4}{x^2-4x-21}$ $=\frac{(x-4)(x+3)}{(x-4)(x+4)}\cdot\frac{(x+4)(2x-1)}{(x+3)(x-7)}$ Apply the basic rule of fractions. Three pairs of factors divide out. $\frac{(x-4)(x+3)}{(x-4)(x+4)}\cdot\frac{(x+4)(2x-1)}{(x+3)(x-7)}$ $=\frac{\cancel{(x-4)}\cancel{(x+3)}}{\cancel{(x-4)}\cancel{(x+4)}}\cdot\frac{\cancel{(x+4)}(2x-1)}{\cancel{(x+3)}(x-7)}$ $=\frac{(2x-1)}{(x-7)}$	**2.** Multiply. $\frac{3x^2+34x+63}{2x^2+7x-15}\cdot\frac{x^2+9x+20}{x^2+13x+36}$

Vocabulary Answers: 1. greatest common factor 2. reciprocals 3. multiply 4. divide

Example	Student Practice
3. Multiply. $\dfrac{x^4-16}{x^3+4x}\cdot\dfrac{2x^2-8x}{4x^2+2x-12}$	**4.** Multiply. $\dfrac{x^3+x}{2x^4-2}\cdot\dfrac{4x^3-4x}{6x^3+38x^2+40x}$

Factor each numerator and denominator. Factoring out the greatest common factor first is very important.

$$\frac{x^4-16}{x^3+4x}\cdot\frac{2x^2-8x}{4x^2+2x-12}$$
$$=\frac{(x^2+4)(x^2-4)}{x(x^2+4)}\cdot\frac{2x(x-4)}{2(2x^2+x-6)}$$
$$=\frac{(x^2+4)(x+2)(x-2)}{x(x^2+4)}\cdot\frac{2x(x-4)}{2(x+2)(2x-3)}$$
$$=\frac{\cancel{(x^2+4)}\cancel{(x+2)}(x-2)}{\cancel{x}\cancel{(x^2+4)}}\cdot\frac{\cancel{2}\cancel{x}(x-4)}{\cancel{2}\cancel{(x+2)}(2x-3)}$$
$$=\frac{(x-2)(x-4)}{2x-3}\text{ or }\frac{x^2-6x+8}{2x-3}$$

Example	Student Practice
5. Divide. $\dfrac{6x+12y}{2x-6y}\div\dfrac{9x^2-36y^2}{4x^2-36y^2}$	**6.** Divide. $\dfrac{5x^2+20x-105}{4x^2-44x+96}\div\dfrac{10x^2+75x+35}{16x^2-140x+96}$

$$=\frac{6x+12y}{2x-6y}\cdot\frac{4x^2-36y^2}{9x^2-36y^2}$$
$$=\frac{6(x+2y)}{2(x-3y)}\cdot\frac{4(x^2-9y^2)}{9(x^2-4y^2)}$$
$$=\frac{(3)(2)(x+2y)}{2(x-3y)}\cdot\frac{(2)(2)(x+3y)(x-3y)}{(3)(3)(x+2y)(x-2y)}$$
$$=\frac{\cancel{(3)}\cancel{(2)}\cancel{(x+2y)}}{\cancel{2}\cancel{(x-3y)}}\cdot\frac{(2)(2)(x+3y)\cancel{(x-3y)}}{\cancel{(3)}(3)\cancel{(x+2y)}(x-2y)}$$
$$=\frac{(2)(2)(x+3y)}{3(x-2y)}$$
$$=\frac{4(x+3y)}{3(x-2y)}$$

Example	Student Practice
7. Divide. $\frac{15-3x}{x+6} \div (x^2-9x+20)$ Note that $x^2-9x+20$ can be written as $\frac{x^2-9x+20}{1}$. $\frac{15-3x}{x+6} \div (x^2-9x+20)$ $=\frac{15-3x}{x+6} \cdot \frac{1}{x^2-9x+20}$ $=\frac{-3(-5+x)}{x+6} \cdot \frac{1}{(x-5)(x-4)}$ $=\frac{-3\cancel{(-5+x)}}{x+6} \cdot \frac{1}{\cancel{(x-5)}(x-4)}$ $=\frac{-3}{(x+6)(x-4)}$ or $-\frac{3}{(x+6)(x-4)}$ or $\frac{3}{(x+6)(4-x)}$	**8.** Divide. $\frac{x+4}{x-2} \div (-x^2+x+20)$

Extra Practice

1. Multiply. $\frac{32x^3}{8x^2-8} \cdot \frac{4x-4}{16x^2}$

2. Multiply. $\frac{2x^2-3x}{4x^3+4x^2-9x-9} \cdot \frac{2x^2+5x+3}{5x+7}$

3. Divide. $\frac{9x^2-49}{3x^2+8x-35} \div (3x^2+x-14)$

4. Divide. $\frac{9x^2-16}{9x^2+24x+16} \div \frac{3x^2-x-4}{5x^2-5}$

Concept Check

Explain how you would divide $\frac{21x-7}{9x^2-1} \div \frac{1}{3x+1}$.

Name: ______________________________ Date: ________________

Instructor: ____________________________ Topic: ________________

Module 14 Rational Expressions and Equations
Topic 3 Adding and Subtracting Rational Expressions

Vocabulary

least common denominator • denominator • different • factor

1. If rational expressions have the same ____________, they can be combined in the same way as arithmetic fractions.

2. If two rational expressions have ____________ denominators, we first change them to equivalent rational expressions with the least common denominator.

3. The ____________ is a product containing each different factor for each denominator of rational expressions.

4. The first step to find the LCD of two or more rational expressions is to ____________ each denominator completely.

Example	Student Practice
1. Add. $\dfrac{5a}{a+2b}+\dfrac{6a}{a+2b}$ Note that the denominators are the same. Only add the numerators. Do not change the denominator. $\dfrac{5a}{a+2b}+\dfrac{6a}{a+2b}=\dfrac{5a+6a}{a+2b}=\dfrac{11a}{a+2b}$	**2.** Add. $\dfrac{4x+3}{3x+4}+\dfrac{5-4x}{3x+4}$
3. Subtract. $\dfrac{3x}{(x+y)(x-2y)}-\dfrac{8x}{(x+y)(x-2y)}$ Write as one fraction and simplify. $\dfrac{3x}{(x+y)(x-2y)}-\dfrac{8x}{(x+y)(x-2y)}$ $=\dfrac{3x-8x}{(x+y)(x-2y)}=\dfrac{-5x}{(x+y)(x-2y)}$	**4.** Subtract. $\dfrac{4x+6}{(3x-2)(x+9)}-\dfrac{2x-5}{(3x-2)(x+9)}$

Vocabulary Answers: 1. denominator 2. different 3. least common denominator 4. factor

Example	**Student Practice**

5. Find the LCD. $\dfrac{5}{2x-4}, \dfrac{6}{3x-6}$

Factor each denominator.

$$2x-4=2(x-2)$$
$$3x-6=3(x-2)$$

The factors are 2, 3, and $(x-2)$. The LCD is the product of these factors.

$$\text{LCD}=(2)(3)(x-2)=6(x-2)$$

6. Find the LCD. $\dfrac{5}{16x+20}, \dfrac{11}{28x+35}$

7. Find the LCD.

$$\frac{5}{12ab^2c}, \frac{13}{18a^3bc^4}$$

The LCD will contain each factor repeated the greatest number of times that it occurs in any one denominator.

$$\begin{aligned} 12ab^2x &= 2\cdot2\cdot3\cdot a\cdot b\cdot b\cdot c \\ 18a^3bc^4 &= 2\cdot3\cdot3\cdot a\cdot a\cdot a\cdot b\cdot c\cdot c\cdot c\cdot c \\ \text{LCD} &= 2\cdot2\cdot3\cdot3\cdot a\cdot a\cdot a\cdot b\cdot b\cdot c\cdot c\cdot c\cdot c \\ \text{LCD} &= 2^2\cdot3^2\cdot a^3\cdot b^2\cdot c^4=36a^3b^2c^4 \end{aligned}$$

8. Find the LCD.

$$\frac{7}{24ab^2c^3}, \frac{13}{36a^2b^4c}$$

9. Add. $\dfrac{5}{xy}+\dfrac{2}{y}$

Find the LCD. The two factors are x and y. Observe that the LCD is xy.

$$\begin{aligned} \frac{5}{xy}+\frac{2}{y} &= \frac{5}{xy}+\frac{2}{y}\cdot\frac{x}{x} \\ &= \frac{5}{xy}+\frac{2x}{xy} \\ &= \frac{5+2x}{xy} \end{aligned}$$

10. Add. $\dfrac{4}{xyz}+\dfrac{2}{y}$

Example	Student Practice
11. Add. $\dfrac{3x}{x^2-y^2}+\dfrac{5}{x+y}$ Factor the first denominator so that $x^2-y^2=(x+y)(x-y)$. Thus, the factors of the denominators are $(x+y)$ and $(x-y)$. Observe that the LCD$=(x+y)(x-y)$. $\dfrac{3x}{x^2-y^2}+\dfrac{5}{x+y}$ $=\dfrac{3x}{(x+y)(x-y)}+\dfrac{5}{(x+y)}\cdot\dfrac{x-y}{x-y}$ $=\dfrac{3x}{(x+y)(x-y)}+\dfrac{5x-5y}{(x+y)(x-y)}$ $=\dfrac{8x-5y}{(x+y)(x-y)}$	**12.** Add. $\dfrac{4}{2x-5}+\dfrac{3x+2}{4x^2-25}$
13. Add. $\dfrac{5}{x^2-y^2}+\dfrac{3x}{x^3+x^2y}$ $\dfrac{5}{x^2-y^2}+\dfrac{3x}{x^3+x^2y}$ $=\dfrac{5}{(x+y)(x-y)}+\dfrac{3x}{x^2(x+y)}$ $=\dfrac{5}{(x+y)(x-y)}\cdot\dfrac{x^2}{x^2}+\dfrac{3x}{x^2(x+y)}\cdot\dfrac{x-y}{x-y}$ $=\dfrac{5x^2}{x^2(x+y)(x-y)}+\dfrac{3x^2-3xy}{x^2(x+y)(x-y)}$ $=\dfrac{x(8x-3y)}{x^2(x+y)(x-y)}$ $=\dfrac{8x-3y}{x(x+y)(x-y)}$	**14.** Add. $\dfrac{6y}{3xy-y^2}+\dfrac{4y}{9x^2-y^2}$

Example	Student Practice
15. Subtract. $\frac{3x+4}{x-2}-\frac{x-3}{2x-4}$	**16.** Subtract. $\frac{x+5}{2x+7}-\frac{x-2}{6x+21}$

Factor the second denominator.

$$\frac{3x+4}{x-2}-\frac{x-3}{2x-4}=\frac{3x+4}{x-2}-\frac{x-3}{2(x-2)}$$
$$=\frac{2}{2}\cdot\frac{3x+4}{x-2}-\frac{x-3}{2(x-2)}$$
$$=\frac{2(3x+4)-(x-3)}{2(x-2)}$$
$$=\frac{6x+8-x+3}{2(x-2)}$$
$$=\frac{5x+11}{2(x-2)}$$

Extra Practice

1. Find the LCD. Do not combine fractions.
$\frac{7}{2x^2-11x+12}, \frac{13}{2x^2+7x-15}$

2. Perform the operation indicated. Be sure to simplify. $\frac{5x}{x+1}-\frac{2x-5}{x+1}$

3. Perform the operation indicated. Be sure to simplify. $\frac{7}{x^2-y^2}+\frac{2y}{xy^2+y^3}$

4. Perform the operation indicated. Be sure to simplify. $\frac{1}{x^2+3x+2}-\frac{2}{x^2-x-6}$

Concept Check

Explain how to find the LCD of the fractions $\frac{3}{10xy^2z}$ and $\frac{6}{25x^2yz^3}$.

Name: ______________________________ Date: ________________
Instructor: ___________________________ Topic: ________________

Module 14 Rational Expressions and Equations
Topic 4 Simplifying Complex Rational Expressions

Vocabulary
complex rational expression • complex fraction • numerator • denominator • LCD

1. A complex rational expression is also called a(n) ____________.
2. A(n) ____________ has a fraction in the numerator or in the denominator, or both.
3. A complex rational expression may contain two or more fractions in the ____________ and denominator.
4. To simplify complex rational expressions, multiply the numerator and denominator of the complex fraction by the ____________ of all the denominators appearing in the complex fraction.

Example	Student Practice
1. Simplify. $\dfrac{\frac{1}{x}}{\frac{2}{y^2}+\frac{1}{y}}$	**2.** Simplify. $\dfrac{\frac{1}{n^2}}{\frac{1}{m}+\frac{3}{4}}$
Add the two fractions in the denominator.	
$\dfrac{\frac{1}{x}}{\frac{2}{y^2}+\frac{1}{y}\cdot\frac{y}{y}}=\dfrac{\frac{1}{x}}{\frac{2+y}{y^2}}$	
Divide the fraction in the numerator by the fraction in the denominator.	
$\frac{1}{x}\div\frac{2+y}{y^2}=\frac{1}{x}\cdot\frac{y^2}{2+y}$ $=\frac{y^2}{x(2+y)}$	

Vocabulary Answers: 1. complex fraction 2. complex rational expression 3. numerator 4. LCD

Example	Student Practice
3. Simplify. $\dfrac{\frac{1}{x}+\frac{1}{y}}{\frac{3}{a}-\frac{2}{b}}$	**4.** Simplify. $\dfrac{\frac{1}{m}+\frac{1}{n}}{\frac{x}{5}-\frac{y}{3}}$
Observe that the LCD of the fractions in the numerator is xy. The LCD of the fractions in the denominator is ab. $\dfrac{\frac{1}{x}+\frac{1}{y}}{\frac{3}{a}-\frac{2}{b}}=\dfrac{\frac{1}{x}\cdot\frac{y}{y}+\frac{1}{y}\cdot\frac{x}{x}}{\frac{3}{a}\cdot\frac{b}{b}-\frac{2}{b}\cdot\frac{a}{a}}=\dfrac{\frac{y+x}{xy}}{\frac{3b-2a}{ab}}$ $=\dfrac{y+x}{xy}\cdot\dfrac{ab}{3b-2a}=\dfrac{ab(y+x)}{xy(3b-2a)}$	
5. Simplify. $\dfrac{\frac{1}{x^2-1}+\frac{2}{x+1}}{x}$	**6.** Simplify. $\dfrac{\frac{3x}{x^2+8x+12}-\frac{2}{x+6}}{5}$
We need to factor x^2-1. $\dfrac{\frac{1}{x^2-1}+\frac{2}{x+1}}{x}$ $=\dfrac{\frac{1}{(x+1)(x-1)}+\frac{2}{x+1}\cdot\frac{x-1}{x-1}}{x}$ $=\dfrac{\frac{1+2x-2}{(x+1)(x-1)}}{x}$ $=\dfrac{2x-1}{(x+1)(x-1)}\cdot\dfrac{1}{x}$ $=\dfrac{2x-1}{x(x+1)(x-1)}$	

Example	Student Practice
7. Simplify. $\dfrac{\dfrac{3}{a+b}-\dfrac{3}{a-b}}{\dfrac{5}{a^2-b^2}}$	**8.** Simplify. $\dfrac{\dfrac{5}{x+y}+\dfrac{2}{x-y}}{\dfrac{7}{x^2-y^2}}$
$=\dfrac{\dfrac{3}{a+b}\cdot\dfrac{a-b}{a-b}-\dfrac{3}{a-b}\cdot\dfrac{a+b}{a+b}}{\dfrac{5}{a^2-b^2}}$	
$=\dfrac{\dfrac{3a-3b}{(a+b)(a-b)}-\dfrac{3a+3b}{(a+b)(a-b)}}{\dfrac{5}{a^2-b^2}}$	
$=\dfrac{\dfrac{-6b}{(a+b)(a-b)}}{\dfrac{5}{(a+b)(a-b)}}$	
$=\dfrac{-6b}{(a+b)(a-b)}\cdot\dfrac{(a+b)(a-b)}{5}$	
$=\dfrac{-6b}{5}$ or $-\dfrac{6b}{5}$	
9. Simplify by multiplying by the LCD. $\dfrac{\dfrac{5}{ab^2}-\dfrac{2}{ab}}{3-\dfrac{5}{2a^2b}}$	**10.** Simplify by multiplying by the LCD. $\dfrac{\dfrac{4}{y}-\dfrac{7}{4x}}{5-\dfrac{9}{xy^2}}$
$=\dfrac{2a^2b^2\left(\dfrac{5}{ab^2}-\dfrac{2}{ab}\right)}{2a^2b^2\left(3-\dfrac{5}{2a^2b}\right)}$	
$=\dfrac{2a^2b^2\left(\dfrac{5}{ab^2}\right)-2a^2b^2\left(\dfrac{2}{ab}\right)}{2a^2b^2(3)-2a^2b^2\left(\dfrac{5}{2a^2b}\right)}$	
$=\dfrac{10a-4ab}{6a^2b^2-5b}$	

Example	Student Practice
11. Simplify by multiplying by the LCD. $\dfrac{\dfrac{3}{a+b}-\dfrac{3}{a-b}}{\dfrac{5}{a^2-b^2}}$	**12.** Simplify by multiplying by the LCD. $\dfrac{\dfrac{5}{x+y}+\dfrac{2}{x-y}}{\dfrac{7}{x^2-y^2}}$

The LCD of all individual fractions in the complex fraction is $(a+b)(a-b)$.

$$=\frac{(a+b)(a-b)\left(\frac{3}{a+b}\right)-(a+b)(a-b)\left(\frac{3}{a-b}\right)}{(a+b)(a-b)\left(\frac{5}{(a+b)(a-b)}\right)}$$

$$=\frac{3(a-b)-3(a+b)}{5}$$

$$=\frac{3a-3b-3a-3b}{5}$$

$$=-\frac{6b}{5}$$

Extra Practice

1. Simplify. $\dfrac{\dfrac{4}{a}+\dfrac{5}{b}}{\dfrac{2}{ab}}$

2. Simplify. $\dfrac{\dfrac{2}{x}+\dfrac{2}{y}}{x+y}$

3. Simplify. $\dfrac{\dfrac{2x}{x^2-25}}{\dfrac{4}{x+5}-\dfrac{3}{x-5}}$

4. Simplify. $\dfrac{\dfrac{3}{4x}-\dfrac{8}{3y}}{\dfrac{7}{a}+\dfrac{5}{b}}$

Concept Check

To simplify the following complex fraction, explain how you would add the two fractions in the numerator.

$$\frac{\frac{7}{x-3}+\frac{15}{2x-6}}{\frac{2}{x+5}}$$

Name: ______________________________ Date: ________________

Instructor: ____________________________ Topic: ________________

Module 14 Rational Expressions and Equations
Topic 5 Solving Equations Involving Rational Expressions

Vocabulary

extraneous solution • no solution • LCD • exclude

1. The first step to solve an equation containing rational expressions is to determine the ____________ of all the denominators.

2. If all of the apparent solutions of an equation are extraneous solutions, we say that the equation has ____________.

3. A value that makes a denominator in the equation equal to zero is called a(n) ____________.

4. ____________ from your solution any value that would make the LCD equal to zero.

Example	Student Practice
1. Solve for x and check your solution. $\frac{5}{x}+\frac{2}{3}=-\frac{3}{x}$ The LCD is $3x$. Multiply each term by $3x$. $3x\left(\frac{5}{x}\right)+3x\left(\frac{2}{3}\right)=3x\left(-\frac{3}{x}\right)$ $15+2x=-9$ $2x=-9-15$ $2x=-24$ $x=-12$ Check. Replace each x by -12. $\frac{5}{-12}+\frac{2}{3}\stackrel{?}{=}-\frac{3}{-12}$ $-\frac{5}{12}+\frac{8}{23}\stackrel{?}{=}\frac{3}{12}$ $\frac{3}{12}=\frac{3}{12}$	**2.** Solve for x and check your solution. $\frac{5}{x}+\frac{1}{4}=-\frac{6}{x}$

Vocabulary Answers: 1. LCD 2. no solution 3. extraneous solution 4. exclude

Example	Student Practice
3. Solve and check. $\dfrac{6}{x+3}=\dfrac{3}{x}$ Observe that the LCD $= x(x+3)$. $x(x+3)\left(\dfrac{6}{x+3}\right)=x(x+3)\left(\dfrac{3}{x}\right)$ $6x=3(x+3)$ $6x=3x+9$ $3x=9$ $x=3$ Check. Replace each x by 3. $\dfrac{6}{3+3}\stackrel{?}{=}\dfrac{3}{3}$ $\dfrac{6}{6}=\dfrac{3}{3}$	**4.** Solve and check. $\dfrac{2}{x+4}=\dfrac{6}{x}$
5. Solve and check. $\dfrac{3}{x+5}-1=\dfrac{4-x}{2x+10}$ $\dfrac{3}{x+5}-1=\dfrac{4-x}{2(x+5)}$ $2(x+5)\left(\dfrac{3}{x+5}\right)-2(x+5)(1)$ $=2(x+5)\left[\dfrac{4-x}{2(x+5)}\right]$ $2(3)-2(x+5)=4-x$ $6-2x-10=4-x$ $-2x-4=4-x$ $-4=4+x$ $-8=x$ The check is left to the student.	**6.** Solve and check. $\dfrac{3}{x+2}-\dfrac{5}{x+5}=\dfrac{x-2}{x^2+7x+10}$

Example	Student Practice

7. Solve and check. $\frac{y}{y-2}-4=\frac{2}{y-2}$

Observe that the LCD is $y-2$.

$$(y-2)\left(\frac{y}{y-2}\right)-(y-2)(4)$$
$$=(y-2)\left(\frac{2}{y-2}\right)$$
$$y-4(y-2)=2$$
$$y-4y+8=2$$
$$-3y+8=2$$
$$-3y=-6$$
$$\frac{-3y}{-3}=\frac{-6}{-3}$$
$$y=2$$

This equation has no solution. We can see immediately that $y=2$ is not a solution of the original equation. When we substitute 2 for y in a denominator, the denominator is equal to zero and the expression is undefined. Suppose that we tried to check the apparent solution by substituting 2 for y.

$$\frac{y}{y-2}-4=\frac{2}{y-2}$$
$$\frac{2}{2-2}-4\stackrel{?}{=}\frac{2}{2-2}$$
$$\frac{2}{0}-4=\frac{2}{0}$$

These expressions are not defined. There is no such number as $2\div0$. We see that 2 does not check. This equation has no solution.

8. Solve and check. $\frac{3x}{x-3}+5=\frac{3x}{x-3}$

Extra Practice

1. Solve and check. If there is no solution, so indicate. $\frac{2}{3x}+\frac{1}{6}=\frac{6}{x}$

2. Solve and check. If there is no solution, so indicate. $\frac{4}{a^2-1}=\frac{2}{a+1}+\frac{2}{a-1}$

3. Solve and check. If there is no solution, so indicate.

$$\frac{6}{5a+10}-\frac{1}{a-5}=\frac{4}{a^2-3a-10}$$

4. Solve and check. If there is no solution, so indicate.

$$\frac{x-2}{x^2-4x-5}+\frac{x+5}{x^2-25}=\frac{2x+13}{x^2+6x+5}$$

Concept Check

Explain how to find the LCD for the following equation. Do not solve the equation.

$$\frac{x}{x^2-9}+\frac{2}{3x-9}=\frac{5}{2x+6}+\frac{3}{2x^2-18}$$

Name: ______________________________ Date: ________________
Instructor: ___________________________ Topic: ________________

Module 14 Rational Expressions and Equations
Topic 6 Ratio, Proportion, and Other Applied Problems

Vocabulary

ratio • proportion • cross multiplying • similar

1. A ____________ is an equation that states that two ratios are equal.

2. ____________ triangles are triangles that have the same shape, but may be different sizes.

3. If $\frac{a}{b}=\frac{c}{d}$, then $ad = bc$ is sometimes called ____________.

4. A ____________ is a comparison of two quantities.

Example	Student Practice
1. Michael took 5 hours to drive 245 miles on the turnpike. At the same rate, how many hours will it take him to drive a distance of 392 miles? Let $x =$ the number of hours it will take to drive 392 miles. If 5 hours are needed to drive 245 miles, then x hours are needed to drive 392 miles. $\frac{5 \text{ hours}}{245 \text{ miles}}=\frac{x \text{ hours}}{392 \text{ miles}}$ $5(392)=245x$ $\frac{1960}{245}=x$ $8=x$ It will take Michael 8 hours to drive 392 miles. To check, do the computations and see if $\frac{5}{245}=\frac{8}{392}$.	**2.** Rose took 4 hours to drive 264 miles. How far will she drive in 9 hours?

Vocabulary Answers: 1. proportion 2. similar 3. cross multiplying 4. ratio

Example	Student Practice
3. A ramp is 32 meters long and rises up 15 meters. A ramp at the same angle is 9 meters long. How high does the second ramp rise? 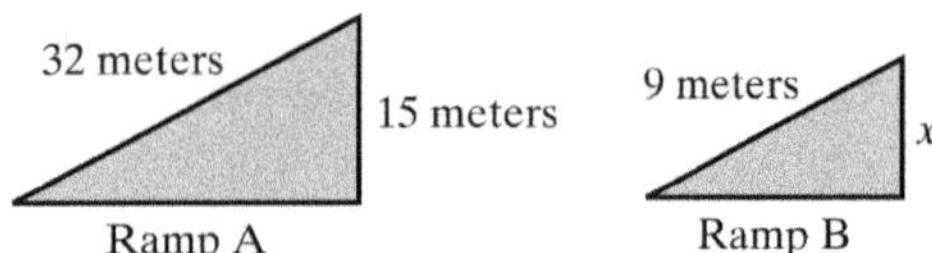 $\frac{32}{9}=\frac{15}{x}$ $32x=(9)(15)$ $32x=135$ $x=\frac{135}{32}$ or $x=4\frac{7}{32}$	**4.** Triangle M is similar to Triangle N. Find the length of side x. Express your answer as a mixed number. 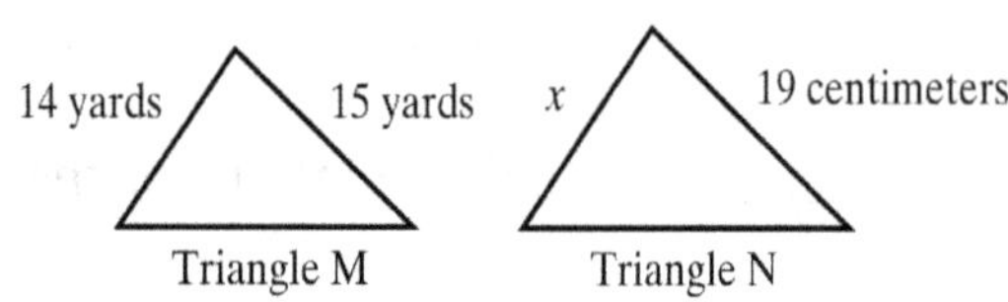
5. A French commuter airline flies from Paris to Avignon. Plane A flies at a speed that is 50 kilometers per hour faster than plane B. Plane A flies 500 kilometers in the amount of time that plane B flies 400 kilometers. Find the speed of each plane. Let $s=$ the speed of plane B in kilometers per hour. Then $s+50=$ the speed of plane A in kilometers per hour. Each plane flies the same amount of time. That is, the time for plane A equals the time for plane B. $\frac{500}{s+50}=\frac{400}{s}$ $500s=(s+50)(400)$ $500s=400s+20,000$ $100s=20,000$ $s=200$ Plane B travels 200 kilometers per hour. Since $s+50=200+50=250,$ plane A travels 250 kilometers per hour.	**6.** Two trains travel in opposite directions for the same amount of time. Train A traveled 200 kilometers, while train B traveled 175 kilometers. Train A traveled 15 kilometers per hour faster than train B. What is the speed of each train?

Example	Student Practice
7. Reynaldo can sort a huge stack of mail on an old sorting machine in 9 hours. His brother Carlos can sort the same amount of mail using a newer sorting machine in 8 hours. How long will it take them to do the job working together? Express your answer in hours and minutes. Round to the nearest minute.	**8.** Two people are painting a side of a house. The first person takes 4 hours to paint the side of the house, while the second person takes 6 hours. How long will it take both of them to paint the side of the house working together? Express your answer in hours and minutes. Round to the nearest minute.

If Reynaldo can do the job in 9 hours, then in 1 hour he could do $\frac{1}{9}$ of the job. If Carlos can do the job in 8 hours, then in 1 hour he could do $\frac{1}{8}$ of the job. Let x = the number of hours it takes Reynaldo and Carlos to do the job together. In 1 hour together they could do $\frac{1}{x}$ of the job. The amount of work Reynaldo can do in 1 hour plus the amount of work Carlos can do in 1 hour must be equal to the amount of work they could do together in 1 hour.

$$\frac{1}{9}+\frac{1}{8}=\frac{1}{x}$$

$$72x\left(\frac{1}{9}\right)+72x\left(\frac{1}{8}\right)=72x\left(\frac{1}{x}\right)$$

$$8x+9x=72$$

$$x=4\frac{4}{17}$$

To change $\frac{4}{17}$ of an hour to minutes,

$$\frac{4}{17}\cancel{\text{hour}}\times\frac{60\text{ minutes}}{1\cancel{\text{hour}}}=\frac{240}{17}\text{ minutes},$$

which is approximately 14.118. Thus doing the job together will take 4 hours and 14 minutes.

Extra Practice

1. Solve. $\frac{3}{10} = \frac{11.4}{x}$

2. The scale on a map of Massachusetts is approximately $\frac{3}{4}$ inch to 25 miles. If the distance from a college to Boston measures 4 inches on the map, how far apart are the two locations?

3. Alicia is 4 feet tall and casts a shadow that is 9 feet long. At the same time of day, a tree casts a shadow that is 36 feet long. How tall is the tree?

4. Toshiro and Mathilda work in the library. Working alone, it takes Toshiro 7 hours to arrange and stack 1000 books, while it takes Mathilda 6.5 hours. To the nearest minute, how long will the job take if they work together to arrange and stack 1000 books?

Concept Check

Mike found that his car used 18 gallons of gas to travel 396 miles. He needs to take a trip of 450 miles and wants to know how many gallons of gas it will take. He set up the equation $\frac{18}{x} = \frac{450}{396}$. Explain what error he made and how he should correctly solve the problem.

MATH COACH

Mastering the skills you need to do well on the test.

Watch the MATH COACH videos in MyMathLab® or on YouTube™ while you work the problems below. These helpful hints will help you avoid making common errors on test problems.

Dividing Rational Expressions—Problem 5 $\dfrac{2a^2-3a-2}{a^2+5a+6} \div \dfrac{a^2-5a+6}{a^2-9}$

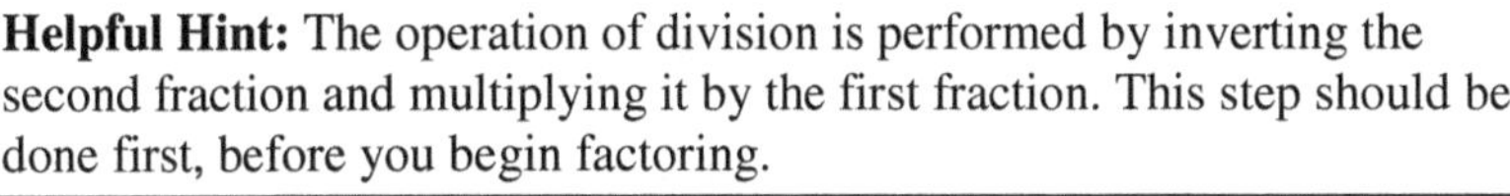

Helpful Hint: The operation of division is performed by inverting the second fraction and multiplying it by the first fraction. This step should be done first, before you begin factoring.

Did you keep the first fraction the same as it is written and then multiply it by $\dfrac{a^2-9}{a^2-5a+6}$? Yes ____ No ____

If you answered No, stop and make this correction to your work.

Were you able to factor each expression so that you obtained $\dfrac{(2a+1)(a-2)}{(a+2)(a+3)} \cdot \dfrac{(a+3)(a-3)}{(a-2)(a-3)}$? Yes ____ No ____

If you answered No, go back and check each factoring step to see if you can get the same result. Complete the problem by dividing out any common factors.

If you answered Problem 5 incorrectly, go back and rework the problem using these suggestions.

Subtracting Rational Expressions with Different Denominators—Problem 8 $\dfrac{3x}{x^2-3x-18} - \dfrac{x-4}{x-6}$

Helpful Hint: First factor the trinomial in the denominator so that you can determine the LCD of these two fractions. Then multiply by what is needed in the second fraction so that it becomes an equivalent fraction with the LCD as the denominator. When subtracting, it is a good idea to place brackets around the second numerator to avoid sign errors.

Did you factor the first denominator into $(x-6)(x+3)$?
Yes ____ No ____

Did you then determine that the LCD is $(x-6)(x+3)$?
Yes ____ No ____

If you answered No to these questions, stop and review how to factor the trinomial in the first denominator and how to find the LCD when working with polynomials as denominators.

Did you multiply the numerator and denominator of the second fraction by $(x+3)$ and place brackets around the product to obtain $\dfrac{3x}{(x-6)(x+3)} - \dfrac{[(x-4)(x+3)]}{(x-6)(x+3)}$?

Yes ____ No ____

Did you obtain $3x - x^2 - x - 12$ in the numerator? Yes ____ No ____

If you answered Yes, you forgot to distribute the negative sign. Rework the problem and make this correction.

As your final step, remember to combine like terms and then factor the new numerator before dividing out common factors.

Now go back and rework the problem using these suggestions.

Simplifying Complex Rational Expressions—Problem 10 Simplify. $\dfrac{\frac{6}{b}-4}{\frac{5}{bx}-\frac{10}{3x}}$

Helpful Hint: There are two ways to simplify this expression:
1)combine the numerators and the denominators separately, or
2)multiply the numerator and denominator of the complex fraction by the LCD of all the denominators. Consider both methods and choose the one that seems easiest to you. We will show the steps of the first method for this particular problem.

Did you multiply 4 by $\frac{b}{b}$ and then subtract $\frac{6}{b}-\frac{4b}{b}$?

Yes ____ No ____

If you answered No, remember that 4 can be written as $\frac{4}{1}$, and to find the LCD, you must multiply numerator and denominator by the variable b.

In the denominator of the complex fraction, did you obtain $3bx$ as the LCD and multiply the first fraction by $\frac{3}{3}$ and the second fraction by $\frac{b}{b}$ before subtracting the two fractions?

Yes ____ No ____

If you answered No, stop and carefully change these two fractions into equivalent fractions with $3bx$ as the denominator. Then subtract the numerators and keep the common denominator.

Were you able to rewrite the problem as follows: $\frac{6-4b}{b} \div \frac{15-10b}{3bx}$?

Yes ____ No ____

If you answered No, examine your steps carefully. Once you have one fraction in the numerator and one fraction in the denominator, you can rewrite the division as multiplication by inverting the fraction in the denominator.

If you answered Problem 10 incorrectly, go back and rework the problem using these suggestions.

Solving Equations Involving Rational Expressions—Problem 14

Solve for x. $\frac{x-3}{x-2}=\frac{2x^2-15}{x^2+x-6}-\frac{x+1}{x+3}$

Helpful Hint: First factor any denominators that need to be factored so that you can determine the LCD of all denominators. Verify that you are solving an equation, then multiply each term of the equation by the LCD. Solve the resulting equation. Check your solution.

Did you factor x^2+x-6 to get $(x+3)(x-2)$ and identify the LCD as $(x+3)(x-2)$? Yes ____ No ____
If you answered No, go back and complete these steps again.

Did you notice that we are solving an equation and that we can multiply each term of the equation by the LCD and then multiply the binomials to obtain the equation $x^2-9=(2x^2-15)-(x^2-x-2)$? Yes ____ No ____

If you answered No, look at the step $-[(x+1)(x-2)]$.

After you multiplied the binomials, did you distribute the negative sign and change the sign of all terms inside the grouping symbols?
Yes ____ No ____

If you answered No, review how to multiply binomials and subtract polynomial expressions. Then combine like terms and solve the equation for x.

Now go back and rework the problem using these suggestions.

Name: ______________________________ Date: ________________
Instructor: ____________________________ Topic: ________________

Module 15 Rational Exponents and Radicals
Topic 1 Rational Exponents

Vocabulary

rational exponents • denominator • quotient rule • exponential factor

1. You need to change fractional exponents to equivalent fractional exponents with the same ____________ when the rules of exponents require you to add or subtract them.

2. Exponents that are fractions are called ____________.

3. When factoring an expression, if the terms contain exponents, we look for the same ____________ in each term.

Example	Student Practice
1. Simplify. $\left(\dfrac{5xy^{-3}}{2x^{-4}y}\right)^{-2}$	**2.** Simplify. $\left(\dfrac{4a^{-3}b}{3a^{-2}b^{-5}}\right)^{-3}$

Apply the rules of exponents.

$$\left(\frac{5xy^{-3}}{2x^{-4}y}\right)^{-2}=\frac{\left(5xy^{-3}\right)^{-2}}{\left(2x^{-4}y\right)^{-2}} \qquad \left(\frac{x}{y}\right)^{n}=\frac{x^n}{y^n}$$

$$=\frac{5^{-2}x^{-2}\left(y^{-3}\right)^{-2}}{2^{-2}\left(x^{-4}\right)^{-2}y^{-2}} \qquad (xy)^n=x^ny^n$$

$$=\frac{5^{-2}x^{-2}y^{6}}{2^{-2}x^{8}y^{-2}} \qquad \left(x^m\right)^n=x^{mn}$$

$$=\frac{5^{-2}}{2^{-2}}\cdot\frac{x^{-2}}{x^{8}}\cdot\frac{y^{6}}{y^{-2}}$$

$$=\frac{2^{2}}{5^{2}}\cdot\frac{x^{-2}}{x^{8}}\cdot\frac{y^{6}}{y^{-2}} \qquad \frac{x^{-n}}{y^{-m}}=\frac{y^m}{x^n}$$

$$=\frac{2^{2}}{5^{2}}\cdot x^{-2-8}\cdot y^{6+2} \qquad \frac{x^m}{x^n}=x^{m-n}$$

$$=\frac{4}{25}x^{-10}y^{8}$$

Vocabulary Answers: 1. denominator 2. rational exponents 3. exponential factor

Example	Student Practice
3. Simplify.	**4.** Simplify.
(a) $\left(x^{2/3}\right)^4$ $\left(x^{2/3}\right)^4 = x^{(2/3)(4/1)} = x^{8/3}$	**(a)** $\left(x^3\right)^{4/5}$
(b) $\dfrac{x^{5/6}}{x^{1/6}}$ $\dfrac{x^{5/6}}{x^{1/6}} = x^{5/6-1/6} = x^{4/6} = x^{2/3}$	**(b)** $\dfrac{x^{7/8}}{x^{1/8}}$
(c) $x^{2/3} \cdot x^{-1/3}$ $x^{2/3} \cdot x^{-1/3} = x^{2/3-1/3} = x^{1/3}$	**(c)** $6^{5/13} \cdot 6^{4/13}$
5. Simplify. Express your answers with positive exponents only.	**6.** Simplify. Express your answers with positive exponents only.
(a) $\left(2x^{1/2}\right)\left(3x^{1/3}\right)$ $\left(2x^{1/2}\right)\left(3x^{1/3}\right) = 6x^{1/2+1/3}$ $= 6x^{3/6+2/6} = 6x^{5/6}$	**(a)** $\left(-3x^{2/5}\right)\left(4x^{1/2}\right)$
(b) $\dfrac{18x^{1/4}y^{-1/3}}{-6x^{-1/2}y^{1/6}}$ $\dfrac{18x^{1/4}y^{-1/3}}{-6x^{-1/2}y^{1/6}} = -3x^{1/4-(-1/2)}y^{-1/3-1/6}$ $= -3x^{1/4+2/4}y^{-2/6-1/6}$ $= -3x^{3/4}y^{-3/6}$ $= -3x^{3/4}y^{-1/2}$ $= -\dfrac{3x^{3/4}}{y^{1/2}}$	**(b)** $\dfrac{25x^{3/4}y^{-5/12}}{5x^{-1/8}y^{1/4}}$

Example	Student Practice
7. Multiply and simplify. $-2x^{5/6}\left(3x^{1/2}-4x^{-1/3}\right)$ We will need to be very careful when we add the exponents for x as we use the distributive property. $-2x^{5/6}\left(3x^{1/2}-4x^{-1/3}\right)$ $=-6x^{5/6+1/2}+8x^{5/6-1/3}$ $=-6x^{5/6+3/6}+8x^{5/6-2/6}$ $=-6x^{8/6}+8x^{3/6}$ $=-6x^{4/3}+8x^{1/2}$	**8.** Multiply and simplify. $4x^{2/3}\left(-2x^{1/2}+6x^{-1/6}\right)$
9. Evaluate. **(a)** $(25)^{3/2}$ $(25)^{3/2}=\left(5^2\right)^{3/2}=5^{2/1\cdot 3/2}=5^3=125$ **(b)** $(27)^{2/3}$ $(27)^{2/3}=\left(3^3\right)^{2/3}=3^{3/1\cdot 2/3}=3^2=9$	**10.** Evaluate. **(a)** $(16)^{3/2}$ **(b)** $(64)^{2/3}$
11. Write as one fraction with positive exponents. $2x^{-1/2}+x^{1/2}$ $2x^{-1/2}+x^{1/2}=\dfrac{2}{x^{1/2}}+\dfrac{x^{1/2}\cdot x^{1/2}}{x^{1/2}}$ $=\dfrac{2}{x^{1/2}}+\dfrac{x^1}{x^{1/2}}$ $=\dfrac{2+x}{x^{1/2}}$	**12.** Write as one fraction with positive exponents. $5x^{-3/4}+x^{1/4}$

Example	Student Practice
13. Factor out the common factor $2x$. $2x^{3/2}+4x^{5/2}$ Rewrite the exponent of each term so that each term contains the factor $2x$ or $2x^{2/2}$. $2x^{3/2}+4x^{5/2}$ $=2x^{2/2+1/2}+4x^{2/2+3/2}$ $=2\left(x^{2/2}\right)\left(x^{1/2}\right)+4\left(x^{2/2}\right)\left(x^{3/2}\right)$ $=2x\left(x^{1/2}+2x^{3/2}\right)$	**14.** Factor out the common factor $6z$. $18z^{5/3}-24z^{4/3}$

Extra Practice

1. Simplify. Express your answers with positive exponents only.

$\left(x^{4/5}\right)^4$

2. Simplify. Express your answers with positive exponents only.

$\left(a^{5/4}b^{2/3}\right)\left(a^{-1/4}b^{2/3}\right)$

3. Write as one fraction with positive exponents.

$y^{-2/3}+3y^{1/3}$

4. Factor out the common factor $5x$.

$10x^{7/4}+25x^{9/8}$

Concept Check

Explain how you would simplify the following. $x^{9/10}\cdot x^{-1/5}$

Name: ______________________________ Date: ________________
Instructor: ____________________________ Topic: ______________

Module 15 Rational Exponents and Radicals
Topic 2 Radical Expressions and Functions

Vocabulary
square root • principal square root • radical sign • radical expression • radicand
index • higher-order roots • cube root • fourth root • square root function

1. Because the symbol $\sqrt{x}$ represents exactly one real number for all real numbers x that are nonnegative, we can use it to define the ____________, $f(x)=\sqrt{x}$.

2. The positive square root is the ____________.

3. A(n) ____________ of a number is a value that when cubed is equal to the original number.

4. In the radical expression $\sqrt[n]{x}$, x is the radicand and n is the ____________.

Example	Student Practice
1. Find the indicated function values of the function $f(x)=\sqrt{2x+4}$. Round your answers to the nearest tenth when necessary.	**2.** Find the indicated function values of the function $f(x)=\sqrt{5x-4}$. Round your answers to the nearest tenth when necessary.
(a) $f(-2)$ Substitute -2 for x in the function. $f(-2)=\sqrt{2(-2)+4}=\sqrt{-4+4}$ $=\sqrt{0}=0$	**(a)** $f(4)$
(b) $f(6)$ $f(6)=\sqrt{2(6)+4}=\sqrt{12+4}$ $=\sqrt{16}=4$	**(b)** $f(1)$
(c) $f(3)$ $f(3)=\sqrt{2(3)+4}=\sqrt{6+4}$ $=\sqrt{10}\approx 3.2$	**(c)** $f(3)$

Vocabulary Answers: 1. square root function 2. principal square root 3. cube root 4. index

Example	**Student Practice**
3. Find the domain of the function. $f(x)=\sqrt{3x-6}$ The expression $3x-6$ must be nonnegative. $3x-6\geq 0$ $3x\geq 6$ $x\geq 2$ Thus, the domain is all real numbers x, where $x\geq 2$.	**4.** Find the domain of the function. $f(x)=\sqrt{\frac{2}{3}x-6}$
5. Change $\sqrt[4]{x^4}$ to rational exponents and simplify. Assume that all variables are nonnegative real numbers. If n is a positive integer and x is a nonnegative real number, then $x^{1/n}=\sqrt[n]{x}$. $\sqrt[4]{x^4}=\left(x^4\right)^{1/4}=x^{4/4}=x^1=x$	**6.** Change $\sqrt[6]{7^6}$ to rational exponents and simplify. Assume that all variables are nonnegative real numbers.
7. Replace all radicals with rational exponents. **(a)** $\sqrt[3]{x^2}$ $\sqrt[3]{x^2}=\left(x^2\right)^{1/3}$ $=x^{2/3}$ **(b)** $\left(\sqrt[5]{w}\right)^7$ $\left(\sqrt[5]{w}\right)^7=\left(w^{1/5}\right)^7$ $=w^{7/5}$	**8.** Replace all radicals with rational exponents. **(a)** $\sqrt[5]{z^3}$ **(b)** $\left(\sqrt[3]{a}\right)^8$

Example	Student Practice
9. Change to radical form. **(a)** $w^{-2/3}$ For positive integers m and n and any real number x where $x \neq 0$ and for which $x^{1/n}$ is defined, then $x^{-m/n} = \frac{1}{x^{m/n}} = \frac{1}{\left(\sqrt[n]{x}\right)^m} = \frac{1}{\sqrt[n]{x^m}}$. $w^{-2/3} = \frac{1}{w^{2/3}} = \frac{1}{\left(\sqrt[3]{w}\right)^2} = \frac{1}{\sqrt[3]{w^2}}$ **(b)** $(3x)^{3/4}$ For positive integers m and n and any real number x for which $x^{1/n}$ is defined, then $x^{m/n} = \left(\sqrt[n]{x}\right)^m = \sqrt[n]{x^m}$. $(3x)^{3/4} = \sqrt[4]{(3x)^3} = \sqrt[4]{27x^3}$ or $(3x)^{3/4} = \left(\sqrt[4]{3x}\right)^3$	**10.** Change to radical form. **(a)** $(ab)^{-5/3}$ **(b)** $(5x)^{1/5}$
11. Change to radical form and evaluate. **(a)** $(-16)^{5/2}$ $(-16)^{5/2} = \left(\sqrt{-16}\right)^5$; however, $\sqrt{-16}$ is not a real number. Thus, $(-16)^{5/2}$ is not a real number. **(b)** $144^{-1/2}$ $144^{-1/2} = \frac{1}{144^{1/2}} = \frac{1}{\sqrt{144}} = \frac{1}{12}$	**12.** Change to radical form and evaluate. **(a)** $64^{-1/3}$ **(b)** $(-81)^{1/4}$

Example	Student Practice
13. Simplify. Assume that x and y may be any real numbers.	**14.** Simplify. Assume that x and y may be any real numbers.
(a) $\sqrt{49x^2}$ Since the index is even, take the absolute value. $\sqrt{49x^2} = 7\lvert x\rvert$	**(a)** $\sqrt{25a^2}$
(b) $\sqrt[4]{81y^{16}}$ $\sqrt[4]{81y^{16}} = 3\lvert y^4\rvert$ Since any number raised to the fourth is positive we can write it as $\sqrt[4]{81y^{16}} = 3y^4$.	**(b)** $\sqrt[3]{64z^6}$
(c) $\sqrt[3]{27x^6y^9}$ Note that the index is odd, thus we do not need the absolute value. $\sqrt[3]{27x^6y^9} = \sqrt[3]{3^3\left(x^2\right)^3\left(y^3\right)^3} = 3x^2y^3$	**(c)** $\sqrt[4]{81x^8y^4}$

Extra Practice

1. Evaluate if possible. $\sqrt[15]{(7)^{15}}$

2. Replace all radicals with rational exponents. Assume that variables represent positive real numbers.
$\sqrt[8]{(4m-3n)^5}$

3. Change to radical form. $4^{-4/7}$

4. Simplify. $\sqrt{49a^6b^{16}}$

Concept Check

Explain how you would simplify the following. $\sqrt[4]{81x^8y^{16}}$

Name: ______________________________ Date: ________________
Instructor: ____________________________ Topic: _______________

Module 15 Rational Exponents and Radicals
Topic 3 Simplifying, Adding, and Subtracting Radicals

Vocabulary
product rule • radicand • index • like radicals • distributive property

1. Like radicals have the same ____________ and index.

2. The____________ for radicals states that for all nonnegative real numbers a and b and positive integers n, $\sqrt[n]{a}\sqrt[n]{b}=\sqrt[n]{ab}.$

3. Only ____________ can be added or subtracted.

Example	**Student Practice**
1. Simplify. $\sqrt{32}$ Recall that for all nonnegative real numbers a and b and positive integers n, $\sqrt[n]{a}\sqrt[n]{b}=\sqrt[n]{ab}$. Factor the radicand into the largest perfect square factors possible, then simplify. $\sqrt{32}=\sqrt{16\cdot 2}=\sqrt{16}\sqrt{2}=4\sqrt{2}$	**2.** Simplify. $\sqrt{50}$
3. Simplify. $\sqrt{48}$ $\sqrt{48}=\sqrt{16\cdot 3}=\sqrt{16}\sqrt{3}=4\sqrt{3}$	**4.** Simplify. $\sqrt{108}$
5. Simplify. $\sqrt[3]{-81}$ Factor the radicand into the largest perfect cube factors possible, then simplify. $\sqrt[3]{-81}=\sqrt[3]{-27}\sqrt[3]{3}=-3\sqrt[3]{3}$	**6.** Simplify. $\sqrt[3]{-162}$

Vocabulary Answers: 1. radicand 2. product rule 3. like radicals

Example	Student Practice
7. Simplify. **(a)** $\sqrt{27x^3y^4}$ Factor out the perfect squares. $\sqrt{27x^3y^4} = \sqrt{9\cdot 3\cdot x^2\cdot x\cdot y^4}$ $= \sqrt{9x^2y^4}\sqrt{3x} = 3xy^2\sqrt{3x}$ **(b)** $\sqrt[3]{16x^4y^3z^6}$ Factor out the perfect cubes. $\sqrt[3]{16x^4y^3z^6} = \sqrt[3]{8\cdot 2\cdot x^3\cdot x\cdot y^3\cdot z^6}$ $= \sqrt[3]{8x^3y^3z^6}\sqrt[3]{2x}$ $= 2xyz^2\sqrt[3]{2x}$	**8.** Simplify. **(a)** $\sqrt{32a^5b^4}$ **(b)** $\sqrt[3]{192x^2y^4z^5}$
9. Combine. $2\sqrt{5}+3\sqrt{5}-4\sqrt{5}$ All three radicals are like radicals because they have the same radicand and index. Thus, they may be combined. $2\sqrt{5}+3\sqrt{5}-4\sqrt{5} = (2+3-4)\sqrt{5}$ $= 1\sqrt{5} = \sqrt{5}$	**10.** Combine. $6\sqrt{3a}-3\sqrt{3a}+\sqrt{3a}$.
11. Combine. $5\sqrt{3}-\sqrt{27}+2\sqrt{48}$ Sometimes when simplifying radicands, you may find like radicals. $5\sqrt{3}-\sqrt{27}+2\sqrt{48}$ $= 5\sqrt{3}-\sqrt{9}\sqrt{3}+2\sqrt{16}\sqrt{3}$ $= 5\sqrt{3}-3\sqrt{3}+2(4)\sqrt{3}$ $= 5\sqrt{3}-3\sqrt{3}+8\sqrt{3}$ $= 10\sqrt{3}$	**12.** Combine. $6\sqrt{5}+\sqrt{80}-3\sqrt{20}$

Example	Student Practice
13. Combine. $6\sqrt{x}+4\sqrt{12x}-\sqrt{75x}+3\sqrt{x}$ $6\sqrt{x}+4\sqrt{12x}-\sqrt{75x}+3\sqrt{x}$ $=6\sqrt{x}+4\sqrt{4}\sqrt{3x}-\sqrt{25}\sqrt{3x}+3\sqrt{x}$ $=6\sqrt{x}+8\sqrt{3x}-5\sqrt{3x}+3\sqrt{x}$ $=6\sqrt{x}+3\sqrt{x}+8\sqrt{3x}-5\sqrt{3x}$ $=9\sqrt{x}+3\sqrt{3x}$	**14.** Combine. $3\sqrt{a}-2\sqrt{32a}+\sqrt{50a}+2\sqrt{a}$
15. Combine. $2\sqrt[3]{81x^3y^4}+3xy\sqrt[3]{24y}$ $2\sqrt[3]{81x^3y^4}+3xy\sqrt[3]{24y}$ $=2\sqrt[3]{27x^3y^3}\sqrt[3]{3y}+3xy\sqrt[3]{8}\sqrt[3]{3y}$ $=2(3xy)\sqrt[3]{3y}+3xy(2)\sqrt[3]{3y}$ $=6xy\sqrt[3]{3y}+6xy\sqrt[3]{3y}$ $=12xy\sqrt[3]{3y}$	**16.** Combine. $\sqrt[3]{40x^6z^4}+3x^2\sqrt[3]{320z^4}$

Extra Practice

1. Simplify. Assume that all variables are nonnegative real numbers. $\sqrt{25x^3}$

2. Simplify. Assume that all variables are nonnegative real numbers. $\sqrt[4]{625ab^{19}}$

3. Combine. $6\sqrt{7}-4\sqrt{5}+4\sqrt{7}$

4. Combine. Assume that all variables represent nonnegative real numbers.
$6\sqrt{48x^2}-2\sqrt{27x^2}-\sqrt{3x^2}$

Concept Check

Explain how you would simplify the following. $\sqrt[4]{16x^{13}y^{16}}$

Name: ______________________________ Date: ________________

Instructor: ______________________________ Topic: ________________

Module 15 Rational Exponents and Radicals
Topic 4 Multiplying and Dividing Radicals

Vocabulary

product rule for radicals • FOIL method • quotient rule for radicals
rationalizing the denominator • conjugates

1. The expressions $a+b$ and $a-b$, where a and b represent any algebraic term, are called ___________.

2. The ___________ states that for all nonnegative real numbers a, all positive numbers b, and positive integers n, $\frac{\sqrt[n]{a}}{\sqrt[n]{b}} = \sqrt[n]{\frac{a}{b}}$.

3. ___________ is the process of transforming a fraction with one or more radicals in the denominator into an equivalent fraction without a radical in the denominator.

Example	Student Practice
1. Multiply. $\left(3\sqrt{2}\right)\left(5\sqrt{11x}\right)$ To multiply, use the product rule for radicals, $\sqrt[n]{a}\sqrt[n]{b} = \sqrt[n]{ab}$. $\left(3\sqrt{2}\right)\left(5\sqrt{11x}\right) = (3)(5)\left(\sqrt{2\cdot 11x}\right)$ $= 15\sqrt{2\cdot 11x} = 15\sqrt{22x}$	**2.** Multiply. $\left(3\sqrt{7}\right)\left(-4\sqrt{3z}\right)$
3. Multiply. $\left(\sqrt{2}+3\sqrt{5}\right)\left(2\sqrt{2}-\sqrt{5}\right)$ To multiply two binomials containing radicals, we can use the distributive property or the FOIL method. For this problem, we use the FOIL method. $\left(\sqrt{2}+3\sqrt{5}\right)\left(2\sqrt{2}-\sqrt{5}\right)$ $= 2\sqrt{4}-\sqrt{10}+6\sqrt{10}-3\sqrt{25}$ $= 4+5\sqrt{10}-15 = -11+5\sqrt{10}$	**4.** Multiply. $\left(4\sqrt{3}-3\sqrt{5}\right)\left(2\sqrt{3}-\sqrt{5}\right)$

Vocabulary Answers: 1. conjugates 2. quotient rule for radicals 3. rationalizing the denominator

Example	Student Practice
5. Multiply. $\left(7-3\sqrt{2}\right)\left(4-\sqrt{3}\right)$ Use the FOIL method. $\left(7-3\sqrt{2}\right)\left(4-\sqrt{3}\right)$ $=28-7\sqrt{3}-12\sqrt{2}+3\sqrt{6}$	**6.** Multiply. $\left(6+4\sqrt{3}\right)\left(2+3\sqrt{2}\right)$
7. Multiply. $\left(\sqrt{7}+\sqrt{3x}\right)^2$ Use $(a+b)^2=a^2+2ab+b^2$, where $a=\sqrt{7}$ and $b=\sqrt{3x}$. $\left(\sqrt{7}+\sqrt{3x}\right)^2$ $=\left(\sqrt{7}\right)^2+2\sqrt{7}\sqrt{3x}+\left(\sqrt{3x}\right)^2$ $=7+2\sqrt{21x}+3x$	**8.** Multiply. $\left(\sqrt{11}-\sqrt{2a}\right)^2$
9. Divide. **(a)** $\sqrt[3]{\frac{125}{8}}$ For all nonnegative real numbers a, all positive numbers b, and positive integers n, $\frac{\sqrt[n]{a}}{\sqrt[n]{b}}=\sqrt[n]{\frac{a}{b}}$. $\sqrt[3]{\frac{125}{8}}=\frac{\sqrt[3]{125}}{\sqrt[3]{8}}=\frac{5}{2}$ **(b)** $\frac{\sqrt{28x^5y^3}}{\sqrt{7x}}$ $\frac{\sqrt{28x^5y^3}}{\sqrt{7x}}=\sqrt{\frac{28x^5y^3}{7x}}=\sqrt{4x^4y^3}$ $=2x^2y\sqrt{y}$	**10.** Divide. **(a)** $\sqrt[3]{\frac{8}{27}}$ **(b)** $\frac{\sqrt{72a^3b^7}}{\sqrt{2b^3}}$

Example	Student Practice
11. Simplify. $\dfrac{3}{\sqrt{12x}}$	**12.** Simplify. $\dfrac{5}{\sqrt{32z}}$

Simplify the radical in the denominator, then multiply in order to rationalize the denominator.

$$\frac{3}{\sqrt{12x}}=\frac{3}{\sqrt{4}\sqrt{3x}}$$
$$=\frac{3}{2\sqrt{3x}}\cdot\frac{\sqrt{3x}}{\sqrt{3x}}$$
$$=\frac{\cancel{3}\sqrt{3x}}{2(\cancel{3}x)}=\frac{\sqrt{3x}}{2x}$$

Example	Student Practice
13. Simplify. $\sqrt[3]{\dfrac{2}{3x^2}}$	**14.** Simplify. $\sqrt[3]{\dfrac{54}{2a}}$

Multiply the numerator and denominator by a value that will make the radicand in the denominator a perfect cube (i.e., rationalize the denominator). In this case, multiply the numerator and denominator by $9x$ because $9x\cdot 3x^2=27x^3$ which is a perfect cube.

$$\sqrt[3]{\frac{2}{3x^2}}=\sqrt[3]{\frac{2}{3x^2}\cdot\frac{9x}{9x}}$$
$$=\sqrt[3]{\frac{18x}{27x^3}}$$

Rewrite with the quotient rule and simplify.

$$\sqrt[3]{\frac{18x}{27x^3}}=\frac{\sqrt[3]{18x}}{\sqrt[3]{27x^3}}$$
$$=\frac{\sqrt[3]{18x}}{3x}$$

Example	Student Practice
15. Simplify. $\dfrac{5}{3+\sqrt{2}}$ Multiply the numerator and denominator by the conjugate of $3+\sqrt{2}$. $\dfrac{5}{3+\sqrt{2}}=\dfrac{5}{3+\sqrt{2}}\cdot\dfrac{3-\sqrt{2}}{3-\sqrt{2}}$ $=\dfrac{15-5\sqrt{2}}{3^2-\left(\sqrt{2}\right)^2}$ $=\dfrac{15-5\sqrt{2}}{9-2}=\dfrac{15-5\sqrt{2}}{7}$	**16.** Simplify. $\dfrac{\sqrt{5}+3\sqrt{7}}{2\sqrt{5}-\sqrt{7}}$

Extra Practice

1. Multiply and simplify. Assume that all variables are nonnegative numbers.
   $\left(3a\sqrt{a}\right)\left(5\sqrt{b}\right)$

2. Multiply and simplify. Assume that all variables are nonnegative numbers.
   $\left(5\sqrt{7}+3\sqrt{10}\right)^2$

3. Divide and simplify. Assume that all variables represent positive numbers.
   $\sqrt{\dfrac{18x}{25y^4}}$

4. Simplify by rationalizing the denominator. $\dfrac{2x}{\sqrt{5}-\sqrt{3}}$

Concept Check

Explain how you would rationalize the denominator of the following. $\dfrac{5\sqrt{3}-3\sqrt{2}}{3\sqrt{2}-2\sqrt{3}}$

Name: ______________________________ Date: ________________
Instructor: ____________________________ Topic: ________________

Module 15 Rational Exponents and Radicals
Topic 5 Radical Equations

Vocabulary
radical equation • isolating the radical term • zero factor • extraneous solution

1. If the result of squaring both sides of a radical equation is a quadratic equation, collect all terms on one side and use the ____________ method to solve.

2. A(n) ____________ is an equation with a variable in one or more of the radicals.

3. Getting one radical expression alone on one side of the equation is referred to as ____________.

Example	Student Practice
1. Solve. $\sqrt{2x+9}=x+3$	**2.** Solve. $\sqrt{3x+7}=x-1$

Square each side and simplify.

$$\left(\sqrt{2x+9}\right)^2=(x+3)^2$$
$$2x+9=x^2+6x+9$$

Collect all terms on one side and factor.

$$2x+9=x^2+6x+9$$
$$0=x^2+4x$$
$$0=x(x+4)$$

Set each factor equal to zero and solve.

$x=0$ or $x+4=0$

$x=0$ $\quad$ $x=-4$

Check.

$x=0$: $\sqrt{2(0)+9}\stackrel{?}{=}(0)+3$

$3=3$

$x=-4$: $\sqrt{2(-4)+9}\stackrel{?}{=}(-4)+3$

$1\neq-1$

Thus, 0 is the only solution.

Vocabulary Answers: 1. zero factor 2. radical equation 3. isolating the radical

Example	Student Practice
3. Solve. $\sqrt{10x+5}-1=2x$	**4.** Solve. $3x=-1+\sqrt{30x+10}$

Isolate the radical term. Then square each side and simplify.

$$\sqrt{10x+5}-1=2x$$
$$\sqrt{10x+5}=2x+1$$
$$\left(\sqrt{10x+5}\right)^2=(2x+1)^2$$
$$10x+5=4x^2+4x+1$$

Collect all terms on one side and factor.

$$10x+5=4x^2+4x+1$$
$$0=4x^2-6x-4$$
$$0=2\left(2x^2-3x-2\right)$$
$$0=2(2x+1)(x-2)$$

Set each factor equal to zero and solve.

$$2x+1=0 \qquad x-2=0$$
$$2x=-1 \qquad x=2$$
$$x=-\frac{1}{2}$$

Check.

$$x=-\frac{1}{2}:\quad \sqrt{10\left(-\frac{1}{2}\right)+5}-1\stackrel{?}{=}2\left(-\frac{1}{2}\right)$$
$$\sqrt{-5+5}-1\stackrel{?}{=}-1$$
$$\sqrt{0}-1\stackrel{?}{=}-1$$
$$-1=-1$$

$$x=2:\quad \sqrt{10(2)+5}-1\stackrel{?}{=}2(2)$$
$$\sqrt{25}-1\stackrel{?}{=}4$$
$$4=4$$

Both answers check. Thus, the solutions are $-\frac{1}{2}$ and 2.

Example	Student Practice
5. Solve. $\sqrt{5x+1}-\sqrt{3x}=1$	**6.** Solve. $\sqrt{x+4}-2=\sqrt{x-4}$

Isolate one of the radicals.

$$\sqrt{5x+1}-\sqrt{3x}=1$$
$$\sqrt{5x+1}=1+\sqrt{3x}$$

Square each side.

$$\sqrt{5x+1}=1+\sqrt{3x}$$
$$\left(\sqrt{5x+1}\right)^2=\left(1+\sqrt{3x}\right)^2$$
$$5x+1=\left(1+\sqrt{3x}\right)\left(1+\sqrt{3x}\right)$$
$$5x+1=1+2\sqrt{3x}+3x$$

Isolate the remaining radical.

$$5x+1=1+2\sqrt{3x}+3x$$
$$2x=2\sqrt{3x}$$
$$x=\sqrt{3x}$$

Square each side.

$$x=\sqrt{3x}$$
$$(x)^2=\left(\sqrt{3x}\right)^2$$
$$x^2=3x$$

Collect all terms on one side and factor.

$$x^2=3x$$
$$x^2-3x=0$$
$$x(x-3)=0$$
$$x=0 \quad \text{or} \quad x-3=0$$
$$x=0 \qquad\qquad x=3$$

The check is left to the student. Verify that both 0 and 3 are valid solutions.

Example	Student Practice
7. Solve. $\sqrt{2y+5}-\sqrt{y-1}=\sqrt{y+2}$	**8.** Solve. $\sqrt{y+6}-\sqrt{y-2}=\sqrt{2y-2}$

$$\sqrt{2y+5}-\sqrt{y-1}=\sqrt{y+2}$$
$$\left(\sqrt{2y+5}-\sqrt{y-1}\right)^2=\left(\sqrt{y+2}\right)^2$$

Expand both sides of the equation.

$$2y+5-2\sqrt{(y-1)(2y+5)}+y-1=y+2$$

Isolate the radical and solve for y.

$$-2\sqrt{(y-1)(2y+5)}=-2y-2$$
$$\sqrt{(y-1)(2y+5)}=y+1$$
$$\left(\sqrt{(y-1)(2y+5)}\right)^2=(y+1)^2$$
$$2y^2+3y-5=y^2+2y+1$$
$$y^2+y-6=0$$
$$(y+3)(y-2)=0$$
$$y=-3 \quad\text{or}\quad y=2$$

The check is left to the student. Verify that 2 is a valid solution but -3 is not.

Extra Practice

1. Solve the radical equation. Check your solution. $\sqrt{10x-5}=5$

2. Solve the radical equation. Check your solution. $\sqrt{x+4}+8=3$

3. Solve the radical equation. This will usually involve squaring each side twice. Check your solution. $\sqrt{3x+4}=2+\sqrt{x}$

4. Solve the radical equation. This will usually involve squaring each side twice. Check your solution. $\sqrt{2x+6}-\sqrt{x+4}=\sqrt{x-4}$

Concept Check

When you try to solve the equation $2+\sqrt{x+10}=x$, you obtain the values $x=-1$ and $x=6$. Explain how you would determine if either of these values is a solution of the radical equation.

Name: ______________________________ Date: ______________

Instructor: ____________________________ Topic: ______________

Module 15 Rational Exponents and Radicals
Topic 6 Complex Numbers

Vocabulary

imaginary number • complex number • real part • imaginary part • conjugates

1. The complex numbers $a+bi$ and $a-bi$ are called __________.

2. The __________ i is defined as $i=\sqrt{-1}$ and $i^2=-1$.

3. A number that can be written in the form $a+bi$, where a and b are real numbers, is a __________.

4. In the complex number $a+bi$, a is the __________ and bi is the imaginary part.

Example	Student Practice
1. Simplify. **(a)** $\sqrt{-36}$ For all positive real numbers a, $\sqrt{-a}=\sqrt{-1}\sqrt{a}=i\sqrt{a}$. $\sqrt{-36}=\sqrt{-1}\sqrt{36}=i(6)=6i$ **(b)** $\sqrt{-17}$ $\sqrt{-17}=\sqrt{-1}\sqrt{17}=i\sqrt{17}$	**2.** Simplify. **(a)** $\sqrt{-41}$ **(b)** $\sqrt{-25}$
3. Multiply. $\sqrt{-16}\cdot\sqrt{-25}$ Rewrite using the definition $\sqrt{-1}=i$. $\left(\sqrt{-16}\right)\left(\sqrt{-25}\right)=\left(i\sqrt{16}\right)\left(i\sqrt{25}\right)$ $=i^2(4)(5)$ $=-1(20)=-20$	**4.** Multiply. $\sqrt{-36}\cdot\sqrt{-9}$

Vocabulary Answers: 1. conjugates 2. imaginary number 3. complex number 4. real part

Example	Student Practice
5. Find the real numbers x and y if $x+3i\sqrt{7}=-2+yi.$ Two complex numbers $a+bi$ and $c+di$ are equal if and only if $a=c$ and $b=d$. The real parts must be equal, so x must be -2. The imaginary parts must also be equal, so y must be $3\sqrt{7}$.	**6.** Find the real numbers x and y if $-3+3yi=x+12i\sqrt{2}.$
7. Subtract. $(6-2i)-(3-5i)$ Subtract the real parts and subtract the imaginary parts. $(6-2i)-(3-5i)$ $=(6-3)+\left[-2-(-5)\right]i$ $=3+(-2+5)i$ $=3+3i$	**8.** Subtract. $(-5+6i)-(-2+4i)$
9. Multiply. $(7-6i)(2+3i)$ Use FOIL. $(7-6i)(2+3i)$ $=(7)(2)+(7)(3i)-(6i)(2)-(6i)(3i)$ $=14+21i-12i-18i^2$ Since $i^2=-1,\ 14+21i-12i-18i^2$ can be written as $14+21i-12i-18(-1)$. Simplify this expression. $14+21i-12i-18(-1)$ $=14+21i-12i+18$ $=32+9i$ Thus, $(7-6i)(2+3i)=32+9i.$	**10.** Multiply. $(6-5i)(3-2i)$

Example	Student Practice
11. Multiply. $3i(4-5i)$ Use the distributive property. Recall that $i^2 = 1$. $3i(4-5i) = (3)(4)i + (3)(-5)i^2$ $= 12i - 15i^2$ $= 12i - 15(-1)$ $= 15 + 12i$	**12.** Multiply. $-4i(3-6i)$
13. Evaluate. **(a)** i^{36} Some values for i^n are shown below. $i = i \quad i^5 = i \quad i^9 = i$ $i^2 = -1 \quad i^6 = -1 \quad i^{10} = -1$ $i^3 = -i \quad i^7 = -i \quad i^{11} = -i$ $i^4 = +1 \quad i^8 = +1 \quad i^{12} = +1$ Divide the exponent by 4 since i^4 raised to any power will be 1. Then use the values from the table above to evaluate the remainder. $i^{36} = \left(i^4\right)^9 = (1)^9 = 1$ **(b)** i^{27} $i^{27} = \left(i^{24+3}\right) = \left(i^{24}\right)\left(i^3\right)$ $= \left(i^4\right)^6\left(i^3\right)$ $= (1)^6(-i)$ $= -i$	**14.** Evaluate. **(a)** i^{22} **(b)** i^{64}

Example	Student Practice
15. Divide. $\frac{3-2i}{4i}$	**16.** Divide. $\frac{6-5i}{4+3i}$

Multiply the numerator and denominator by the conjugate of the denominator. The conjugate of $0+4i$ is $0-4i$, or simply $-4i$.

$$\frac{3-2i}{4i}=\frac{3-2i}{4i}\cdot\frac{-4i}{-4i}$$
$$=\frac{-12i+8i^2}{-16i^2}$$
$$=\frac{-12i+8(-1)}{-16(-1)}$$
$$=\frac{-8-12i}{16}$$
$$=\frac{\cancel{4}(-2-3i)}{\cancel{4}\cdot 4}$$
$$=\frac{-2-3i}{4}\text{ or }-\frac{1}{2}-\frac{3}{4}i$$

Extra Practice

1. Multiply and simplify. Place in i notation before doing any other operations.
$\left(6+\sqrt{-2}\right)\left(3-\sqrt{-2}\right)$

2. Evaluate. $i^{82}-i^{31}$

3. Multiply and simplify.
$(3+2i)(2+i)$

4. Divide. $\frac{6}{5-2i}$

Concept Check

Explain how you would simplify the following. $(3+5i)^2$

Name: ______________________________ Date: ________________
Instructor: ____________________________ Topic: ________________

Module 15 Rational Exponents and Radicals
Topic 7 Variation

Vocabulary

direct variation • inverse variation • combined variation • constant of variation

1. ____________ is where one variable is a constant multiple of the reciprocal of the other.

2. In the direct variation equation $y = kx$, k is the ____________.

3. ____________ is where one variable is a constant multiple of the other.

4. If a quantity depends on the variation of two or more variables, it is called joint or ____________.

Example	Student Practice
1. The time of a pendulum's period varies directly with the square root of its length. If the pendulum is 1 foot long when the time is 0.2 second, find the time when its length is 4 feet.	**2.** A car's stopping distance varies directly with the square of its speed. A car that is traveling 35 miles per hour can stop in 49 feet. What distance will it take to stop if it is traveling 60 miles per hour?

1. The time of a pendulum's period varies directly with the square root of its length. If the pendulum is 1 foot long when the time is 0.2 second, find the time when its length is 4 feet.

Let $t =$ the time and $L =$ the length. This gives us the equation $t = k\sqrt{L}$. Substitute $L = 1$ and $t = 0.2$ into the equation to find k.

$$t = k\sqrt{L}$$
$$(0.2) = k\left(\sqrt{1}\right)$$
$$0.2 = k$$

We can rewrite the equation as $t = 0.2\sqrt{L}$. Find the value of t when $L = 4$.

$$t = 0.2\sqrt{L}$$
$$= 0.2\sqrt{4} = (0.2)(2) = 0.4 \text{ seconds}$$

Vocabulary Answers: 1. inverse variation 2. constant of variation 3. direct variation 4. combined variation

Example	Student Practice
3. If y varies inversely with x and $y = 12$ when $x = 5$, find the value of y when $x = 14$. Write the equation as $y = \frac{k}{x}$. Substitute $y = 12$ and $x = 5$ to find k. $12 = \frac{k}{5}$ $60 = k$ Rewrite the equation as $y = \frac{60}{x}$. Find the value of y when $x = 14$. $y = \frac{60}{x} = \frac{60}{14} = \frac{30}{7}$	**4.** If y varies inversely with x and $y = 12$ when $x = 5$, find the value of y when $x = 14$.
5. The amount of light from a light source varies inversely with the square of the distance from the light source. If an object receives 6.25 lumens when the light source is 8 meters away, how much light will the object receive if the light source is 4 meters away? Let $L =$ light and $d =$ distance, which gives the equation $L = \frac{k}{d^2}$. Substitute $L = 6.25$ and $d = 8$ to find k. $6.25 = \frac{k}{8^2}$ $400 = k$ Rewrite the original equation with $k = 400$. Then find L when $d = 4$. $L = \frac{400}{d^2} = \frac{400}{4^2} = \frac{400}{16} = 25$ lumens The check is left to the student.	**6.** The amount of light from a light source varies inversely with the square of the distance from the light source. If an object receives 9.31 lumens when the light source is 10 meters away, how much light will the object receive if the light source is 7 meters away?

Example	Student Practice
7. y varies directly with x and z and inversely with d^2. When $x=7$, $z=3$, and $d=4$, the value of y is 20. Find the value of y when $x=5$, $z=6$, and $d=2$.	**8.** y varies directly with m^4 and n and inversely with p^2. When $m=2$, $n=3$, and $p=5$, the value of y is 63. Find the value of y when $m=1$, $n=8$, and $p=3$.

Write the equation as $y=\frac{kxz}{d^2}$.

Substitute the known values and solve for k.

$$y=\frac{kxz}{d^2}$$
$$20=\frac{k(7)(3)}{4^2}$$
$$20=\frac{21k}{16}$$
$$320=21k$$
$$\frac{320}{21}=k$$

Substitute this value of k into the original equation.

$$y=\frac{\frac{320}{21}xz}{d^2} \quad \text{or} \quad y=\frac{320xz}{21d^2}$$

Now find y when $x=5$, $z=3$, and $d=2$.

$$y=\frac{320xz}{21d^2}$$
$$y=\frac{320(5)(6)}{21(2)^2}=\frac{9600}{84}=\frac{800}{7}$$

Thus, the value of y is $\frac{800}{7}$.

Extra Practice

1. The distance a spring stretches varies directly with the weight of the object hung on the spring. If a 12-pound weight stretches a spring 15 inches, how far will a 32-pound weight stretch this spring?

2. If y varies inversely with the cube of x, and $y = 50$ when $x = 3$, find y when $x = 7$. Round to the nearest tenth.

3. The speed of a car varies inversely with the amount of time it takes to cover a certain distance. At 50 mph, a car travels a certain distance in 16 seconds. What is the speed of the car that travels the same distance in 10 seconds?

4. y varies directly with x, and inversely with z. $y = 40$ when $x = 6$ and $z = 15$. Find the value of y when $x = 0.5$ and $z = 12$. Round to the nearest tenth.

Concept Check

If y varies directly with the square root of x and $y = 50$ when $x = 5$, explain how you would find the constant of variation k.

MATH COACH

Mastering the skills you need to do well on the test.

Watch the MATH COACH videos in MyMathLab® or on YouTube™ while you work the problems below. These helpful hints will help you avoid making common errors on test problems.

Simplifying Expressions with Rational Exponents—

Problem 7 Evaluate. $16^{5/4}$

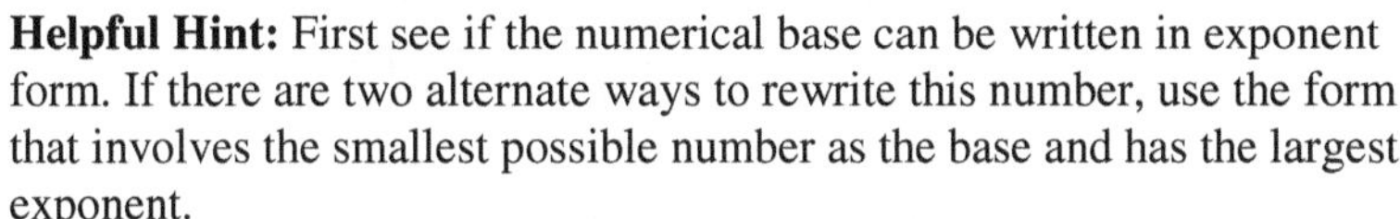

Helpful Hint: First see if the numerical base can be written in exponent form. If there are two alternate ways to rewrite this number, use the form that involves the smallest possible number as the base and has the largest exponent.

Did you write 16 as 2^4?

Yes ____ No ____

If you answered No, think of the different ways that 16 can be written in exponent form: 16^1, 4^2, or 2^4.

Your choice is most likely between 4^2 and 2^4. The problem works out more easily with the smallest possible base, which is 2. See if you can redo this step correctly.

Do you see that if we use the rule $\left(x^m\right)^n = x^{mn}$, we obtain the expression $\left(2^4\right)^{5/4} = 2^{4/1 \cdot 5/4}$?

Yes ____ No ____

If you answered No, please review the rule about raising a power to a power and see if you can obtain the right result. Now simplify this expression further and evaluate the resulting exponential expression.

If you answered Problem 7 incorrectly, go back and rework the problem using these suggestions.

Adding and Subtracting Radical Expressions—Problem 12

Combine where possible. $\sqrt{40x} - \sqrt{27x} + 2\sqrt{12x}$

Helpful Hint: Remember to simplify each radical expression first. For this problem, remember to keep the variable x inside the radical.

Did you simplify $\sqrt{40x}$ to $2\sqrt{10x}$? Yes ____ No ____

Did you simplify $\sqrt{27x}$ to $3\sqrt{3x}$? Yes ____ No ____

If you answered No to either question, remember that $\sqrt{40x} = \sqrt{4} \cdot \sqrt{10x}$ and $\sqrt{27x} = \sqrt{9} \cdot \sqrt{3x}$. You can take the square roots of both 4 and 9. Try to do that part of the problem again and then simplify to see if you obtain the same results.

Did you simplify $2\sqrt{12x}$ to $4\sqrt{3x}$? Yes ____ No ____

If you answered No, remember that $2\sqrt{12x} = 2 \cdot \sqrt{4} \cdot \sqrt{3x}$. Go back and see if you can obtain the correct answer to that step. Remember that in your final step, you can only add radical expressions if they have the same radicand.

Now go back and rework the problem using these suggestions.

Rationalizing the Denominator—Problem 18 Rationalize the denominator. $\dfrac{1+2\sqrt{3}}{3-\sqrt{3}}$

Helpful Hint: You will need to multiply both the numerator and the denominator by the conjugate of the denominator. Do this very carefully using the FOIL method.

Did you multiply the numerator and the denominator by $\left(3+\sqrt{3}\right)$?

Yes ____ No ____

If you answered No, review the definition of conjugate. Note that the conjugate of $3-\sqrt{3}$ is $3+\sqrt{3}$.

Did you multiply the binomials in the denominator to get $9+3\sqrt{3}-3\sqrt{3}-\sqrt{9}$?

Yes ____ No ____

Did you multiply the binomials in the numerator to get $3+\sqrt{3}+6\sqrt{3}+2\sqrt{9}$?

Yes ____ No ____

If you answered No to either question, go back over each step of multiplying the binomial expressions using the FOIL method. Be careful in writing which numbers go outside the radical and which numbers go inside the radical. Now simplify each expression.

If you answered Problem 18 incorrectly, go back and rework the problem using these suggestions.

Solving a Radical Equation—Problem 20 Solve and check your solution(s). $5+\sqrt{x+15}=x$

Helpful Hint: Always isolate a radical term on one side of the equation before squaring each side.

Did you isolate the radical term first to obtain the equation $\sqrt{x+15}=x-5$?

Yes ____ No ____

After squaring each side, did you obtain $x+15=x^2-10x+25$?

Yes ____ No ____

If you answered No to either question, review the process for solving a radical equation. Remember that $\left(\sqrt{x+15}\right)^2=x+15$ and $(x-5)^2=x^2-10x+25$. Go back and try to complete these steps again.

Did you collect all like terms on one side to obtain the equation $0=x^2-11x+10$?

Yes ____ No ____

If you answered No, remember to add $-x-15$ to both sides of the equation.

Now factor the quadratic equation and set each factor equal to 0 to find the solution(s). Make sure to check your possible solutions and discard any solution that does not check when substituted back into the original equation.

If you answered Problem 20 incorrectly, go back and rework the problem using these suggestions.

Name: ______________________________ Date: ________________
Instructor: ___________________________ Topic: ________________

Module 16 Quadratic Equations
Topic 1 Quadratic Equations

Vocabulary
quadratic equation • standard form • square root property • complete the square

1. The __________ states that if $x^2 = a$, then $x = \pm\sqrt{a}$ for all real numbers a.

2. When we __________ we are changing the polynomial to a perfect square trinomial.

3. $ax^2 + bx + c = 0$ is the __________ of a quadratic equation.

4. An equation written in the form $ax^2 + bx + c = 0$, where a, b, and c are real numbers and $a \neq 0$, is called a __________.

Example	Student Practice
1. Solve and check. $x^2 - 36 = 0$ If we add 36 to each side, we have $x^2 = 36$. $x = \pm\sqrt{36}$ $x = \pm 6$ The two roots are 6 and –6. Check. $(6)^2 - 36 \stackrel{?}{=} 0 \quad (-6)^2 - 36 \stackrel{?}{=} 0$ $36 - 36 \stackrel{?}{=} 0 \quad 36 - 36 \stackrel{?}{=} 0$ $0 = 0 \quad 0 = 0$	**2.** Solve and check. $x^2 - 25 = 0$
3. Solve. $x^2 = 48$ $x = \pm\sqrt{48}$ $x = \pm\sqrt{16 \cdot 3} = \pm 4\sqrt{3}$ The two roots are $4\sqrt{3}$ and $-4\sqrt{3}$.	**4.** Solve. $x^2 = 24$

Vocabulary Answers: 1. square root property 2. complete the square 3. standard form 4. quadratic equation

Example	Student Practice
5. Solve and check. $3x^2+2=77$ $3x^2+2=77$ $3x^2=75$ $x^2=25$ $x=\pm\sqrt{25}$ $x=\pm 5$ The roots are 5 and -5. The check is left to the student.	**6.** Solve and check. $6x^2-3=93$
7. Solve and check. $4x^2=-16$ $4x^2=-16$ $x^2=-4$ $x=\pm\sqrt{-4}$ $x=\pm 2i$ The roots are $2i$ and $-2i$. Check. $4(2i)^2\stackrel{?}{=}-16 \quad 4(-2i)^2\stackrel{?}{=}-16$ $4(-4)\stackrel{?}{=}-16 \quad 4(-4)\stackrel{?}{=}-16$ $-16=-16 \quad -16=-16$	**8.** Solve and check. $2x^2=-50$
9. Solve. $(4x-1)^2=5$ $(4x-1)^2=5$ $4x-1=\pm\sqrt{5}$ $4x=1\pm\sqrt{5}$ $x=\dfrac{1\pm\sqrt{5}}{4}$ The roots are $\dfrac{1+\sqrt{5}}{4}$ and $\dfrac{1-\sqrt{5}}{4}$.	**10.** Solve. $(3x+8)^2=6$

Example	Student Practice
11. Solve by completing the square and check. $x^2+6x+1=0$	**12.** Solve by completing the square and check. $x^2+4x+1=0$

First rewrite the equation in the form $ax^2+bx=c$. Thus, we obtain $x^2+6x=-1$.

Next verify that the coefficient of the quadratic term (x^2) is 1. We want to add a constant term to x^2+6x so that we get a perfect square trinomial. We do this by taking half the coefficient of x and squaring it.

$\left(\frac{6}{2}\right)^2=3^2=9$

Adding 9 to x^2+6x gives the perfect square trinomial x^2+6x+9, which we factor as $(x+3)^2$. Add 9 to both the left and right side of our equation. We now have $x^2+6x+9=-1+9$.

Factor the left side then use the square root property.

$(x+3)^2=8$

$x+3=\pm\sqrt{8}$

$x+3=\pm 2\sqrt{2}$

Next we solve for x by subtracting 3 from each side of the equation.

$x=-3\pm 2\sqrt{2}$

The roots are $-3+2\sqrt{2}$ and $-3-2\sqrt{2}$.

The check is left to the student. Be sure to check the solution in the original equation, and not the perfect square trinomial.

Example	Student Practice
13. Solve by completing the square. $3x^2-8x+1=0$	**14.** Solve by completing the square. $5x^2-4x-3=0$

$$3x^2-8x+1=0$$
$$3x^2-8x=-1$$

Divide each term by 3 so that the coefficient of the quadratic term is 1.

$$\frac{3x^2}{3}-\frac{8x}{3}=-\frac{1}{3}$$
$$x^2-\frac{8}{3}x+\frac{16}{9}=-\frac{1}{3}+\frac{16}{9}$$
$$\left(x-\frac{4}{3}\right)^2=\frac{13}{9}$$
$$x-\frac{4}{3}=\pm\sqrt{\frac{13}{9}}$$
$$x-\frac{4}{3}=\pm\frac{\sqrt{13}}{3}$$
$$x=\frac{4\pm\sqrt{13}}{3}$$

Extra Practice

1. Solve the equation by using the square root property. Express any complex numbers using i notation. $x^2+64=0$

2. Solve the equation by using the square root property. Express any complex numbers using i notation. $(3x+1)^2=15$

3. Solve the equation by completing the square. Express any complex numbers using i notation. $x^2+12x+35=0$

4. Solve the equation by completing the square. Express any complex numbers using i notation. $4x^2+3=x$

Concept Check

Explain how you would decide what to add to each side of the equation to complete the square for the following equation. $x^2+x=1$

Name: ______________________________ Date: ________________
Instructor: ___________________________ Topic: ________________

Module 16 Quadratic Equations
Topic 2 Solutions to Quadratic Equations – Quadratic Formula

Vocabulary
quadratic formula • standard form • discriminant • complex number

1. The expression b^2-4ac is called the ____________.

2. The ____________ says that for all equations $ax^2+bx+c=0$, $x=\dfrac{-b\pm\sqrt{b^2-4ac}}{2a}$.

Example	Student Practice
1. Solve by using the quadratic formula. $x^2+8x=-3$ The standard form is $x^2+8x+3=0$. We substitute $a=1$, $b=8$, and $c=3$. $x=\dfrac{-b\pm\sqrt{b^2-4ac}}{2a}$ $x=\dfrac{-8\pm\sqrt{8^2-4(1)(3)}}{2(1)}$ $x=\dfrac{-8\pm2\sqrt{13}}{2}$ $x=-4\pm\sqrt{13}$	**2.** Solve by using the quadratic formula. $x^2+4x=-7-8x$
3. Solve by using the quadratic formula. $2x^2-48=0$ The standard form is $2x^2-0x-48=0$. Therefore, $a=2$, $b=0$, and $c=-48$. $x=\dfrac{-0\pm\sqrt{(0)^2-4(2)(-48)}}{2(2)}$ $x=\dfrac{\pm8\sqrt{6}}{4}$ $x=\pm2\sqrt{6}$	**4.** Solve by using the quadratic formula. $4x^2-60=0$

Vocabulary Answers: 1. discriminant 2. quadratic formula

Example	Student Practice
5. A small company that manufactures canoes makes a daily profit p according to the equation $p=-100x^2+3400x-26{,}196$, where p is measured in dollars and x is the number of canoes made per day. Find the number of canoes that must be made each day to produce a zero profit for the company. Round your answer to the nearest whole number.	**6.** A company that manufactures guitars makes a daily profit p according to the equation $p=-100x^2+3024x-20{,}240$, where p is measured in dollars and x is the number of guitars made per day. Find the number of guitars that must be made each day to produce a zero profit for the company. Round your answer to the nearest whole number.

Since $p=0$, the equation is $0=-100x^2+3400x-26{,}196$. Thus, $a=-100$, $b=3400$, and $c=-26{,}196$.

$$x=\frac{-3400\pm\sqrt{3400^2-4(-100)(-26{,}196)}}{2(-100)}$$

$$x=\frac{-3400\pm\sqrt{1{,}081{,}600}}{-200}$$

$$x=\frac{-3400\pm1040}{-200}$$

We now obtain two answers.

$$x=\frac{-3400+1040}{-200}$$

$$x=\frac{-2360}{-200}=11.8\approx12$$

$$x=\frac{-3400-1040}{-200}$$

$$x=\frac{-4440}{-200}=22.2\approx22$$

A zero profit is obtained when approximately 12 canoes are produced or when approximately 22 canoes are produced.

Example	Student Practice
7. Solve by using the quadratic formula. $\frac{2x}{x+2}=1-\frac{3}{x+4}$ The LCD is $(x+2)(x+4)$. $\frac{2x}{x+2}=1-\frac{3}{x+4}$ $2x(x+4)=(x+2)(x+4)-3(x+2)$ $x^2+5x-2=0$ $x=\frac{-5\pm\sqrt{5^2-4(1)(-2)}}{2(1)}$ $x=\frac{-5\pm\sqrt{33}}{2}$	**8.** Solve by using the quadratic formula. $\frac{6}{x}+\frac{x}{x-3}=-\frac{4}{5}$
9. Solve and simplify your answer. $8x^2-4x+1=0$ $x=\frac{-(-4)\pm\sqrt{(-4)^2-4(8)(1)}}{2(8)}$ $x=\frac{4\pm 4i}{16}$ $x=\frac{1\pm i}{4}$	**10.** Solve by using the quadratic formula. $6x^2+4x+3=0$
11. What type of solutions does the equation $2x^2-9x-35=0$ have? Do not solve the equation. $a=2,\ b=-9,$ and $c=-35.$ $b^2-4ac=(-9^2)-4(2)(-35)=361$ Since the discriminant is positive, the equation has two real roots. 361 is a perfect square. The equation has two different rational solutions.	**12.** Use the discriminant to find what type of solutions the equation $2x^2-5x+43=0$ has. Do not solve the equation.

Example	Student Practice
13. Find a quadratic equation whose roots are 5 and −2. $x=5 \quad x=-2$ $x-5=0 \quad x+2=0$ $(x-5)(x+2)=0$ $x^2-3x-10=0$	**14.** Find a quadratic equation whose roots are 3 and 4.
15. Find a quadratic equation those solutions are $3i$ and $-3i$. $x-3i=0$ and $x+3i=0$ $(x-3i)(x+3i)=0$ $x^2+3ix-3ix-9i^2=0$ $x^2+9=0$	**16.** Find a quadratic equation whose solutions are $3i\sqrt{5}$ and $-3i\sqrt{5}$

Extra Practice

1. Solve by the quadratic formula. Simplify your answers. $3x^2+12x+7=0$

2. Simplify, then solve by the quadratic formula. Simplify your answers. Use i notation for nonreal complex numbers.
$\frac{1}{x}-\frac{2}{x+2}=\frac{1}{5}$

3. Use the discriminant to find what type of solutions the following equation has. Do not solve the equation.
$x^2-3(2x-3)=0$

4. Write a quadratic equation having the given solutions -3 and $-\frac{1}{2}$.

Concept Check

Explain how you would determine if $2x^2-6x+3=3$ has two rational, two irrational, one rational, or two nonreal complex solutions.

Name: ______________________________ Date: ________________

Instructor: ____________________________ Topic: ________________

Module 16 Quadratic Equations

Topic 3 Solving Equations That Are Quadratic in Form

Vocabulary

quadratic in form • linear term • standard form • substitution

1. Before making a substitution and then factoring, we first must put the equation into ____________.

2. An equation is ____________ if we can substitute a linear term for the variable raised to the lowest power and get an equation of the form $ay^2 + by + c = 0$.

Example	Student Practice
1. Solve. $x^4 - 13x^2 + 36 = 0$ Let $y = x^2$. Then $y^2 = x^4$. Substitute these into the equation. $y^2 - 13y + 36 = 0$ Factor the equation and then solve for x. $(y-4)(y-9) = 0$ $y - 4 = 0$ or $y - 9 = 0$ $y = 4$ $\quad y = 9$ Replace y with x^2. $x^2 = 4$ or $x^2 = 9$ $x = \pm\sqrt{4}$ $\quad x = \pm\sqrt{9}$ $x = \pm 2$ $\quad x = \pm 3$ Thus, there are four solutions to the original equation: $x = +2$, $x = -2$, $x = +3$, and $x = -3$. Check these values to verify that they are solutions. The check is left to the student.	**2.** Solve. $x^4 + 9x^2 - 400 = 0$

Vocabulary Answers: 1. standard form 2. quadratic in form

Example	Student Practice
3. Solve for all real roots. $2x^6 - x^3 - 6 = 0$ Let $y = x^3$. Then $y^2 = x^6$. Thus, we have the following: $2y^2 - y - 6 = 0$ $(2y+3)(y-2) = 0$ $2y+3=0$ or $y-2=0$ $y = -\frac{3}{2}$ $y = 2$ $x^3 = -\frac{3}{2}$ $x^3 = 2$ $x = \sqrt[3]{-\frac{3}{2}}$ $x = \sqrt[3]{2}$ $x = \frac{\sqrt[3]{-12}}{2}$ The check is left to the student.	**4.** Solve for all real roots. $3x^6 - 83x^3 + 54 = 0$
5. Solve and check your solutions. $x^{2/3} - 3x^{1/3} + 2 = 0$ Let $y = x^{1/3}$. Then $y^2 = x^{2/3}$. $y^2 - 3y + 2 = 0$ $(y-2)(y-1) = 0$ $y-2=0$ or $y-1=0$ $y = 2$ $y = 1$ $x^{1/3} = 2$ $x^{1/3} = 1$ $\left(x^{1/3}\right)^3 = (2)^3$ $\left(x^{1/3}\right)^3 = (1)^3$ $x = 8$ $x = 1$ The check is left to the student.	**6.** Solve and check your solutions. $x^{2/3} - 65x^{1/3} + 64 = 0$

Example	Student Practice
7. Solve and check your solutions. $2x^{1/2}=5x^{1/4}+12$	**8.** Solve and check your solutions. $2x^{1/2}=11x^{1/4}+15$

This equation in standard form is $2x^{1/2}-5x^{1/4}-12=0$. Let $y=x^{1/4}$, making $y^2=x^{1/2}$. Then solve.

$$2y^2-5y-12=0$$
$$(2y+3)(y-4)=0$$
$$2y+3=0 \quad \text{or} \quad y-4=0$$
$$y=-\frac{3}{2} \qquad y=4$$
$$x^{1/4}=-\frac{3}{2} \qquad x^{1/4}=4$$
$$\left(x^{1/4}\right)^4=\left(-\frac{3}{2}\right)^4 \qquad \left(x^{1/4}\right)^4=(4)^4$$
$$x=\frac{81}{16} \qquad x=256$$

Check by substituting the solutions into the original equation.

$x=\frac{81}{16}$

$$2\left(\frac{81}{16}\right)^{1/2}-5\left(\frac{81}{16}\right)^{1/4}-12\stackrel{?}{=}0$$
$$\frac{9}{2}-\frac{15}{2}-12\stackrel{?}{=}0$$
$$-15\neq 0$$

$x=256$

$$2(256)^{1/2}-5(256)^{1/4}-12\stackrel{?}{=}0$$
$$32-20-12\stackrel{?}{=}0$$
$$0=0$$

$\frac{81}{16}$ is extraneous and not a valid solution. The only valid solution is 256.

Example	Student Practice
9. Solve and check your solutions. $x^{-2}=5x^{-1}+14$	**10.** Solve and check your solutions. $x^{-2}+x^{-1}=12$

Write the equation in standard form. Let $y=x^{-1}$, making $y^2=x^{-2}$. Then solve.

$$x^{-2}=5x^{-1}+14$$
$$x^{-2}-5x^{-1}-14=0$$
$$y^2-5y-14=0$$
$$(y-7)(y+2)=0$$
$$y-7=0 \quad \text{or} \quad y+2=0$$
$$y=7 \qquad y=-2$$
$$x^{-1}=7 \qquad x^{-1}=-2$$
$$x=\frac{1}{7} \qquad x=-\frac{1}{2}$$

The check is left to the student.

Extra Practice

1. Solve. Express any nonreal complex numbers with *i* notation.
   $x^4+10x^2+21=0$

2. Solve for all real roots. $x^6+27x^3=0$

3. Solve for all real roots.
   $6x^{2/5}+18x^{1/5}+12=0$

4. Solve for all real roots. $3x^{-2}+3x^{-1}=168$

Concept Check

Explain how you would solve the following. $x^8-6x^4=0$

Name: ______________________________ Date: ______________

Instructor: ______________________________ Topic: ______________

Module 16 Quadratic Equations
Topic 4 Formulas and Applications

Vocabulary

hypotenuse • leg • Pythagorean Theorem • area • surface area

1. The ___________ states that if c is the length of the longest side of a right triangle and a and b are the lengths of the other two sides, then $a^2 + b^2 = c^2$.

2. The longest side of a right triangle is called the ___________.

Example	Student Practice
1. The surface area of a sphere is given by $A = 4\pi r^2$. Solve this equation for r. (You do not need to rationalize the denominator.) $A = 4\pi r^2$ $\frac{A}{4\pi} = r^2$ $\pm\sqrt{\frac{A}{4\pi}} = r$ $\pm\frac{1}{2}\sqrt{\frac{A}{\pi}} = r$ Since the radius of a sphere must be a positive value, we use only the principal root. $r = \frac{1}{2}\sqrt{\frac{A}{\pi}}$	**2.** The volume of a sphere is given by $V = \frac{4}{3}\pi r^3$. Solve this equation for r. (You do not need to rationalize the denominator.)
3. Solve for y. $y^2 - 2yz - 15z^2 = 0$ Factor, then set each factor equal to 0. $y^2 - 2yz - 15z^2 = 0$ $(y+3z)(y-5z) = 0$ $y + 3z = 0$ or $y - 5z = 0$ $y = -3z$ $\quad$ $y = 5z$	**4.** Solve for x. $x^2 - 3xy - 10y^2 = 0$

Vocabulary Answers: 1. Pythagorean Theorem 2. hypotenuse

Example	Student Practice
5. Solve for x. $2x^2+3wx-4z=0$ We use the quadratic formula where the variable is considered to be x, and w and z are considered constants. Thus, $a=2$, $b=3w$, and $c=-4z$. $x=\frac{-b\pm\sqrt{b^2-4ac}}{2a}$ $x=\frac{-3w\pm\sqrt{(3w)^2-4(2)(-4z)}}{2(2)}$ $x=\frac{-3w\pm\sqrt{9w^2+32z}}{4}$	**6.** Solve for x. $4x^2-6yx+9z=0$
7. Complete parts **(a)** and **(b)**. **(a)** Solve the Pythagorean Theorem $a^2+b^2=c^2$ for a. $a^2+b^2=c^2$ $a^2=c^2-b^2$ $a=\pm\sqrt{c^2-b^2}$ a, b, and c must be positive numbers because they represent lengths, only use the positive root, $a=\sqrt{c^2-b^2}$. **(b)** Find the value of a if $c=13$ and $b=5$. $a=\sqrt{c^2-b^2}$ $a=\sqrt{(13)^2-(5)^2}$ $a=\sqrt{144}$ $a=12$	**8.** Complete parts **(a)** and **(b)**. **(a)** Solve the Pythagorean Theorem $a^2+b^2=c^2$ for c. **(b)** Find the value of c if $a=15$ and $b=20$.

Example	Student Practice
9. The perimeter of a triangular piece of land is 12 miles. One leg of the triangle is 1 mile longer than the other leg. Find the length of each boundary of the land if the triangle is a right triangle.	**10.** The perimeter of a triangular piece of land is 60 miles. One leg of the triangle is 4 miles longer than twice the other leg. Find the length of each boundary of the land if the triangle is a right triangle.

We are given that the perimeter is 12 miles, so $x+(x+1)+c=12$.

Thus, $c=-2x+11$.

By the Pythagorean Theorem, $x^2+(x+1)^2=(-2x+11)^2$.

$$x^2+(x+1)^2=(-2x+11)^2$$
$$x^2+x^2+2x+1=4x^2-44x+121$$
$$0=2x^2-46x+120$$
$$0=x^2-23x+60$$

Use the quadratic formula to solve.

$$x=\frac{-(23)\pm\sqrt{(-23)^2-4(1)(60)}}{2(1)}$$
$$x=\frac{23\pm\sqrt{289}}{2}$$
$$x=\frac{23\pm17}{2}$$
$$x=\frac{40}{2}=20 \text{ or } x=\frac{6}{2}=3.$$

20 is too large, so the only answer that makes sense is $x=3$.

Thus, the sides of the triangle are $x=3$, $x+1=4$, and $-2x+11=-2(3)+11=5$.

The longest boundary of this triangular piece of land is 5 miles. The other two boundaries are 4 miles and 3 miles.

Example	Student Practice
11. A triangular sign marks the edge of the rocks in Rockport Harbor. The sign has an area of 35 square meters. Find the base and altitude of this triangular sign if the base is 3 meters shorter than the altitude. The area of the triangle is given by $A=\frac{1}{2}ab$. Let $x=$ the length in meters of the altitude. Then $x-3=$ the length in meters of the base. Solve for x. $35=\frac{1}{2}x(x-3)$ $0=x^2-3x-70$ $0=(x-10)(x+7)$ $x=10$ or $x=-7$ We disregard -7. Thus altitude $=x=10$ meters and base $=x-3=7$ meters. The check is left to the student.	**12.** The length of a rectangle is 2 yards shorter than twice the width. The area of the rectangle is 84 square yards. Find the dimensions of the rectangle.

Extra Practice

1. Solve $s=\frac{1}{2}gt^2$ for t. Assume that all other variables are nonzero.

2. Solve $(a+2)x^2-7x+3y=0$ for x. Assume that all other variables are nonzero.

3. $c=10$ and $b=3a$; use the Pythagorean Theorem to find a and b.

4. Andrew drove at a constant speed on a dirt road for 75 miles. He then traveled 20 mph faster on a paved road for 180 miles. If he drove for 7 hours, find the car's speed for each part of the trip.

Concept Check

In a right triangle the hypotenuse measures 12 meters. One of the legs of the triangle is three times as long as the other. Explain how you would find the length of each leg.

MATH COACH

Mastering the skills you need to do well on the test.

Watch the MATH COACH videos in MyMathLab® or on YouTube™ while you work the problems below. These helpful hints will help you avoid making common errors on test problems.

Solving Quadratic Equations Involving Fractions—Problem 6

Solve the quadratic equation. $\dfrac{2x}{2x+1}-\dfrac{6}{4x^2-1}=\dfrac{x+1}{2x-1}$

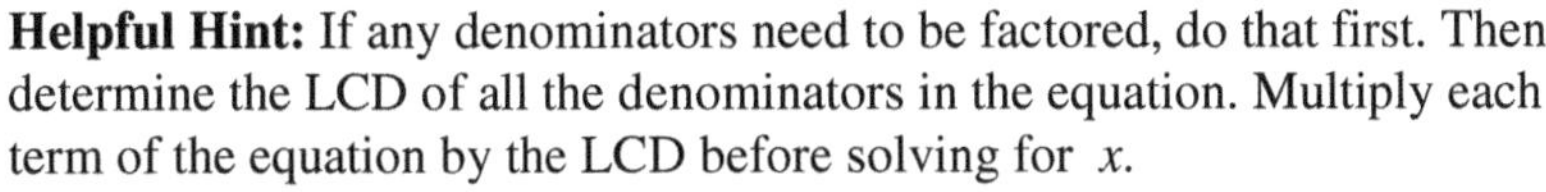

Helpful Hint: If any denominators need to be factored, do that first. Then determine the LCD of all the denominators in the equation. Multiply each term of the equation by the LCD before solving for x.

Did you factor $4x^2-1$ as $(2x+1)(2x-1)$?

Yes ____ No ____

Did you identify the LCD to be $(2x+1)(2x-1)$?

Yes ____ No ____

If you answered No to these questions, review how to factor the difference of two squares and how to find the LCD of polynomial denominators.

Did you multiply the LCD by each term of the equation and remove parentheses to obtain $2x^2-5x-7=0$?

Yes ____ No ____

Did you then use the quadratic formula and substitute for a, b, and c to get $x=\dfrac{-(-5)\pm\sqrt{(-5)^2-4(2)(-7)}}{2(2)}$?

Yes ____ No ____

If you answered No to these questions, remember that your final equation should be in the form $ax^2+bx+c=0$. Then use $a=2$, $b=-5$, and $c=-7$ in the quadratic formula and simplify your result.

If you answered Problem 6 incorrectly, go back and rework the problem using these suggestions.

Solving Equations That Are Quadratic in Form—Problem 9

Solve for any valid real roots. $x^4-11x^2+18=0$

Helpful Hint: Let $y=x^2$ and then let $y^2=x^4$. Write the new quadratic equation after these replacements have been made.

After making the necessary replacements, did you obtain the equation $y^2-11y+18=0$? Yes ____ No ____

Did you solve the quadratic equation using any method to result in $y=9$ and $y=2$? Yes ____ No ____

If you answered No to these questions, stop and complete these steps again.

If $y=9$ and $y=2$, can you conclude that $x^2=9$ and $x^2=2$? Yes ____ No ____

If you take the square root of each side of each equation, can you obtain $x=\pm 3$ and $x=\pm\sqrt{2}$? Yes ____ No ____

If you answered No to these questions, remember that when you take the square root of each side of the equation there are two sign possibilities. Note that your final solution should consist of four values for x.

Now go back and rework the problem using these suggestions.

Solving a Quadratic Equation with Several Variables—Problem 13

Solve for the variable specified. $5y^2+2by+6w=0$; for y

Helpful Hint: Think of the equation being written as $ay^2+by+c=0$. The quantities for a, b, or c may contain variables. Use the quadratic formula to solve for y.

Did you determine that $a=5$, $b=2b$, and $c=6w$?
Yes ____ No ____

Did you substitute these values into the quadratic formula and simplify to obtain $y=\dfrac{-2b\pm\sqrt{4b^2-120w}}{10}$?

Yes ____ No ____

If you answered No to these questions, review the Helpful Hint again to make sure that you find the correct values for a, b, and c. Carefully substitute these values into the quadratic formula and simplify.

Were you able to simplify the radical expression by factoring out a 4 to obtain $\sqrt{4\left(b^2-30w\right)}$? Yes ____ No ____

Did you simplify this expression further to get $2\sqrt{b^2-30w}$? Yes ____ No ____

If you answered No to these questions, remember to always simplify radicals whenever possible. Make sure to write your solution as a simplified rational expression.

Now go back and rework the problem using these suggestions.

Name: ______________________________ Date: ________________
Instructor: ____________________________ Topic: ________________

Module 17 The Conic Sections
Topic 1 The Distance Formula and the Circle

Vocabulary

conic section • distance formula • circle • radius • center

1. A ____________ is defined as the set of all points in a plane that are a fixed distance from a point in that plane.

2. The ____________ states that the distance between two points (x_1, y_1) and (x_2, y_2) is $d = \sqrt{(x_2 - x_1)^2 + (y_2 - y_1)^2}$.

3. A ____________ is a shape formed by slicing a cone with a plane.

4. The point from which the set of points in a circle are a fixed distance from is called the ____________.

Example	Student Practice
1. Find the distance between $(3,-4)$ and $(-2,-5)$. To use the distance formula, we arbitrarily let $(x_1, y_1) = (3,-4)$ and $(x_2, y_2) = (-2,-5)$. Substitute the appropriate values into the formula. $d = \sqrt{(x_2 - x_1)^2 + (y_2 - y_1)^2}$ $= \sqrt{(-2-3)^2 + [-5-(-4)]^2}$ $= \sqrt{(-5)^2 + (-5+4)^2}$ $= \sqrt{(-5)^2 + (-1)^2}$ $= \sqrt{25+1}$ $= \sqrt{26}$	**2.** Find the distance between $(5,10)$ and $(3,5)$.

Vocabulary Answers: 1. circle 2. distance formula 3. conic section 4. center

Example	Student Practice
3. Find the center and radius of the circle $(x-2)^2+(y-3)^2=25$. Then sketch its graph. From the equation of a circle, $(x-h)^2+(y-k)^2=r^2$, we see that $(h,k)=(2,3)$. Thus, the center of the circle is at $(2,3)$. Since $r^2=25$, the radius of the circle is $r=5$. To graph the circle, start by graphing the center. Then, use the radius to graph the circle. The graph is shown below. 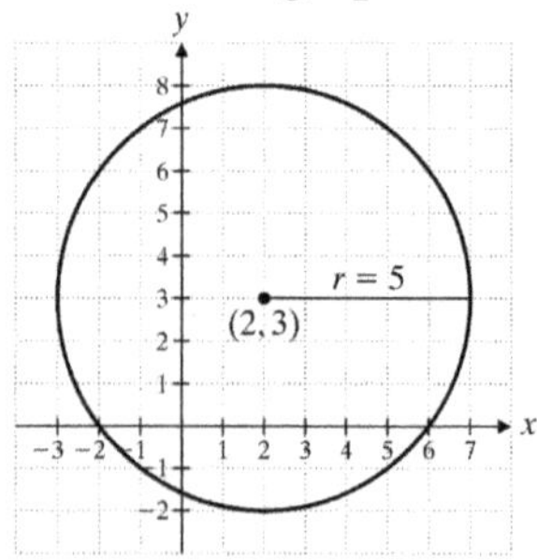	**4.** Find the center and radius of the circle $(x-1)^2+(y-4)^2=16$. Then sketch its graph. 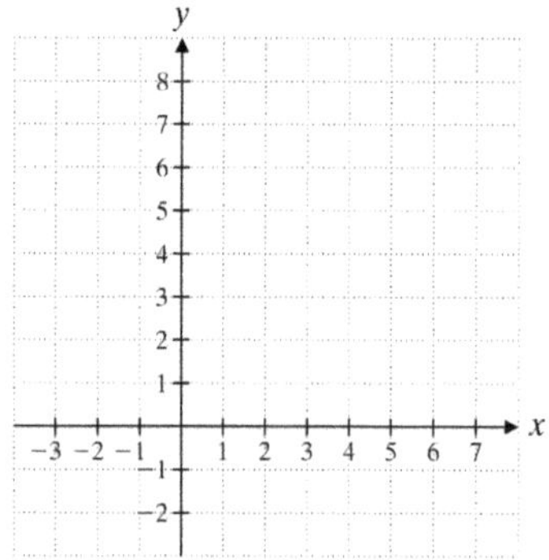
5. Write the equation of the circle with center $(-1,3)$ and radius $\sqrt{5}$. Put your answer in standard form. We are given that $(h,k)=(-1,3)$ and $r=\sqrt{5}$. Substitute the values into the standard form, $(x-h)^2+(y-k)^2=r^2$. $(x-h)^2+(y-k)^2=r^2$ $[x-(-1)]^2+(y-3)^2=(\sqrt{5})^2$ $(x+1)^2+(y-3)^2=5$ Be careful of the signs. It is easy to make a sign error in these steps.	**6.** Write the equation of the circle with center $(14,-5)$ and radius $\sqrt{7}$. Put your answer in standard form.

Example	Student Practice

7. Write the equation of the circle $x^2+2x+y^2+6y+6=0$ in standard form. Find the radius and center of the circle and sketch its graph.

If we multiply out the terms in the standard form of the equation of a circle, we have the following.

$$(x-h)^2+(y-k)^2=r^2$$
$$(x^2-2hx+h^2)+(y^2-2ky+k^2)=r^2$$

Comparing this with the given equation, $(x^2+2x)+(y^2+6y)=-6$, suggests we can complete the squares to put the equations in standard form.

$$x^2+2x+\_\_\_+y^2+6y+\_\_\_=-6$$
$$x^2+2x+1+y^2+6y+9=-6+1+9$$
$$x^2+2x+1+y^2+6y+9=4$$
$$(x+1)^2+(y+3)^2=4$$

Thus, the center is at $(-1,-3)$, and the radius is 2. The sketch of the circle is shown below.

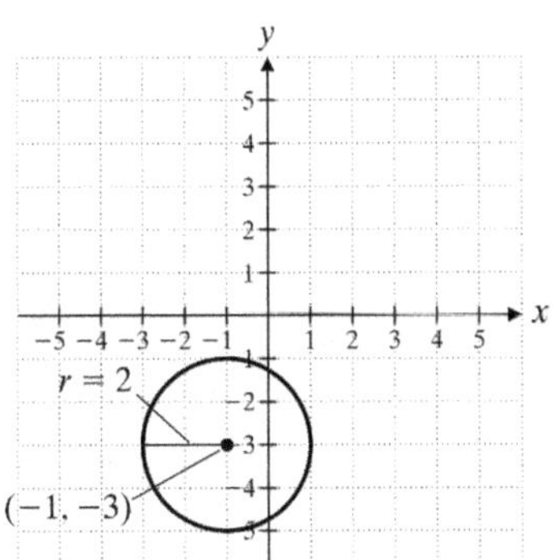

8. Write the equation of the circle $x^2+4x-2+y^2-2y=2$ in standard form. Find the radius and center of the circle and sketch its graph.

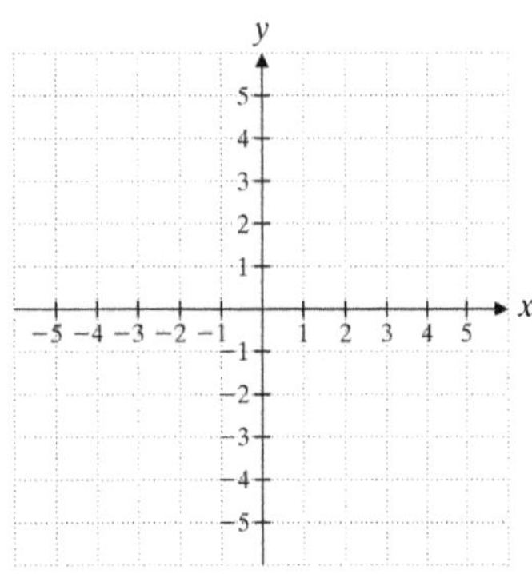

Extra Practice

1. Find the distance between $(-0.5, 8.2)$ and $(3.5, 6.2)$.

2. Write the equation of the circle with center $\left(0, \frac{6}{5}\right)$ and radius $\sqrt{13}$. Put your answer in standard form.

3. Find the center and radius of the circle $x^2 + y^2 = 36$. Then sketch its graph.

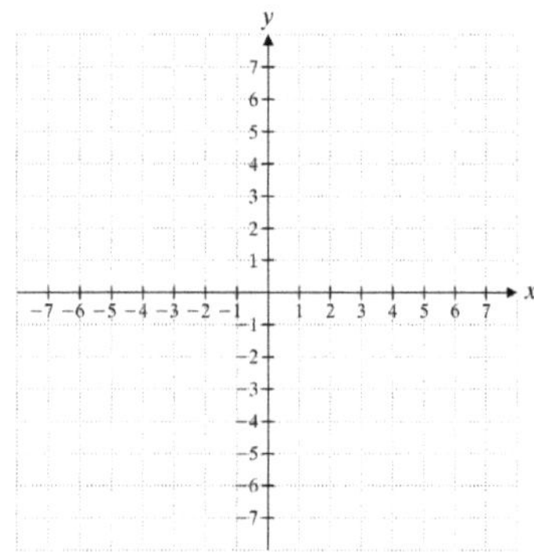

4. Write the equation of the circle $x^2 + 10x + y^2 - 6y + 29 = 0$ in standard form. Find the radius and center of the circle.

Concept Check

Explain how you would find the values of the unknown coordinate x if the distance between $(-6, 8)$ and $(x, 12)$ is 4.

Name: ______________________________ Date: ______________

Instructor: ___________________________ Topic: ______________

Module 17 The Conic Sections
Topic 2 The Parabola

Vocabulary

parabola • directrix • focus • axis of symmetry • vertex

1. The point at which the parabola crosses the axis of symmetry is the __________.

2. A(n) __________ is defined as the set of points in a plane that are a fixed distance from some fixed line and some fixed point that is not on the line.

3. The fixed point that a parabola is a fixed distance from is called a(n) __________.

Example	Student Practice
1. Graph $y=(x-2)^2$. Identify the vertex and the axis of symmetry.	**2.** Graph $y=(x+1)^2$. Identify the vertex and the axis of symmetry.

Make a table of values. Begin with $x=2$ in the middle of the table of values because $(2-2)^2=0$. That is, when $x=2$, $y=0$. Then fill in the x- and y-values above and below $x=2$.

x	4	3	2	1	0
y	4	1	0	1	4

Plot the points and draw the graph.

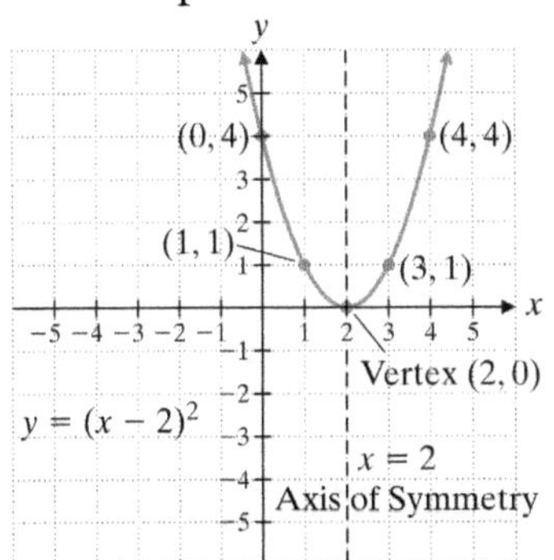

The vertex is $(2,0)$, and the axis of symmetry is the line $x=2$.

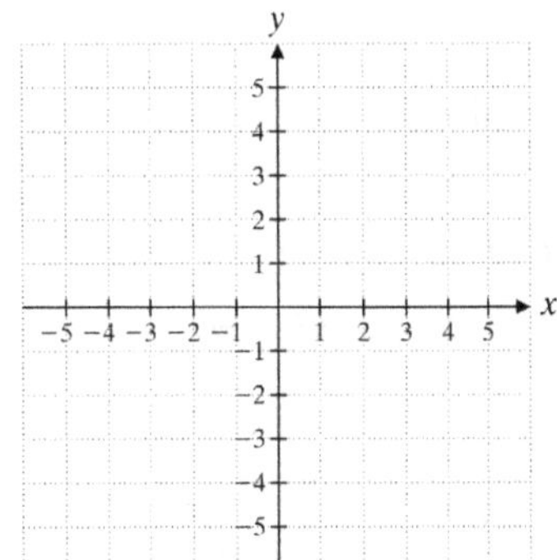

Vocabulary Answers: 1. vertex 2. parabola 3. focus

Example	Student Practice
3. Graph $y=-\frac{1}{2}(x+3)^2-1$.	**4.** Graph $y=-\frac{1}{4}(x+2)^2-4$.

Rewrite the equation in standard form.

$y=-\frac{1}{2}[x-(-3)]^2+(-1)$

Thus, $a=-\frac{1}{2}$, $h=-3$, and $k=-1$, so it is a vertical parabola. The parabola opens downward since $a<0$. The vertex is $(-3,-1)$ and the axis of symmetry is the line $x=-3$. The y-intercept is $(0,-5.5)$. Plot a few points on either side of the axis of symmetry.

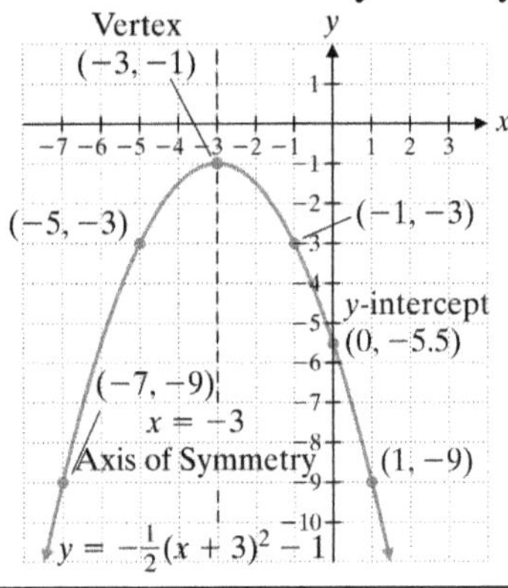

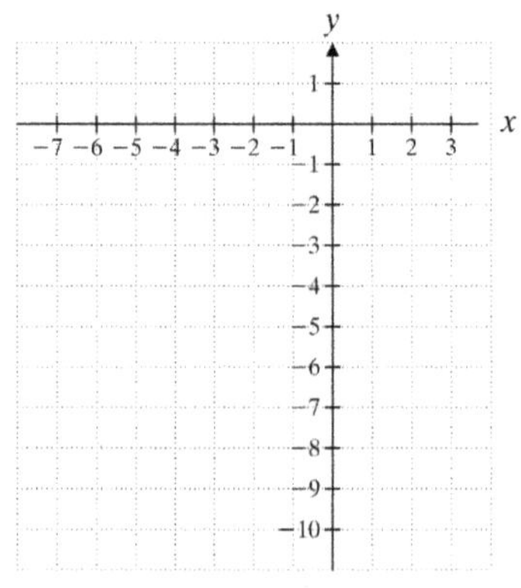

5. Graph $x=-2y^2$.	**6.** Graph $x=-3y^2$.

This is a horizontal parabola because the y term is squared. Make a table of values, but choose values for y instead of x. The graph is shown below.

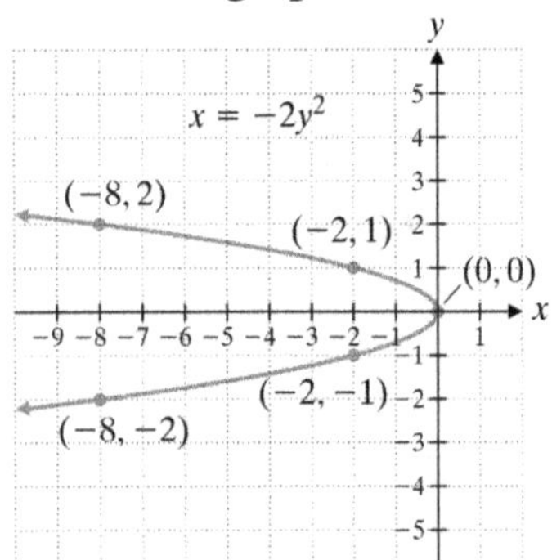

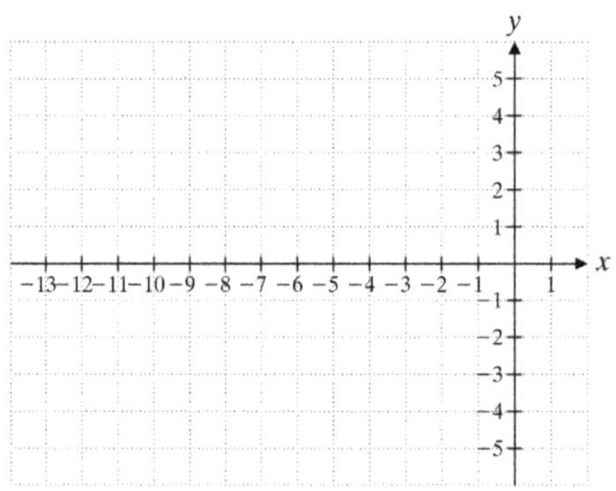

Example	Student Practice
7. Place the equation $x = y^2 + 4y + 1$ in standard form. Then graph it.	**8.** Place the equation $x = y^2 + 6y + 7$ in standard form. Then graph it.

$$x = y^2 + 4y + 1$$
$$= y^2 + 4y + \left(\frac{4}{2}\right)^2 - \left(\frac{4}{2}\right)^2 + 1$$
$$= (y+2)^2 - 3$$

Notice that $a = 1$, $k = -2$, and $h = -3$. Since a is positive, the parabola opens to the right. The vertex is $(-3, -2)$ and the axis of symmetry is $y = -2$. Let $y = 0$, to find the x-intercept, $(1, 0)$.

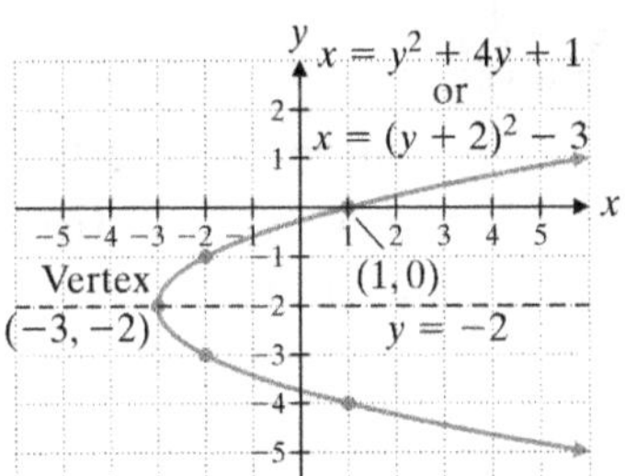

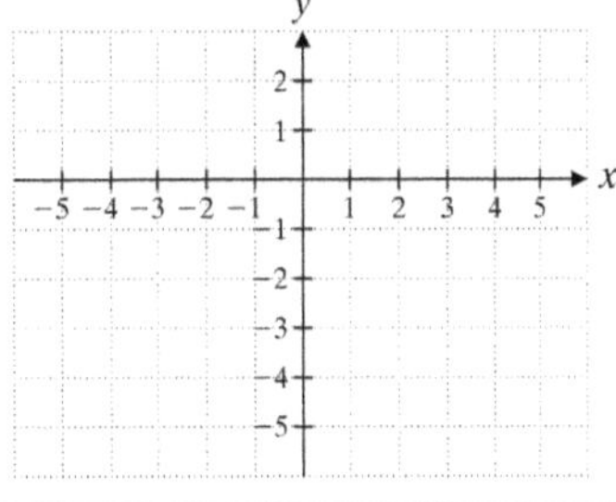

Example	Student Practice
9. Place the equation $y = 2x^2 - 4x - 1$ in standard form. Then graph it.	**10.** Place the equation $y = -2x^2 - 8x - 4$ in standard form. Then graph it.

$$y = 2x^2 - 4x - 1$$
$$= 2\left[x^2 - 2x + (1)^2\right] - 2(1)^2 - 1$$
$$= 2(x-1)^2 - 3$$

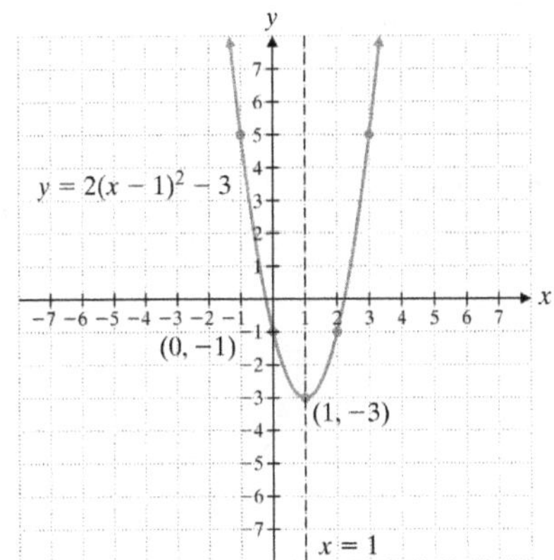

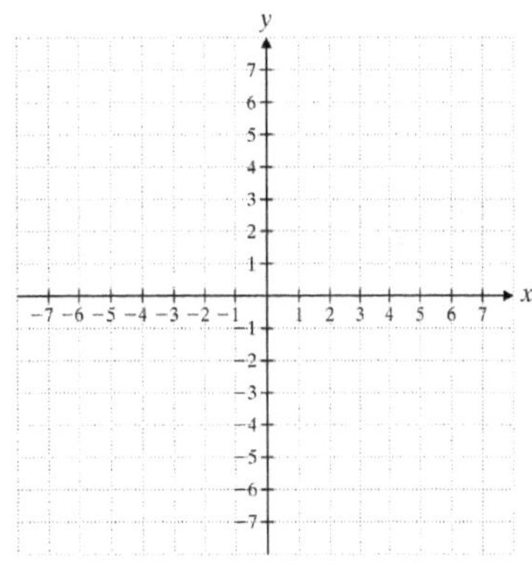

Extra Practice

1. Graph $y = 2(x+4)^2 - 3$ and label the vertex. Find the y-intercept.

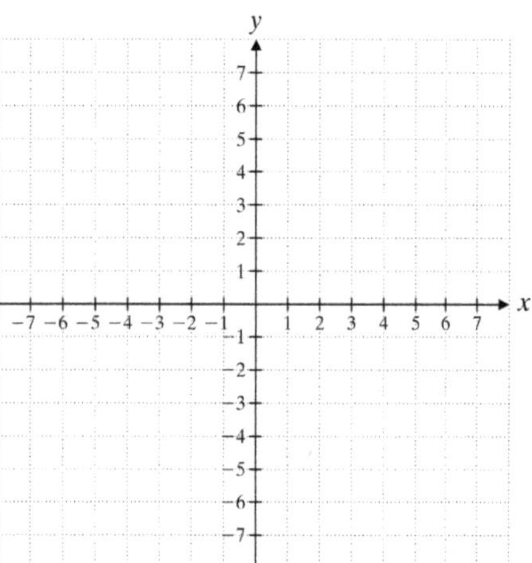

2. Graph $y = -2\left(x - \frac{3}{2}\right)^2 + 3$ and label the vertex. Find the y-intercept.

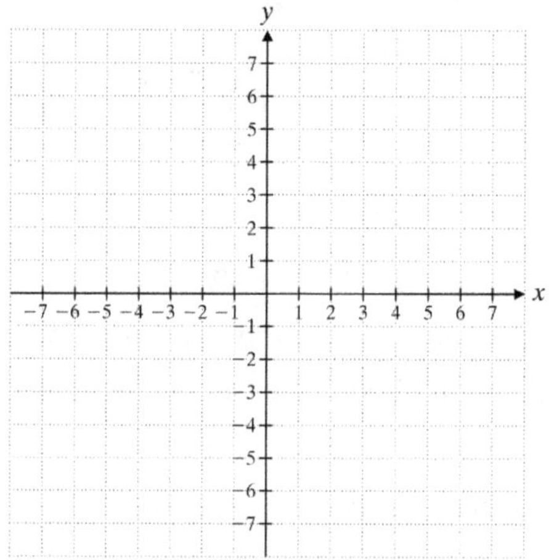

3. Graph $x = 3(y+1)^2 - 3$ and label the vertex. Find the x-intercept.

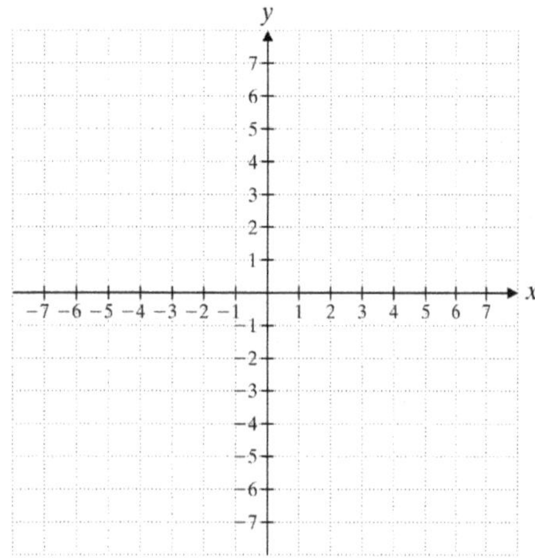

4. Place the equation $x = -3y^2 + 12y + 6$ in standard form. Determine (a) whether the parabola is horizontal or vertical, (b) the direction it opens, and (c) the vertex.

Concept Check

Explain how you can tell which way the parabola opens for these equations: $y = 2x^2$, $y = -2x^2$, $x = 2y^2$, and $x = -2y^2$.

Name: ______________________________ Date: ________________
Instructor: ____________________________ Topic: ________________

Module 17 The Conic Sections
Topic 3 The Ellipse

Vocabulary

ellipse • foci • vertices • center at the origin • center at (h,k)

1. The fixed points in an ellipse are called ____________.

2. We define a(n) ____________ as the set of points in a plane such that for each point in the set, the sum of its distances to two fixed points is constant.

Example	Student Practice
1. Graph $x^2+3y^2=12$. Label the intercepts.	**2.** Graph $9x^2+16y^2=144$. Label the intercepts.

Example 1. Rewrite the equation in standard form.

$$x^2+3y^2=12$$

$$\frac{x^2}{12}+\frac{3y^2}{12}=\frac{12}{12}$$

$$\frac{x^2}{12}+\frac{y^2}{4}=1$$

Thus, we have the following:

$a^2=12$ so $a=2\sqrt{3}$

$b^2=4$ so $b=2$

The x-intercepts are $(-2\sqrt{3},0)$ and $(2\sqrt{3},0)$, and the y-intercepts are $(0,2)$ and $(0,-2)$. We plot these points and draw the ellipse.

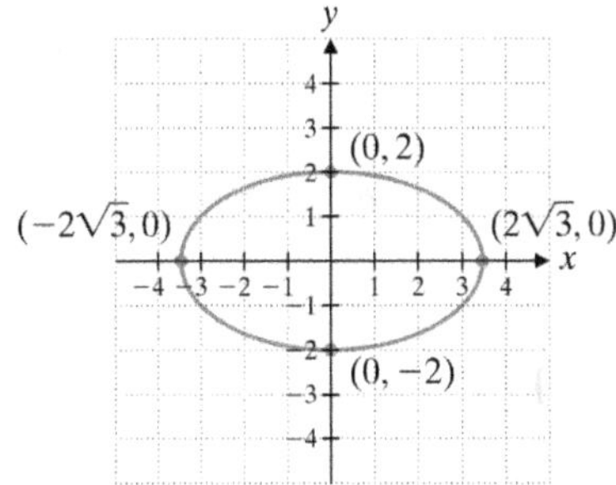

Student Practice 2.

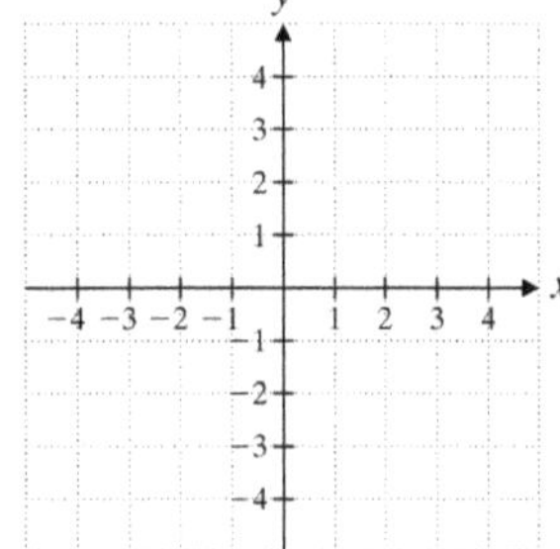

Vocabulary Answers: 1. foci 2. ellipse

Example	Student Practice
3. Graph $\frac{(x-5)^2}{9}+\frac{(y-6)^2}{4}=1.$	**4.** Graph $\frac{(x-2)^2}{16}+\frac{(y-3)^2}{4}=1.$

Notice that this ellipse has the form $\frac{(x-h)^2}{a^2}+\frac{(y-k)^2}{b^2}=1.$

The center is (h,k). Note that a and b are not the x-intercepts and y-intercepts now. You'll see that a is the horizontal distance from the center of the ellipse to a point on the ellipse. Similarly, b is the vertical distance. Hence, when the center of the ellipse is not at the origin, the ellipse may not cross either axis.

The center of the ellipse is $(5,6)$, $a=3$, and $b=2$. Therefore, we begin at $(5,6)$. We plot points 3 units to the left, 3 units to the right, 2 units up, and 2 units down from $(5,6)$. The points we plot are the vertices of the ellipse.

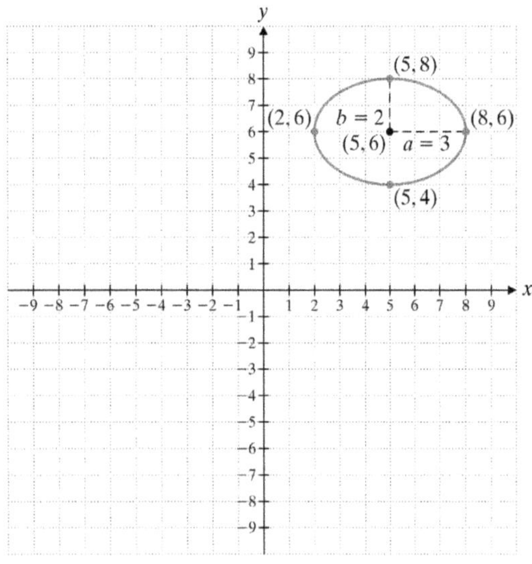

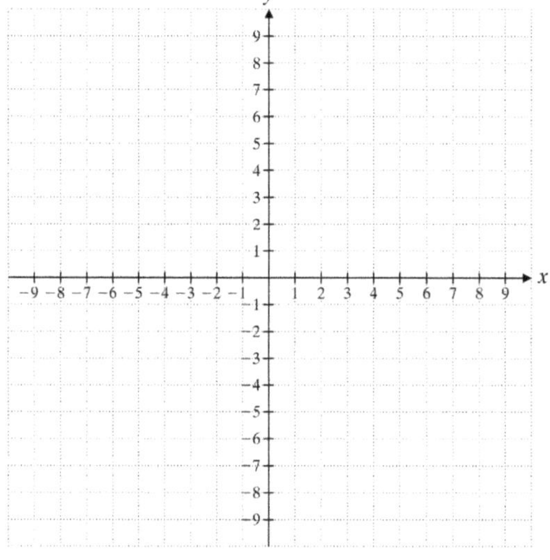

Extra Practice

1. Graph $\dfrac{x^2}{4}+\dfrac{y^2}{36}=1$. Label the intercepts.

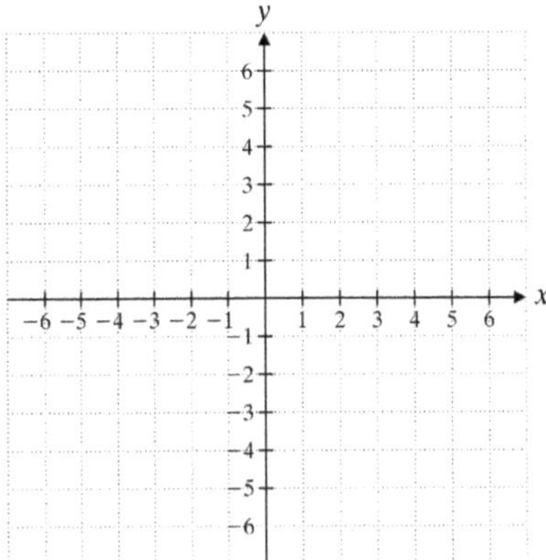

2. Graph $x^2+4y^2=16$. Label the intercepts.

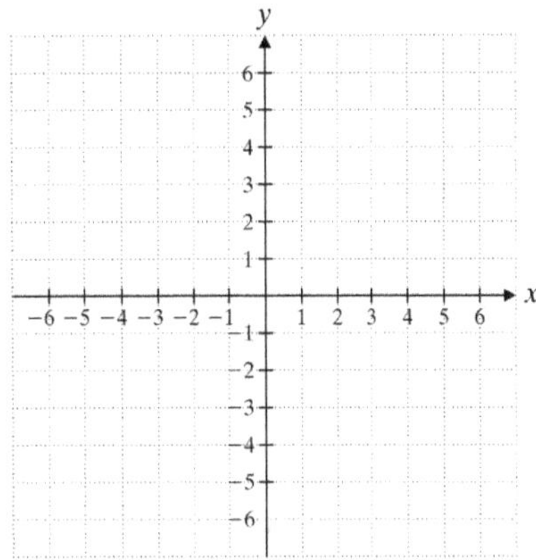

3. Graph $\dfrac{(x+2)^2}{16}+\dfrac{(y-3)^2}{9}=1$. Label the center.

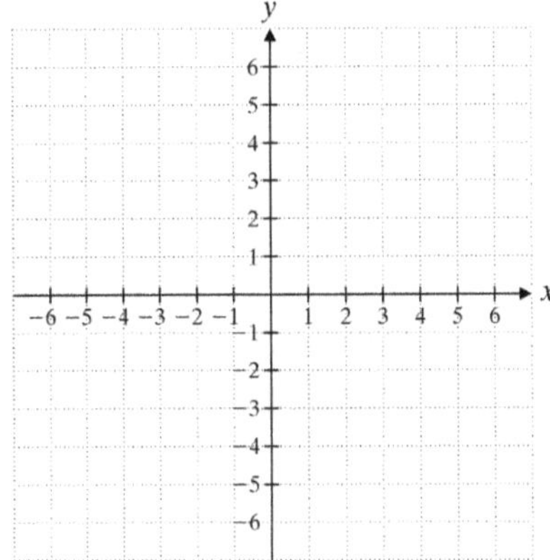

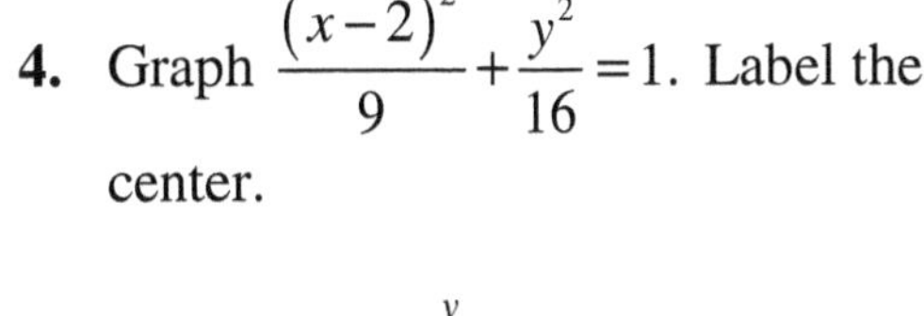

4. Graph $\dfrac{(x-2)^2}{9}+\dfrac{y^2}{16}=1$. Label the center.

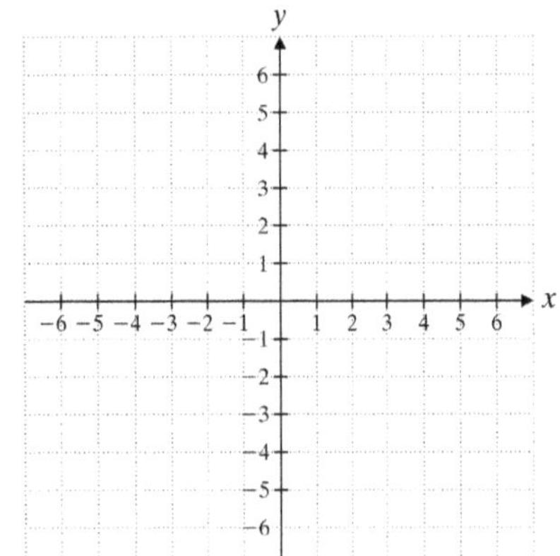

Concept Check

Explain how you would find the four vertices of the ellipse $12x^2+y^2-36=0$.

Name: ______________________________ Date: ________________
Instructor: ____________________________ Topic: ________________

Module 17 The Conic Sections
Topic 4 The Hyperbola

Vocabulary
hyperbola • foci • axis • vertices • asymptotes • fundamental rectangle

1. The points where the hyperbola intersects its axis are called the ___________.

2. We define a(n) ___________ as the set of points in a plane such that for each point in the set, the absolute value of the difference of its distances to two fixed points is constant.

Example	Student Practice
1. Graph $\frac{x^2}{25}-\frac{y^2}{16}=1$.	**2.** Graph $\frac{x^2}{4}-\frac{y^2}{9}=1$.

The equation has the form $\frac{x^2}{a^2}-\frac{y^2}{b^2}=1$, so it is a horizontal hyperbola. $a^2=25$, so $a=5$; $b^2=16$, so $b=4$. Since the hyperbola is horizontal, it has vertices $(a,0)$ and $(-a,0)$ or $(5,0)$ and $(-5,0)$. To draw the asymptotes, we construct a fundamental rectangle with corners at $(5,4)$, $(5,-4)$, $(-5,4)$, and $(-5,-4)$. We draw extended diagonals of the rectangle as the asymptotes. We construct each branch of the curve so that it passes through a vertex and gets closer to the asymptotes as it moves away from the origin.

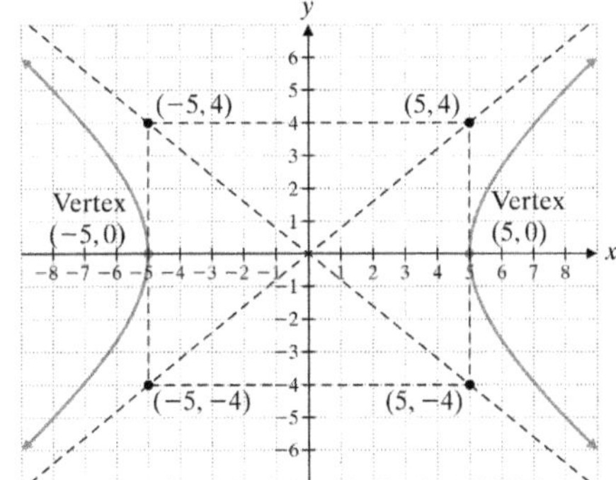

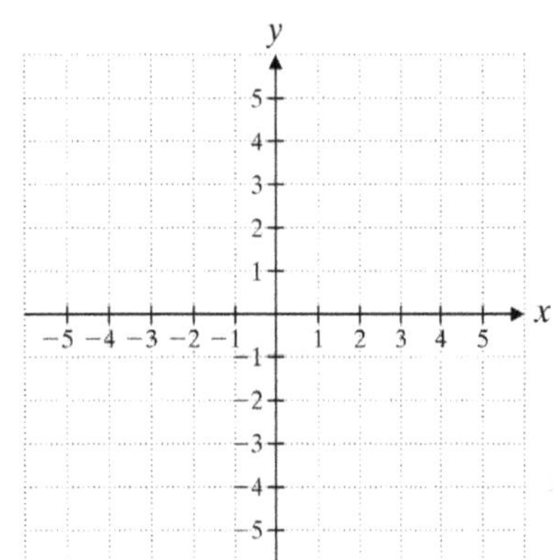

Vocabulary Answers: 1. vertices 2. hyperbola

Example	Student Practice
3. Graph $4y^2 - 7x^2 = 28$.	**4.** Graph $3y^2 - 2x^2 = 18$.

To find the vertices and asymptotes, we must rewrite the equation in standard form. Divide each term by 28.

$$4y^2 - 7x^2 = 28$$
$$\frac{4y^2}{28} - \frac{7x^2}{28} = \frac{28}{28}$$
$$\frac{y^2}{7} - \frac{x^2}{4} = 1$$

Thus, we have the standard form of a vertical hyperbola with center at the origin.

Here $b^2 = 7$, so $b = \sqrt{7}$; $a^2 = 4$, so $a = 2$.

The hyperbola has vertices at $\left(0, \sqrt{7}\right)$ and $\left(0, -\sqrt{7}\right)$. The fundamental rectangle has corners at $\left(2, \sqrt{7}\right)$, $\left(2, -\sqrt{7}\right)$, $\left(-2, \sqrt{7}\right)$, and $\left(-2, -\sqrt{7}\right)$.

To aid us in graphing, we measure the distance $\sqrt{7}$ as approximately 2.6.

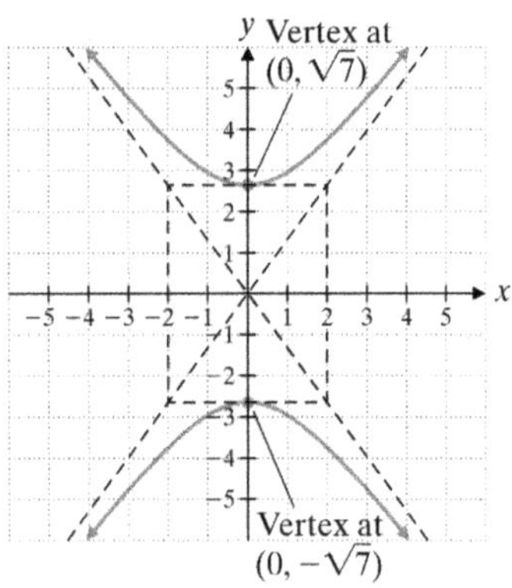

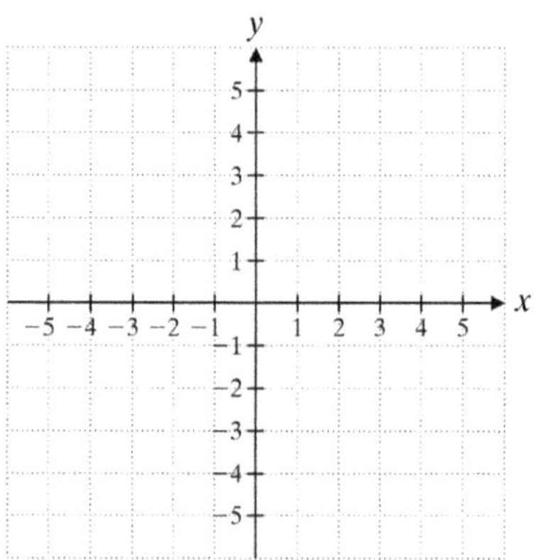

Example	Student Practice
5. Graph $\frac{(x-4)^2}{9}-\frac{(y-5)^2}{4}=1.$	**6.** Graph $\frac{(x-1)^2}{4}-\frac{(y-3)^2}{16}=1.$

The center is at $(4,5)$, and the hyperbola is horizontal. We have $a=3$ and $b=2$, so the vertices are $(4\pm 3,5)$, or $(7,5)$ and $(1,5)$. We can sketch the hyperbola more readily if we draw a fundamental rectangle. Using $(4,5)$ as the center, we construct a rectangle $2a$ units wide and $2b$ units high. We then draw and extend the diagonals of the rectangle. The extended diagonals are the asymptotes for the branches of the hyperbola.

In this example, since $a=3$ and $b=2$, we draw a rectangle $2a=6$ units wide and $2b=4$ units high with a center at $(4,5)$. We draw extended diagonals through the rectangle. From the vertex at $(7,5)$, we draw a branch of the hyperbola opening to the right. From the vertex at $(1,5)$, we draw a branch of the hyperbola opening to the left. The graph of the hyperbola is shown.

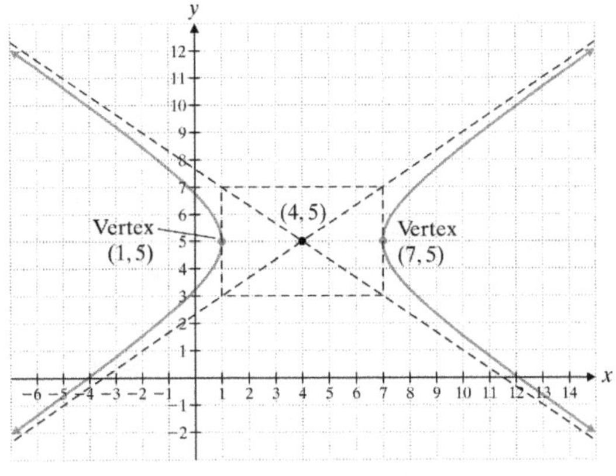

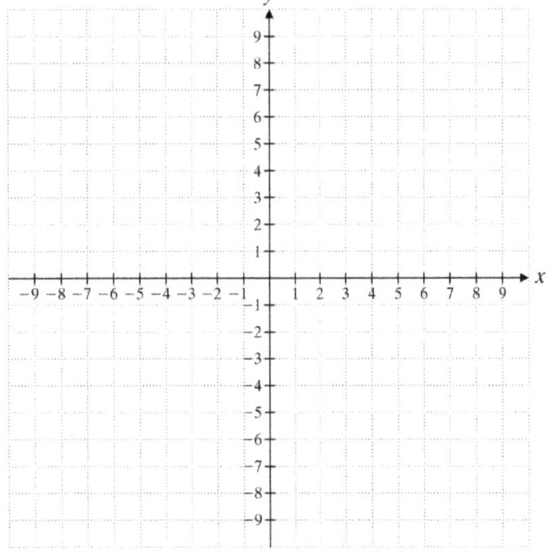

Extra Practice

1. Graph $\dfrac{x^2}{4}-\dfrac{y^2}{36}=1.$

2. Graph $y^2-x^2=16.$

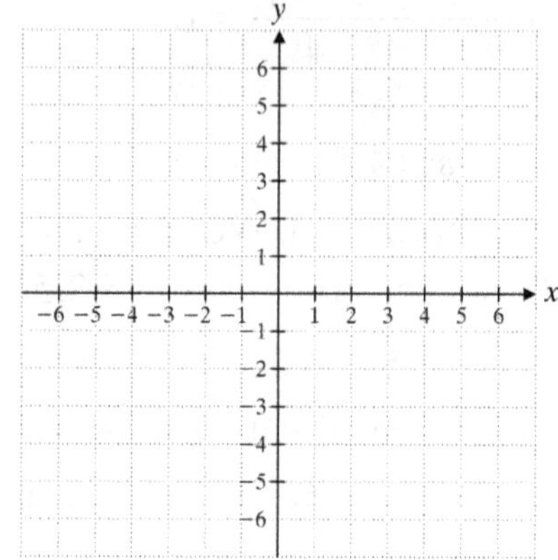

3. Graph $\dfrac{(x+2)^2}{4}-\dfrac{(y+1)^2}{9}=1.$

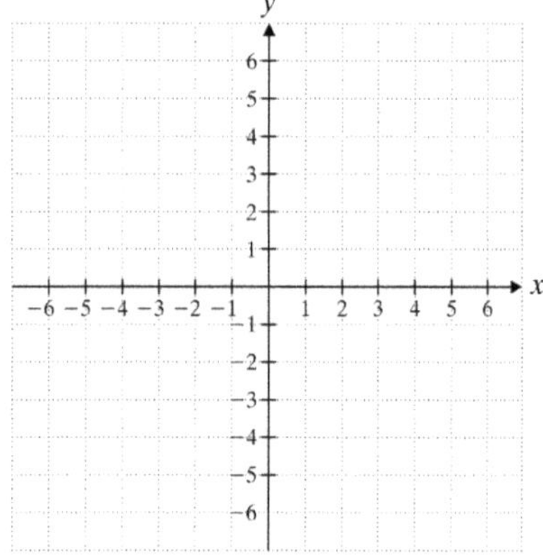

4. Graph $\dfrac{(y+1)^2}{9}-\dfrac{(x-1)^2}{9}=1.$

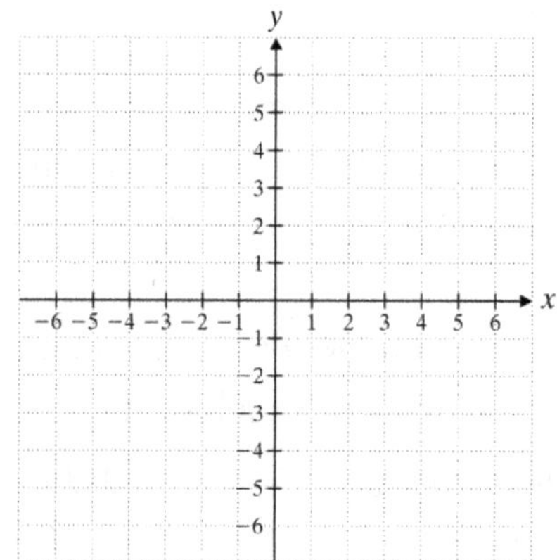

Concept Check

Explain how you would find the equation of one of the asymptotes for the hyperbola $49x^2-4y^2=196.$

Name: ______________________________ Date: ________________

Instructor: ____________________________ Topic: ________________

Module 17 The Conic Sections
Topic 5 Nonlinear Systems of Equations

Vocabulary

nonlinear equation • nonlinear system of equations

1. A ____________ includes at least one nonlinear equation.

2. Any equation that is of second degree or higher is a ____________.

Example	**Student Practice**
1. Solve the following nonlinear system and verify your answer with a sketch.	**2.** Solve the following nonlinear system and verify your answer with a sketch.

Example

1. Solve the following nonlinear system and verify your answer with a sketch.

$$x+y-1=0 \quad (1)$$
$$y-1=x^2+2x \quad (2)$$

We will use the substitution method. Solve for y in equation (1).

$$x+y-1=0$$
$$y=-x+1 \quad (3)$$

Substitute (3) into equation (2). Then solve the resulting quadratic equation.

$$y-1=x^2+2x$$
$$(-x+1)-1=x^2+2x$$
$$-x+1-1=x^2+2x$$
$$0=x^2+3x$$
$$0=x(x+3)$$
$$x=0 \quad \text{or} \quad x=-3$$

Now substitute the values for x in the equation $y=-x+1$.

Continued on the next page.

Student Practice

2. Solve the following nonlinear system and verify your answer with a sketch.

$$x+y-5=0$$
$$y-5=x^2+10x$$

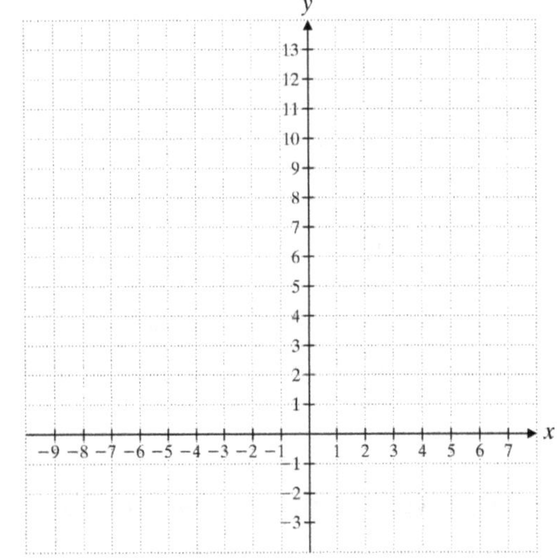

Vocabulary Answers: 1. nonlinear system of equations 2. nonlinear equation

Example	Student Practice
For $x=-3$: $y=-(-3)+1=+3+1=4$ For $x=0$: $y=-(0)+1=+1=1$ Thus, the solutions of the system are $(-3,4)$ and $(0,1)$. Sketch the system. Equation (2) describes a parabola. Write it in the following form. $y=x^2+2x+1=(x+1)^2$ This is a parabola opening upward with its vertex at $(-1,0)$. Equation (1) can be written as $y=-x+1$, which is a straight line with slope $=-1$ and y- intercept $(0,1)$. 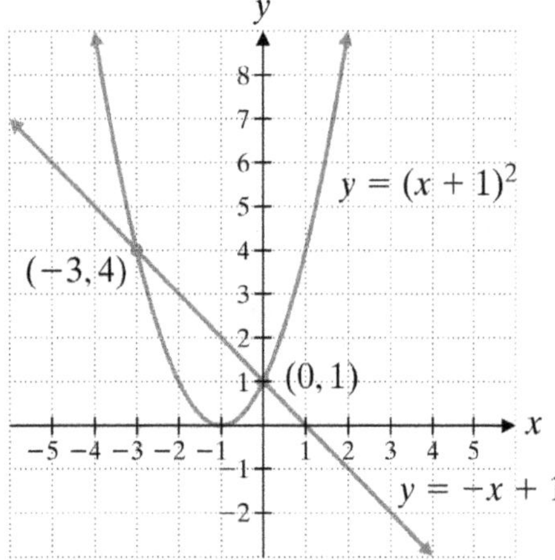 The sketch verifies that the two graphs intersect at $(-3,4)$ and $(0,1)$.	

Example	Student Practice
3. Solve the following nonlinear system and verify your answer with a sketch. $y-2x=0 \quad (1)$ $\frac{x^2}{4}+\frac{y^2}{9}=1 \quad (2)$	**4.** Solve the following nonlinear system and verify your answer with a sketch. $y+2x=0$ $x^2+\frac{y^2}{4}=1$

Solving equation (1) for y yields

$y=2x \quad (3)$.

Substitute (3) into equation (2).

$$\frac{x^2}{4}+\frac{(2x)^2}{9}=1$$

$$36\left(\frac{x^2}{4}\right)+36\left(\frac{(2x)^2}{9}\right)=36(1)$$

$$9x^2+16x^2=36$$

$$x=\pm\sqrt{\frac{36}{25}}$$

$$x=\pm1.2$$

For $x=+1.2$: $y=2(1.2)=2.4$.

For $x=-1.2$: $y=2(-1.2)=-2.4$.

We recognize $\frac{x^2}{4}+\frac{y^2}{9}=1$ as an ellipse and $y=2x$ as a straight line. The sketch shows that the points of intersection at $(1.2, 2.4)$ and $(-1.2, -2.4)$ are reasonable.

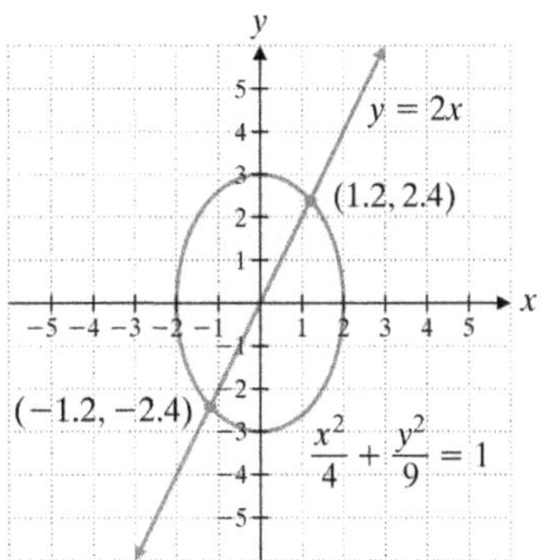

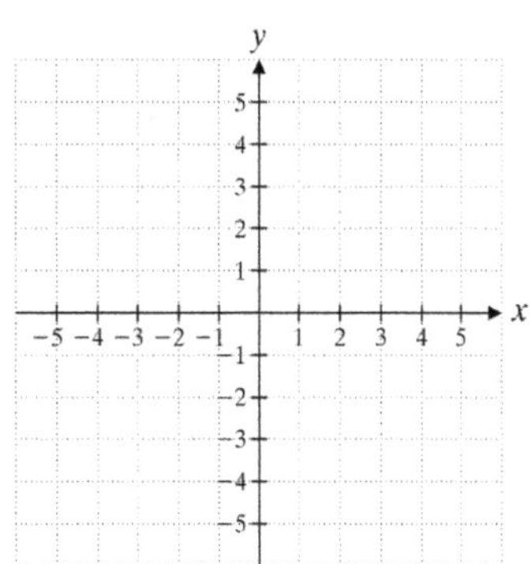

Example	Student Practice
5. Solve the system. $4x^2+y^2=1 \quad (1)$ $x^2+4y^2=1 \quad (2)$	**6.** Solve the system. $9x^2+y^2=1 \quad (1)$ $x^2+9y^2=1 \quad (2)$

Since neither equation is linear, we will use the addition method. Multiply equation (1) by -4 and add to equation (2).

$$\begin{array}{r} -16x^2-4y^2=-4 \\ x^2+4y^2=\ \ 1 \\ \hline -15x^2 \qquad\quad =-3 \end{array}$$

$$x^2=\frac{-3}{-15}$$

$$x^2=\frac{1}{5}$$

$$x=\pm\sqrt{\frac{1}{5}}$$

If $x=+\sqrt{\frac{1}{5}}$, then $x^2=\frac{1}{5}$. Substituting this value into equation (2) gives

$$\frac{1}{5}+4y^2=1$$

$$4y^2=\frac{4}{5}$$

$$y=\pm\sqrt{\frac{1}{5}}$$

Similarly, if $x=-\sqrt{\frac{1}{5}}$, then $y=\pm\sqrt{\frac{1}{5}}$. In this case, we have four solutions. If we rationalize each expression, the four solutions are $\left(\frac{\sqrt{5}}{5},\frac{\sqrt{5}}{5}\right)$, $\left(\frac{\sqrt{5}}{5},-\frac{\sqrt{5}}{5}\right)$, $\left(-\frac{\sqrt{5}}{5},\frac{\sqrt{5}}{5}\right)$, and $\left(-\frac{\sqrt{5}}{5},-\frac{\sqrt{5}}{5}\right)$.

Extra Practice

1. Solve the following nonlinear system and verify your answer with a sketch.

$y = x^2 - 4$

$y = x + 2$

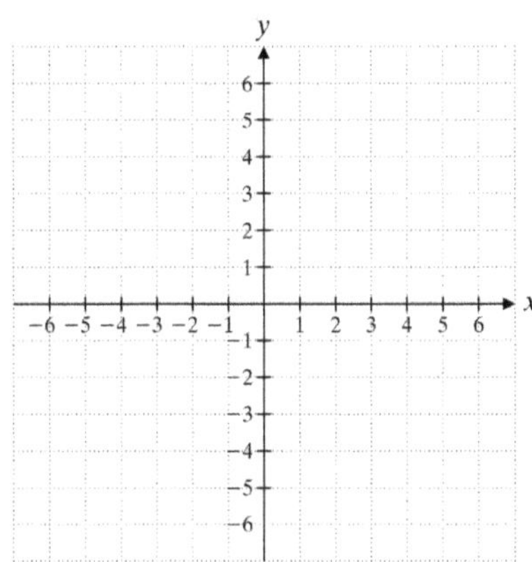

2. Solve the following nonlinear system by the substitution method.

$x^2 - 49y^2 - 25 = 0$

$x + 7y - 2 = 0$

3. Solve the following nonlinear system and verify your answer with a sketch.

$x^2 + y^2 = 25$

$20x^2 - 5y^2 = 100$

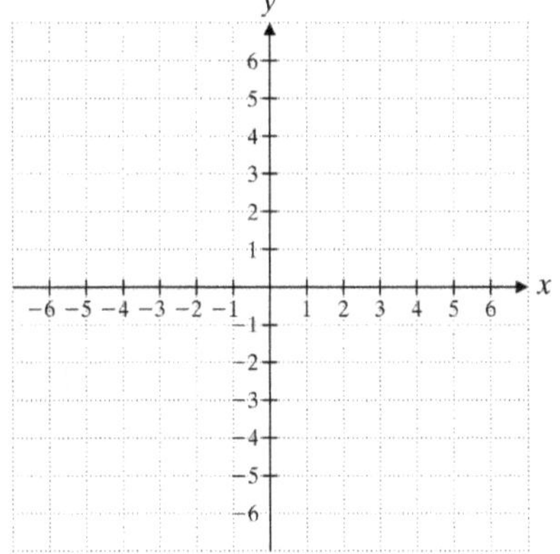

4. Solve the following nonlinear system by the addition method.

$4x^2 = 3y^2 + 24$

$2(x^2 - 15) = -3y^2$

Concept Check

Explain how you would solve the following system.

$y^2 + 2x^2 = 18$

$xy = 4$

MATH COACH

Mastering the skills you need to do well on the test.

Watch the MATH COACH videos in MyMathLab® or on YouTube™ while you work the problems below. These helpful hints will help you avoid making common errors on test problems.

Rewriting and Graphing the Equation of a Parabola—
Problem 2 Rewrite the equation in standard form. Find the center or vertex, plot at least one other point, identify the conic, and sketch the curve.

$$y^2 - 6y - x + 13 = 0$$

> **Helpful Hint:** If the equation of the parabola has a y^2 term, then it can be written in standard form $x = a(y-k)^2 + h$. The vertex is (h,k), and the graph is a horizontal parabola.

Did you first rewrite the equation as $x = y^2 - 6y + 13$?
Yes ____ No ____
Next, did you complete the square to obtain $x = (y-3)^2 + 4$? Yes ____ No ____

If you answered No to these questions, remember that when completing the square, we take half of -6 and square it, which results in 9. We must add both a positive and a negative 9 to the right side of the equation. Go back and try to complete these steps again.

Did you see that since $a = 1$ and $a > 0$, the parabola opens to the right? Yes ____ No ____

Did you identify the vertex as $(4,3)$?
Yes ____ No ____

If you answered No to these questions, review the helpful hint and other information about horizontal parabolas. Stop and perform this step again.

In your final answer, be sure to give the equation in standard form, list the vertex, identify the conic, and graph the equation.

If you answered Problem 2 incorrectly, go back and rework the problem using these suggestions.

Identifying and Graphing a Hyperbola with a Center at the Origin—Problem 5

Identify and graph each conic section. Label the center and/or vertex as appropriate. $\frac{x^2}{10} - \frac{y^2}{9} = 1$

> **Helpful Hint:** The standard form of a hyperbola with the center at the origin and vertices $(-a,0)$ and $(a,0)$ has the equation $\frac{x^2}{a^2} - \frac{y^2}{b^2} = 1$.

Did you realize that this equation represents a horizontal hyperbola with vertices $(-\sqrt{10},0)$ and $(\sqrt{10},0)$?

Yes ____ No ____
If you answered No, review the rules for the standard form of an equation of a hyperbola with its center at the origin and read the Helpful Hint. Then remember that since $a^2 = 10$ and $a > 0$, $a = \sqrt{10}$. Likewise, if $b^2 = 9$ and $b > 0$, then $b = 3$.
Did you obtain a fundamental rectangle with corners at $(\sqrt{10},3)$, $(\sqrt{10},-3)$, $(-\sqrt{10},3)$, and $(-\sqrt{10},-3)$?

Yes ____ No ____

If you answered No, note that drawing a fundamental rectangle with corners at (a,b), $(a,-b)$, $(-a,b)$, and $(-a,-b)$ can help in creating the graph. The extended diagonals of the rectangle become asymptotes.

In your final answer, remember to identify the conic section, create its graph, and label the center and the vertices.

Now go back and rework the problem using these suggestions.

Identifying and Graphing an Ellipse With a Center Not at the Origin—Problem 7

Identify and graph each conic section. Label the center and/or vertex as appropriate. $\frac{(x+2)^2}{16}+\frac{(y-5)^2}{4}=1$

> **Helpful Hint:** When an ellipse has a center that is not at the origin, the center has the coordinates (h,k) and the equation in standard form is $\frac{(x-h)^2}{a^2}+\frac{(y-k)^2}{b^2}=1$, where both a and b are greater than zero.

Did you determine that the center of the ellipse is at $(-2,5)$ with $a=4$ and $b=2$?
Yes ____ No ____

If you answered No, remember that the standard form of the equation involves $x-h$ and $y-k$, so you must be careful in determining the signs of h and k.

Did you start at the center and find points a units to the left, a units to the right, b units up, and b units down to plot the points $(-6,5)$, $(2,5)$, $(-2,7)$, and $(-2,3)$?
Yes ____ No ____

If you answered No, remember that to find these four points you need to find the following: $(h-a,k)$, $(h+a,k)$, $(h,k+b)$, and $(h,k-b)$.

Plot all four points and the center and label these on your graph. Then use the four points to make a sketch of the ellipse. Remember to identify the conic as an ellipse in your final answer.

If you answered Problem 7 incorrectly, go back and rework the problem using these suggestions.

Solving a System of Nonlinear Equations—Problem 16

Solve. $x^2+2y^2=15$
$x^2-y^2=6$

> **Helpful Hint:** When two equations in a system have the form $ax^2+by^2=c$, where a, b, and c are real numbers, then it may be easiest to solve the system by the addition method.

If you multiplied the second equation by 2 and added the result to the first equation, do you get $3x^2=27$?
Yes ____ No ____

Can you solve this equation for x to get $x=3$ and $x=-3$?
Yes ____ No ____

If you answered No to these questions, remember that when you add the two equations together, the y^2 term adds to 0.

If you substitute $x=3$ into the first equation, do you get the equation $9+2y^2=15$? Yes ____ No ____

Can you solve this equation for y to get $y=\sqrt{3}$ and $y=-\sqrt{3}$? Yes ____ No ____

If you answered No to these questions, try substituting $x=3$ into the equation again and be careful to avoid calculation errors. Remember that you must also perform this same step with $x=-3$. Since x is squared, your results for y should be the same.

Your final answer should have a total of four possible ordered pair solutions to this system.

Now go back and rework this problem using these suggestions.

Name: ______________________________ Date: ________________
Instructor: ____________________________ Topic: _______________

Module 18 Additional Properties of Functions
Topic 1 Function Notation

Vocabulary
function notation • free-fall • radius • surface area

1. The surface area of a sphere is a function of ___________.

2. The approximate distance an object in ___________ travels when there is no initial downward velocity is given by the distance function $d(t)=16t^2$.

Example	**Student Practice**
1. If $g(x)=5-3x$, find the following.	**2.** If $h(x)=6x-7$, find the following.
(a) $g(a)$ $g(a)=5-3a$	**(a)** $h(b)$
(b) $g(a+3)$ $g(a+3)=5-3(a+3)=5-3a-9$ $=-4-3a$	**(b)** $h(b+6)$
(c) $g(a)+g(3)$ This requires us to find each addend separately, then add them together. $g(a)=5-3a$ $g(3)=5-3(3)=5-9=-4$ $g(a)+g(3)=(5-3a)+(-4)$ $=5-3a-4$ $=1-3a$ Notice that $g(a+3)\neq g(a)+g(3)$.	**(c)** $h(b)+h(6)$

Vocabulary Answers: 1. radius 2. free-fall

Example	Student Practice
3. If $r(x)=\frac{4}{x+2}$, find $r(a+3)-r(a)$. Express this result as one fraction.	**4.** If $k(x)=\frac{5}{x+6}$, find $k(a+2)-k(a)$. Express this result as one fraction.
$r(a+3)-r(a)=\frac{4}{a+3+2}-\frac{4}{a+2}$ $=\frac{4}{a+5}-\frac{4}{a+2}$	
To express this as one fraction, we note that the LCD $=(a+5)(a+2)$.	
$r(a+3)-r(a)$ $=\frac{4(a+2)}{(a+5)(a+2)}-\frac{4(a+5)}{(a+2)(a+5)}$ $=\frac{4a+8}{(a+5)(a+2)}-\frac{4a+20}{(a+2)(a+5)}$ $=\frac{4a-4a+8-20}{(a+5)(a+2)}=\frac{-12}{(a+5)(a+2)}$	
5. Let $f(x)=3x-7$. Find $\frac{f(x+h)-f(x)}{h}$.	**6.** Let $f(x)=5x+8$. Find $\frac{f(x+h)-f(x)}{h}$.
First find $f(x+h)$, then subtract $f(x)$.	
$f(x+h)=3(x+h)-7=3x+3h-7$	
$f(x+h)-f(x)$ $=(3x+3h-7)-(3x-7)$ $=3x+3h-7-3x+7$ $=3h$	
Therefore, $\frac{f(x+h)-f(x)}{h}=\frac{3h}{h}=3$.	

Example	Student Practice
7. The surface area of a sphere is given by $S=4\pi r^2$ where r is the radius. If we use $\pi=3.14$ as an approximation, this becomes $S=4(3.14)r^2$, or $S=12.56r^2$.	**8.** The surface area of a sphere is given by $S=4\pi r^2$ where r is the radius. If we use $\pi=3.14$ as an approximation, this becomes $S=4(3.14)r^2$, or $S=12.56r^2$.
(a) Find the surface area of a sphere with a radius of 3 centimeters. Write surface area as a function of r and solve for $S(r)$ when $r=3$. $S(3)=12.56(3)^2=113.04\text{ cm}^2$	**(a)** Find the surface area of a sphere with a radius of 6 centimeters.
(b) Suppose that an error is made and the radius is calculated to be $(3+e)$ centimeters. Find an expression for the surface area as a function of the error e. $S(e)=12.56(3+e)^2$ $=113.04+75.36e+12.56e^2$	**(b)** Suppose that an error is made and the radius is calculated to be $(6+e)$ centimeters. Find an expression for the surface area as a function of the error e.
(c) Evaluate the surface area for $r=(3+e)$ cm when $e=0.2$. Round your answer to the nearest hundredth of a cm. What is the difference in the surface area due to the error in measurement? An error in measurement was made, so use the function found in part **(b)**. $S(0.2)$ $=113.04+75.36(0.2)+12.56(0.2)^2$ $=128.6144$ $\approx 128.61\text{ cm}^2$ If the radius of 3 cm was incorrectly calculated as 3.2 cm, the surface area would be too large by approximately $128.61-113.04=15.57\text{ cm}^2$.	**(c)** Evaluate the surface area for $r=(6+e)$ cm when $e=0.3$. Round your answer to the nearest hundredth of a cm. What is the difference in the surface area due to the error in measurement?

Extra Practice

1. If $g(x)=5x^2-8x+9$, find $g(a+1)$.

2. If $h(x)=\sqrt{x+2}$, find $h\left(4a^2+6\right)$.

3. If $s(x)=\frac{5}{x+2}$, find $s\left(-\frac{4}{3}\right)+s(-3)$.

4. Find $\frac{f(x+h)-f(x)}{h}$ for $f(x)=4x^2$.

Concept Check

Explain how you would find $k(2a-1)$ if $k(x)=\sqrt{3x+1}$.

Name: ______________________________ Date: ________________
Instructor: ___________________________ Topic: ________________

Module 18 Additional Properties of Functions
Topic 2 General Graphing Procedures for Functions

Vocabulary
function • vertical line test • vertical shift • horizontal shift

1. $f(x)+k$ represents the graph of $f(x)$ with a ___________ upwards of k units.

2. A ___________ must have no ordered pairs that have the same first coordinates and different second coordinates.

3. The ___________ states that if any vertical line intersects the graph of a relation more than once, the relation is not a function.

4. $f(x+h)$ represents the graph of $f(x)$ with a ___________ to the left of h units.

Example	**Student Practice**
1. Determine whether the following is the graph of a function.	**2.** Determine whether the following is the graph of a function.

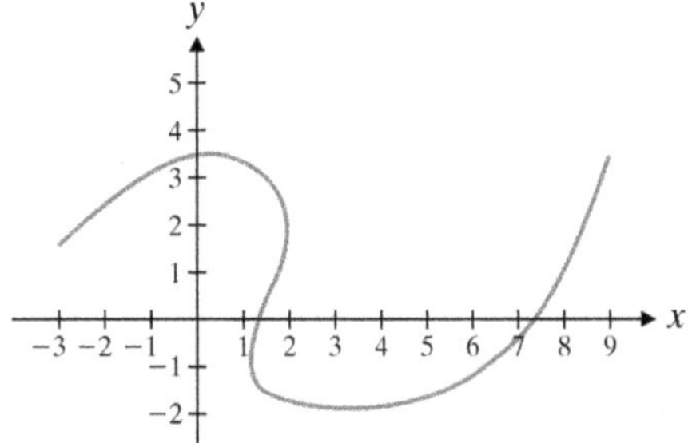

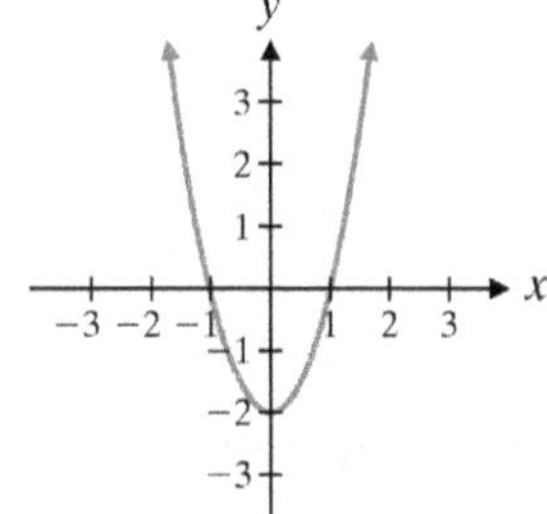

A vertical line intersects the graph more than once, so by the vertical line test, this relation is not a function.

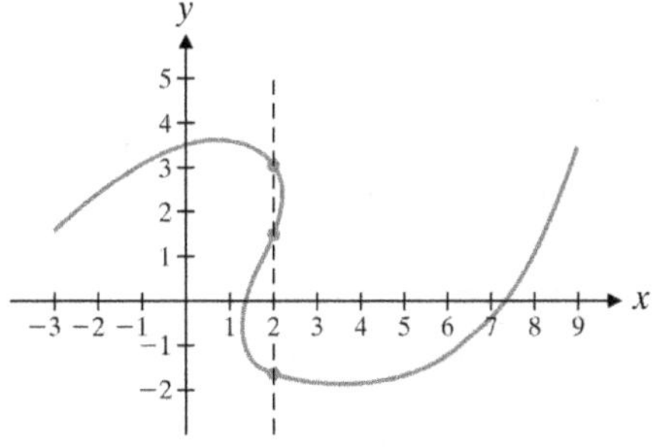

Vocabulary Answers: 1. vertical shift 2. function 3. vertical line test 4. horizontal shift

Example	Student Practice
3. Graph the functions on one coordinate plane. $f(x)=x^2$ and $h(x)=x^2+2$	**4.** Graph the functions on one coordinate plane. $f(x)=x^2$ and $h(x)=x^2-3$

First we make a table of values for $f(x)$ and for $h(x)$.

x	f(x) = x²
−2	4
−1	1
0	0
1	1
2	4

x	h(x) = x² + 2
−2	6
−1	3
0	2
1	3
2	6

Graph on the same coordinate plane.

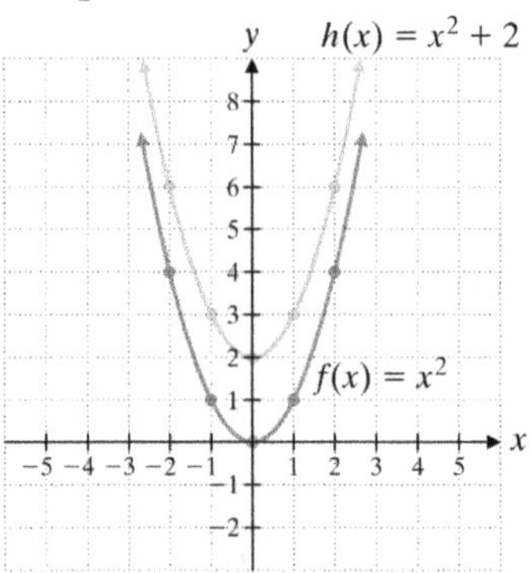

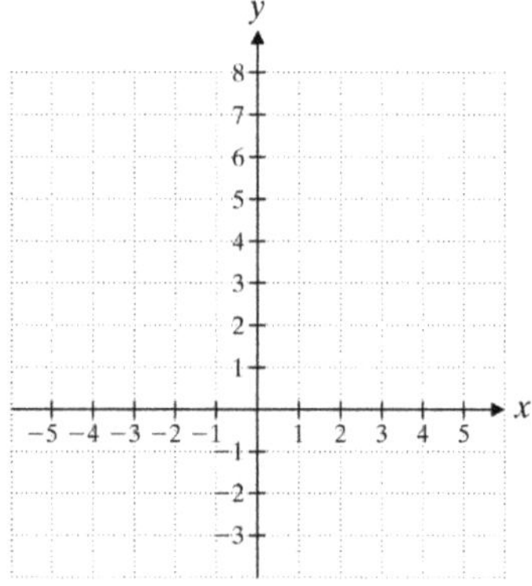

Notice that the graph of $h(x)$ is the graph of $f(x)$ moved 2 units upward.

5. Graph the functions on one coordinate plane. $f(x)=\|x\|$ and $p(x)=\|x-3\|$	**6.** Graph the functions on one coordinate plane. $f(x)=\|x\|$ and $p(x)=\|x+3\|$

Graph each function on the same coordinate plane.

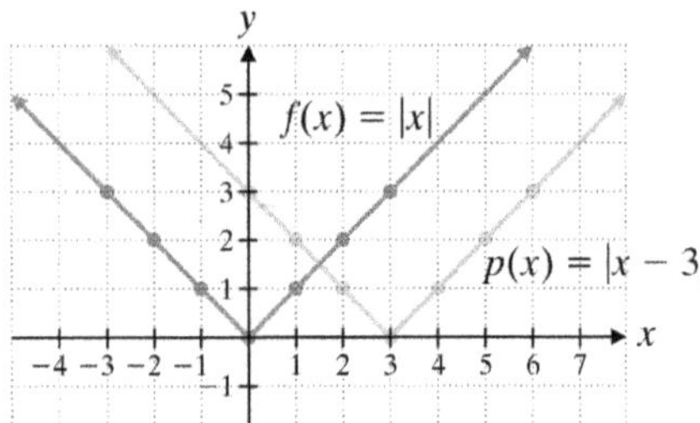

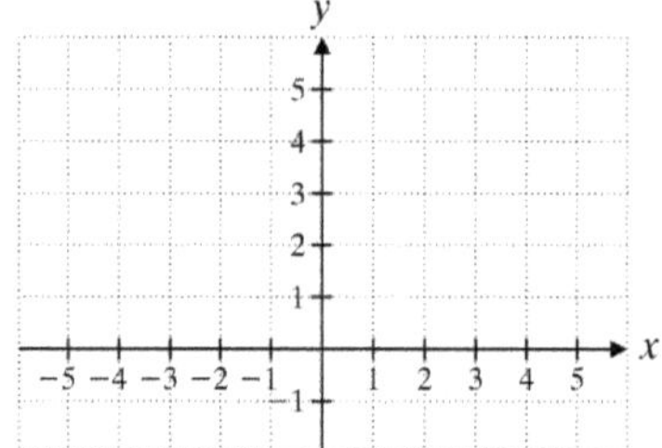

The graph of $p(x)$ is the graph of $f(x)$ shifted 3 units to the right.

Example	Student Practice

7. Graph the functions on one coordinate plane. $f(x)=x^3$ and $h(x)=(x-3)^3-2$

First we make a table of values for $f(x)$ and graph the function.

x	f(x)
−2	−8
−1	−1
0	0
1	1
2	8

Next we recognize that $h(x)$ will have a similar shape, but the curve will be shifted 3 units to the right and 2 units downward. We draw the graph of $h(x)$ using these shifts.

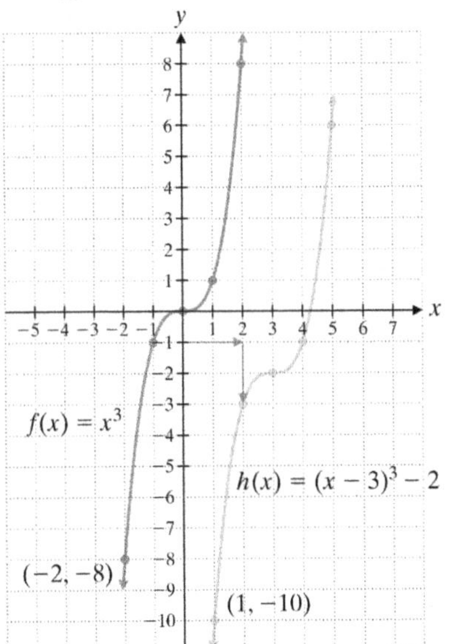

The point $(-2,-8)$ has been shifted 3 units to the right and 2 units down to the point $(-2+3,-8+(-2))$ or $(1,-10)$.

The point $(-1,-1)$ is a point on $f(x)$. Use the same reasoning to find the image of $(-1,-1)$ on the graph of $h(x)$. Verify by checking the graphs.

8. Graph the functions on one coordinate plane.

$f(x)=x^3$ and $h(x)=(x+4)^3-2$

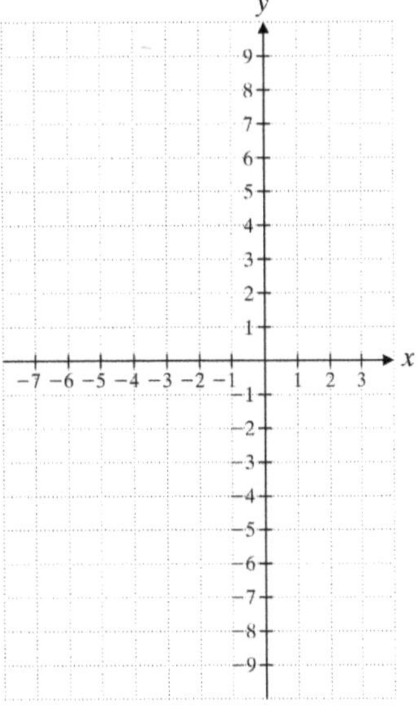

Extra Practice

1. Determine whether the following is the graph of a function.

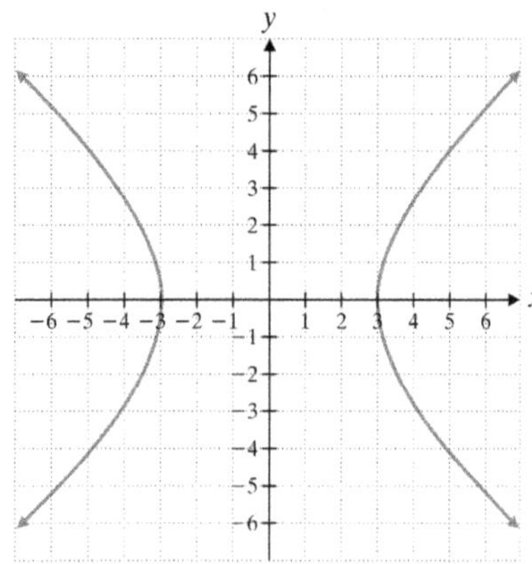

2. Graph the functions on one coordinate plane. $f(x) = x^2$ and $g(x) = (x+3)^2 - 2$

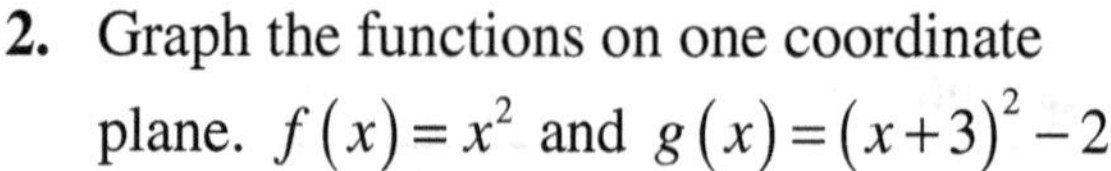

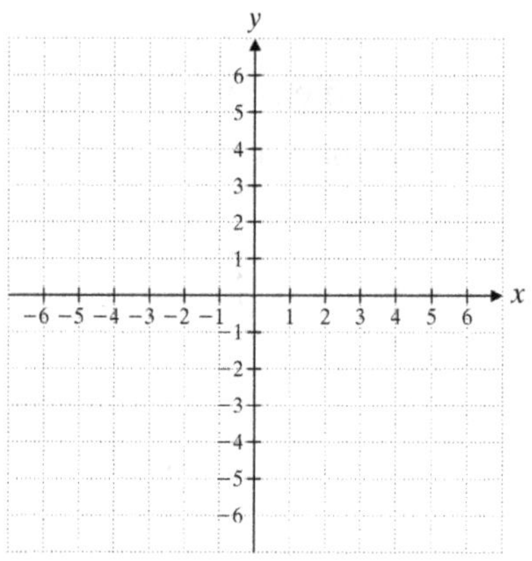

3. Graph the functions on one coordinate plane. $f(x) = x^3$ and $g(x) = (x+2)^3 + 2$

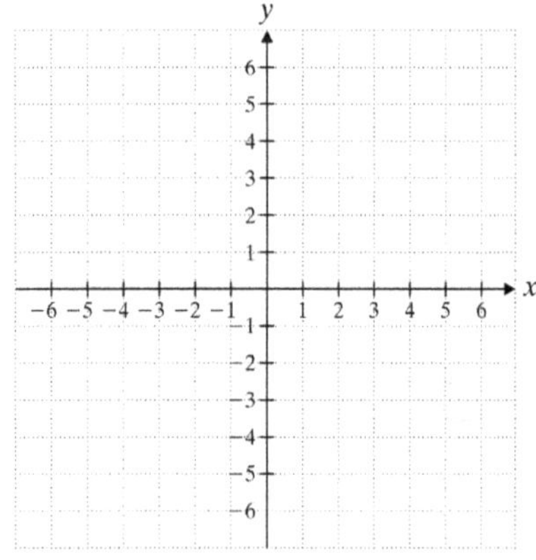

4. Graph the functions on one coordinate plane. $f(x) = \dfrac{3}{x}$ and $g(x) = \dfrac{3}{x-2}$

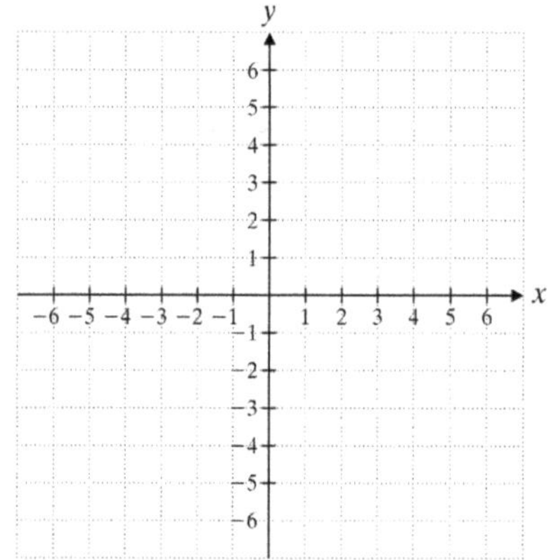

Concept Check

Explain how you can use a vertical line to determine whether a graph represents a function.

Name: ______________________________ Date: ________________
Instructor: ____________________________ Topic: ________________

Module 18 Additional Properties of Functions
Topic 3 Algebraic Operations on Functions

Vocabulary

sum • difference • product • quotient • composition

1. When finding the ____________ of a function, we must be careful to avoid division by zero.

2. The ____________ of the functions f and g, is defined as follows: $(fg)(x) = f(x) \cdot g(x)$.

3. The ____________ of the functions f and g, denoted $f \circ g$, is defined as follows: $(f \circ g)(x) = f\left[g(x)\right]$.

4. The ____________ of the functions f and g, is defined as follows: $(f+g)(x) = f(x) + g(x)$.

Example	Student Practice
1. Suppose that $f(x)=3x^2-3x+5$ and $g(x)=5x-2$. **(a)** Find $(f+g)(x)$. $(f+g)(x)=f(x)+g(x)$ $=(3x^2-3x+5)+(5x-2)$ $=3x^2+2x+3$ **(b)** Evaluate $(f+g)(x)$ when $x=3$. Write $(f+g)(3)$ and use the formula obtained in **(a)**. $(f+g)(x)=3x^2+2x+3$ $(f+g)(3)=3(3)^2+2(3)+3$ $=27+6+3=36$	**2.** Suppose that $f(x)=2x^2+4x-7$ and $g(x)=8x+1$. **(a)** Find $(f+g)(x)$. **(b)** Evaluate $(f+g)(x)$ when $x=4$.

Vocabulary Answers: 1. quotient 2. composition 3. product 4. sum

Example	Student Practice
3. Given $f(x)=x^2-5x+6$ and $g(x)=2x-1$, find the following. **(a)** $(fg)(x)$ $(fg)(x)=f(x)\cdot g(x)$ $=(x^2-5x+6)(2x-1)$ $=2x^3-11x^2+17x-6$ **(b)** $(fg)(-4)$ Use the formula obtained in **(a)**. $(fg)(x)=2x^3-11x^2+17x-6$ $(fg)(-4)$ $=2(-4)^3-11(-4)^2+17(-4)-6$ $=-128-176-68-6=-378$	**4.** Given $f(x)=x^2-3x+8$ and $g(x)=4x-3$, find the following. **(a)** $(fg)(x)$ **(b)** $(fg)(-2)$
5. Given $f(x)=3x+1$, $g(x)=2x-1$, and $h(x)=9x^2+6x+1$, find the following. **(a)** $\left(\frac{f}{g}\right)(x)$ $\left(\frac{f}{g}\right)(x)=\frac{3x+1}{2x-1}$ The denominator of the quotient can never be zero. Since $2x-1\neq 0$, we know that $x\neq\frac{1}{2}$. **(b)** $\left(\frac{f}{h}\right)(x)$ $\left(\frac{f}{h}\right)(x)=\frac{3x+1}{9x^2+6x+1}$ $=\frac{3x+1}{(3x+1)(3x+1)}=\frac{1}{3x+1}$ Since $3x+1\neq 0$, we know $x\neq-\frac{1}{3}$.	**6.** Given $f(x)=2x+3$, $g(x)=x-4$, and $h(x)=4x^2+12x+9$, find the following. **(a)** $\left(\frac{f}{g}\right)(x)$ **(b)** $\left(\frac{f}{h}\right)(x)$

Example	Student Practice
7. Given $f(x)=3x-2$ and $g(x)=2x+5$, find $f[g(x)]$. First substitute $g(x)=2x+5$. Then apply the formula for $f(x)$. Remove the parentheses and simplify. $f[g(x)]=f(2x+5)$ $=3(2x+5)-2$ $=6x+15-2$ $=6x+13$	**8.** Given $f(x)=5x+3$ and $g(x)=3x-4$, find $f[g(x)]$.
9. Given $f(x)=\sqrt{x-4}$ and $g(x)=3x+1$, find the following. **(a)** $f[g(x)]$ Substitute $g(x)=3x+1$. Then apply the formula for $f(x)$. $f[g(x)]=f(3x+1)$ $=\sqrt{(3x+1)-4}$ $=\sqrt{3x+1-4}$ $=\sqrt{3x-3}$ **(b)** $g[f(x)]$ $g[f(x)]=g\left(\sqrt{x-4}\right)$ $=3\left(\sqrt{x-4}\right)+1$ $=3\sqrt{x-4}+1$ We note that $g[f(x)]\neq f[g(x)]$.	**10.** Given $f(x)=\sqrt{x+6}$ and $g(x)=4x-1$, find the following. **(a)** $f[g(x)]$ **(b)** $g[f(x)]$

Example	Student Practice
11. Given $f(x)=2x$ and $g(x)=\frac{1}{3x-4}$, $x\neq\frac{4}{3}$, find the following.	**12.** Given $f(x)=4x$ and $g(x)=\frac{1}{2x-5}$, $x\neq\frac{5}{2}$, find the following.
(a) $(f\circ g)(x)$	**(a)** $(f\circ g)(x)$
$(f\circ g)(x)=f[g(x)]=f\left(\frac{1}{3x-4}\right)$ $=2\left(\frac{1}{3x-4}\right)=\frac{2}{3x-4}$	
(b) $(f\circ g)(2)$	**(b)** $(f\circ g)\left(\frac{1}{2}\right)$
$(f\circ g)(2)=\frac{2}{3(2)-4}=\frac{2}{6-4}=\frac{2}{2}=1$	

Extra Practice

1. Given $f(x)=1.6x^3-2.7x$ and $g(x)=4.6x^2+7.6$, find the following.

(a) $(f+g)(x)$

(b) $(f-g)(x)$

(c) $(f+g)(3)$

(d) $(f-g)(-2)$

2. Given $f(x)=x^2-8x+16$ and $g(x)=x-4$, find the following.

(a) $(fg)(x)$

(b) $\left(\frac{f}{g}\right)(x)$

(c) $(fg)(3)$

(d) $\left(\frac{f}{g}\right)(-3)$

3. Given $f(x)=x-5$ and $g(x)=6-2x$, find $f[g(x)]$.

4. Given $f(x)=\left|\frac{4}{3}x+1\right|$ and $g(x)=-3x-2$, find $f[g(x)]$.

Concept Check

If $f(x)=5-2x^2$ and $g(x)=3x^2-5x+1$, explain how you would find $(f-g)(-4)$.

Name: ______________________________ Date: ________________

Instructor: ____________________________ Topic: ________________

Module 18 Additional Properties of Functions
Topic 4 Inverse of a Function

Vocabulary

inverse function f • one-to-one function • horizontal line test

1. The ____________ states that if any horizontal line intersects the graph of a function more than once, the function is not one-to-one.

2. We call a function f^{-1} that reverses the domain and range of a function f the ____________.

3. A(n) ____________ is a function in which no ordered pairs have the same second coordinate.

Example	**Student Practice**
1. Indicate whether the following functions are one-to-one.	**2.** Indicate whether the following functions are one-to-one.
(a) $M=\{(1,3),(2,7),(5,8),(6,12)\}$ M is a function because no ordered pairs have the same first coordinate. M is also a one-to-one function because no ordered pairs have the same second coordinate.	**(a)** $L=\{(1,6),(4,3),(7,6),(9,16)\}$
(b) $P=\{(1,4),(2,9),(3,4),(4,18)\}$ P is a function, but it is not one-to-one because the ordered pairs $(1,4)$ and $(3,4)$ have the same second coordinate.	**(b)** $N=\{(2,6),(3,4),(6,9),(7,11)\}$

Vocabulary Answers: 1. horizontal line test 2. inverse function f 3. one-to-one function

Example	Student Practice

3. Determine whether the functions graphed are one-to-one functions.

(a)

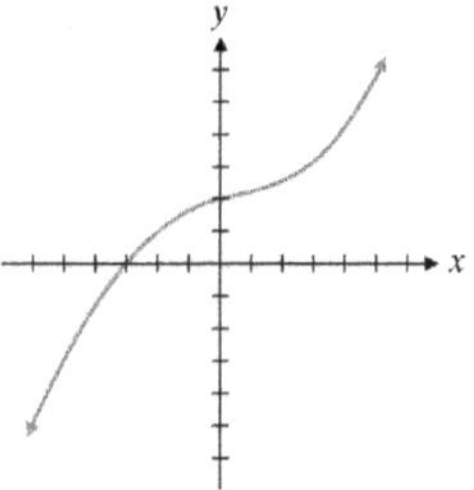

This graph represents a one-to-one function. Horizontal lines cross the graph at most once.

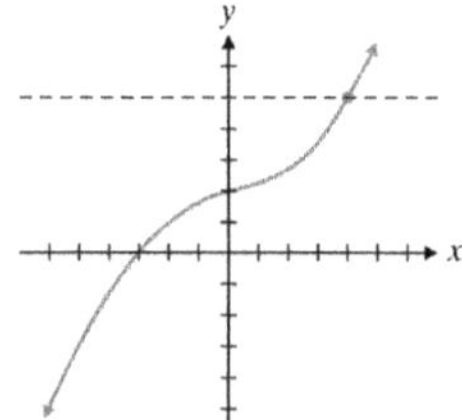

(b)

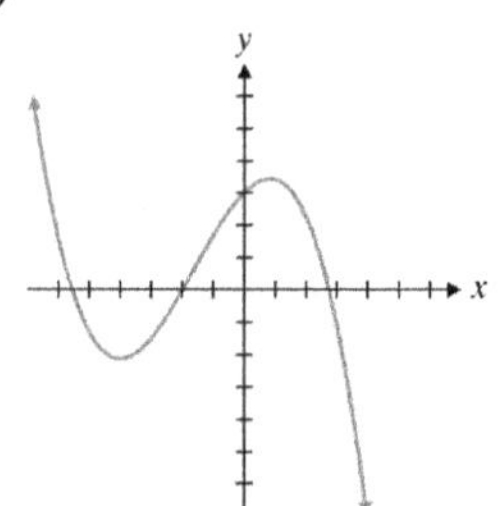

This graph does not represent a one-to-one function. A horizontal line exists that crosses the graph more than once.

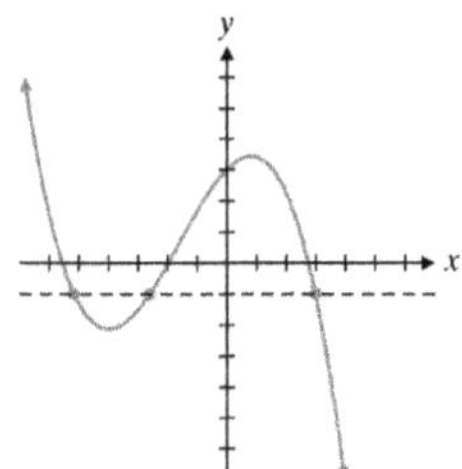

4. Determine whether the functions graphed are one-to-one functions.

(a)

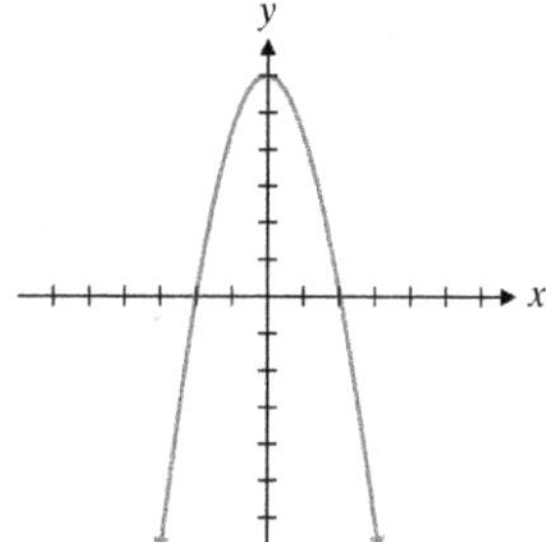

(b)

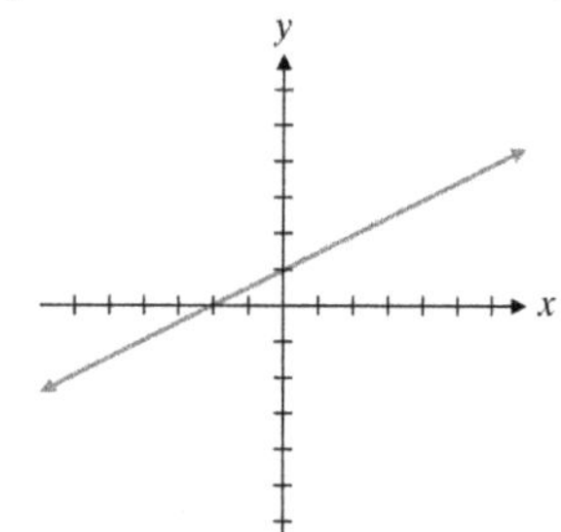

Example	Student Practice
5. Determine the inverse function of $F=\{(6,1),(12,2),(13,5),(14,6)\}$. Since we have a list of ordered pairs, we interchange the coordinates of each ordered pair. The inverse function of F is as follows. $F^{-1}=\{(1,6),(2,12),(5,13),(6,14)\}$	**6.** Determine the inverse function of $Q=\{(2,4),(6,2),(9,7),(11,3)\}$.
7. Find the inverse of $f(x)=7x-4$. Replace $f(x)$ with y. $y=7x-4$ Interchange the variables x and y. $x=7y-4$ Solve for y in terms of x. $x=7y-4$ $x+4=7y$ $\frac{x+4}{7}=y$ Replace y with $f^{-1}(x)$. $f^{-1}(x)=\frac{x+4}{7}$	**8.** Find the inverse of $f(x)=5x+8$.

Example	Student Practice
9. Find the inverse function of $f(x)=\frac{9}{5}x+32$, which converts Celsius temperature (x) into equivalent Fahrenheit temperature. $y=\frac{9}{5}x+32$ $x=\frac{9}{5}y+32$ $5x=9y+160$ $\frac{5x-160}{9}=y$ $f^{-1}(x)=\frac{5x-160}{9}$ Our inverse function $f^{-1}(x)$ will now convert Fahrenheit temperature to Celsius temperature.	**10.** Find the inverse function of $f(x)=150+3(x-25)$, which gives the cost of a rental truck if the company charges a base rate of \$150 plus \$3 for every mile traveled over 25 miles. Here x is the total number of miles traveled in the rental truck.
11. If $f(x)=3x-2$, find $f^{-1}(x)$. Graph f and f^{-1} on the same set of axes. Draw the line $y=x$ as a dashed line for reference. Following the procedure to find $f^{-1}(x)$ yields $f^{-1}(x)=\frac{x+2}{3}$. Now we graph each line. 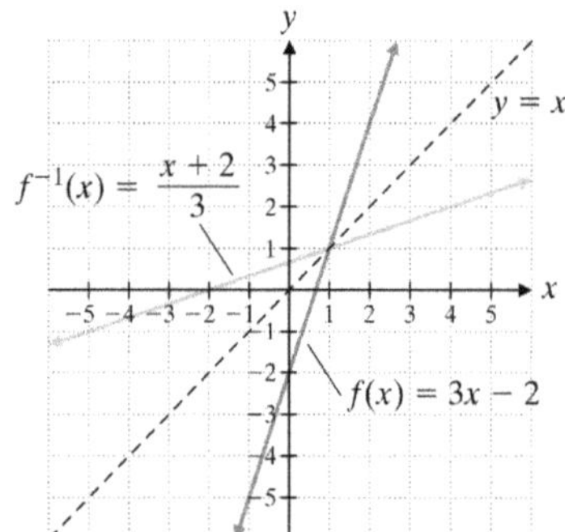 We see that the graphs of f and f^{-1} are symmetric about the line $y=x$.	**12.** If $f(x)=4x+1$, find $f^{-1}(x)$. Graph f and f^{-1} on the same set of axes. Draw the line $y=x$ as a dashed line for reference. 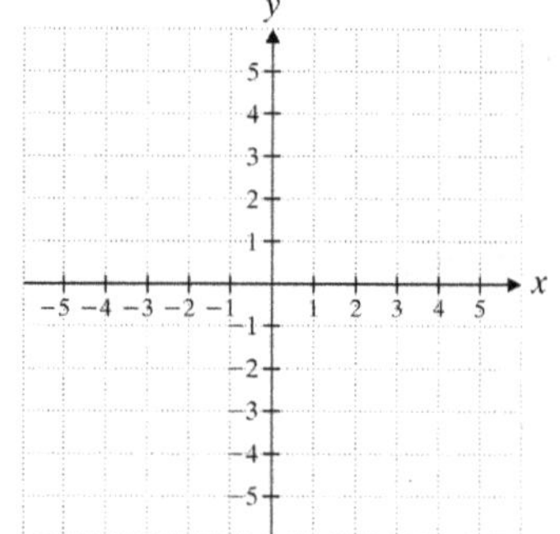

Extra Practice

1. Indicate whether the following function is one-to-one.

   $B=\{(7,9),(9,7),(-7,-9),(-9,-7)\}$

2. Find the inverse of $f(x)=x-4$.

3. Find the inverse of $f(x)=\dfrac{5}{3x-4}$.

4. Find the inverse of $g(x)=-3x-4$. Graph the function and its inverse on the same set of axes. Draw the line $y=x$ as a dashed line for reference.

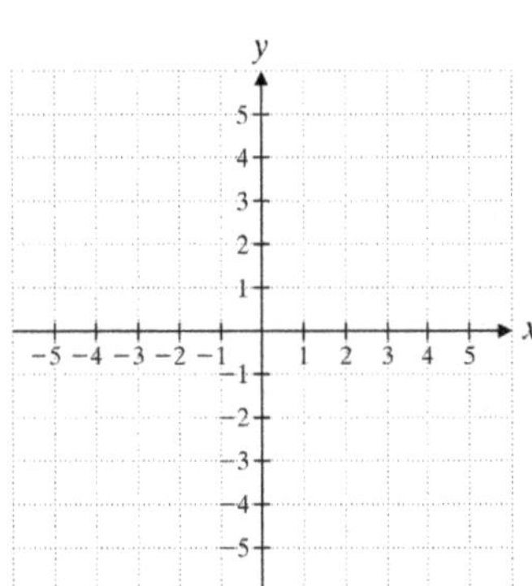

Concept Check

If $f(x)=\dfrac{x-5}{3}$, explain how you would find the inverse function.

MATH COACH

Mastering the skills you need to do well on the test.

Watch the MATH COACH videos in MyMathLab® or on YouTube™ while you work the problems below. These helpful hints will help you avoid making common errors on test problems.

Using Function Notation to Evaluate Expressions—

Problem 5 For the function $f(x)=3x^2-2x+4$, find $f(a+1)$.

> **Helpful Hint:** The key idea is to replace every x with the expression $a+1$ and then simplify the result. Use parentheses around the substitutions to avoid calculation errors.

Did you substitute $a+1$ into the function to get $3(a+1)^2-2(a+1)+4$?
Yes ____ No ____

Did you evaluate $(a+1)^2$ to get a^2+2a+1?
Yes ____ No ____

If you answered No to these questions, remember to replace every x with $a+1$. Use parentheses to avoid calculation errors. Note that when you square a binomial, you must be sure to write down all the terms.

Next, did you simplify further to obtain $3a^2+6a+3-2a-2+4$? Yes ____ No ____

If you answered No, remember to multiply all three terms of a^2+2a+1 by 3. Multiply both terms of $a+1$ by -2.

In your final step, you can collect like terms to write your answer in simplest form.

If you answered Problem 5 incorrectly, go back and rework the problem using these suggestions.

Graphing a Function with a Horizontal and Vertical Shift—Problem 10 Graph each pair of functions on one coordinate plane. $f(x)=x^2$

$$g(x)=(x-1)^2+3$$

> **Helpful Hint:** First graph $f(x)$. The graph of $f(x-h)+k$ is the graph of $f(x)$ shifted h units to the right and k units upward (assuming that $h>0$ and $k>0$).

Can you determine that the graph of $f(x)$ passes through the points $(0,0)$, $(1,1)$, $(-1,1)$, $(2,4)$, and $(-2,4)$?
Yes ____ No ____

If you answered No, try building a table of values in which you replace x with 0, 1, -1, 2, and -2 and find the corresponding values of $f(x)$. Plot those points and connect the points with a curve to form the graph of $f(x)=x^2$.

Do you see that the function $g(x)=(x-1)^2+3$ has the values of $h=1$ and $k=3$ when you apply the Helpful Hint?
Yes ____ No ____

Did you find that the graph of $g(x)$ is the graph of $f(x)$ shifted one unit to the right and 3 units up? Yes ____ No ____

If you answered No to these questions, reread the Helpful Hint carefully. Notice that the values of h and k are both greater than zero.

Finish by graphing $g(x)$ on the same coordinate plane as your graph of $f(x)$.

Now go back and rework the problem using these suggestions.

Finding the Composition of Two Functions—Problem 14(a)

If $f(x)=\frac{1}{2}x-3$ and $g(x)=4x+5$, find $(f\circ g)(x)$.

Helpful Hint: First rewrite $(f\circ g)(x)$ as $f\left[g(x)\right]$. Most students find this expression more logical. Then substitute $g(x)$ for the value of x in $f(x)$.

First, did you rewrite the problem as

$f\left[g(x)\right]=\frac{1}{2}(4x+5)-3$?

Yes ____ No ____

If you answered No, substitute the expression for $g(x)$ for the value of x in the expression for $f(x)$.

Next did you simplify the resulting expression to

$f\left[g(x)\right]=2x+\frac{5}{2}-3$?

Yes ____ No ____

If you answered No, remember that $\frac{1}{2}(4x)=2x$ and $\frac{1}{2}(5)=\frac{5}{2}$. As your final step, combine like terms to write your expression in simplest form.

If you answered Problem 14(a) incorrectly, go back and rework the problem using these suggestions.

Finding and Graphing the Inverse of a Function—Problem 18 Given $f(x)=-3x+2$, find f^{-1}. Graph f and its inverse f^{-1} on one coordinate plane. Graph $y=x$ as a dashed line for reference.

Helpful Hint: Use the following four steps to find the inverse of a function:

1. Replace $f(x)$ with y.
2. Interchange x and y.
3. Solve for y in terms of x.
4. Replace y with $f^{-1}(x)$.

Did you substitute y for $f(x)$ to get $y=-3x+2$ and then interchange x and y to get $x=-3y+2$?

Yes ____ No ____

If you answered No, review the first two steps in the Helpful Hint and perform these steps again.

Did you solve the equation for y to get $y=-\frac{1}{3}x+\frac{2}{3}$?

Yes ____ No ____

If you answered No, remember to add $3y$ to each side. Next add $-x$ to each side and then divide each side of the equation by 3. As your last step in finding the inverse, replace y with $f^{-1}(x)$.

Remember to graph $f(x)$ and $f^{-1}(x)$ on the same coordinate plane. Add the graph of $y=x$ as a dashed line for reference.

Now go back and rework the problem using these suggestions.

Name: ______________________________ Date: ________________
Instructor: ___________________________ Topic: ________________

Module 19 Logarithmic and Exponential Functions
Topic 1 The Exponential Function

Vocabulary

exponential function • *e* • asymptote • property of exponential equations
compound interest • radioactive decay • base

1. The ___________ says that if $b^x = b^y$, then $x = y$ for $b > 0$ and $b \neq 1$.

2. The function $f(x) = b^x$, where $b > 0$, $b \neq 1$, and x is a real number, is called a(n) ___________.

3. For every exponential function, the x- axis is a(n) ___________.

Example	Student Practice
1. Graph $f(x) = 2^x$.	**2.** Graph $f(x) = 5^x$.

Make a table of values for x and $f(x)$.

$f(-1) = 2^{-1} = \frac{1}{2}, f(0) = 2^0 = 1, f(1) = 2^1 = 2$

Continue to evaluate $f(x)$ at different integers to make a table of values for x and $f(x)$. Using these points, draw the graph.

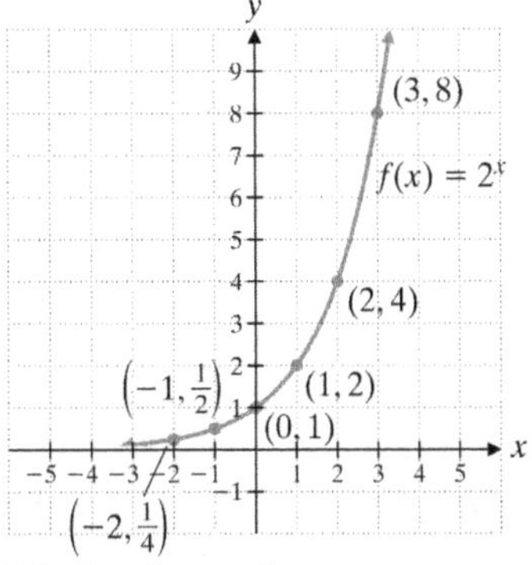

Notice how the curve comes very close to the x- axis but never touches it. The x- axis is an asymptote for every exponential function.

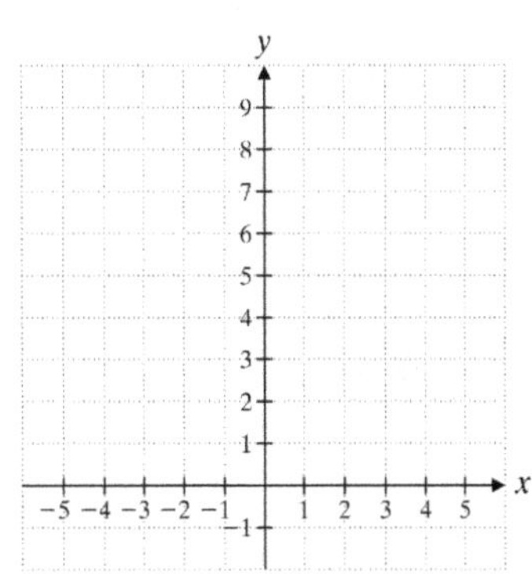

Vocabulary Answers: 1. property of exponential equations 2. exponential function 3. asymptote

Example	Student Practice

3. Graph $f(x)=\left(\frac{1}{2}\right)^x$.

Rewrite the function as follows.

$$f(x)=\left(\frac{1}{2}\right)^x=\left(2^{-1}\right)^x=2^{-x}$$

Evaluate the function for a few values of x. Using these points, draw the graph.

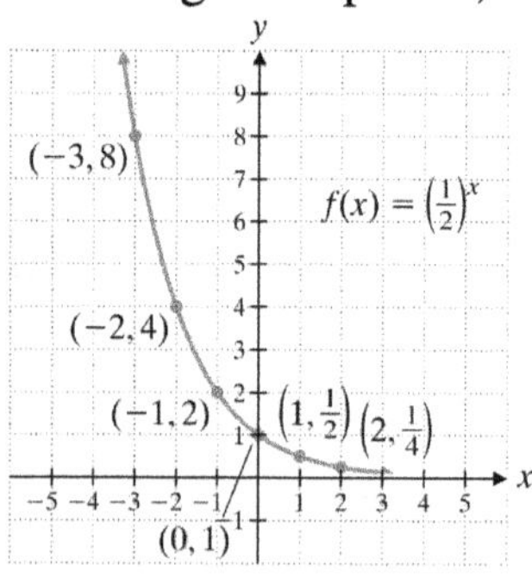

Note that as x increases, $f(x)$ decreases.

4. Graph $f(x)=\left(\frac{2}{3}\right)^x$.

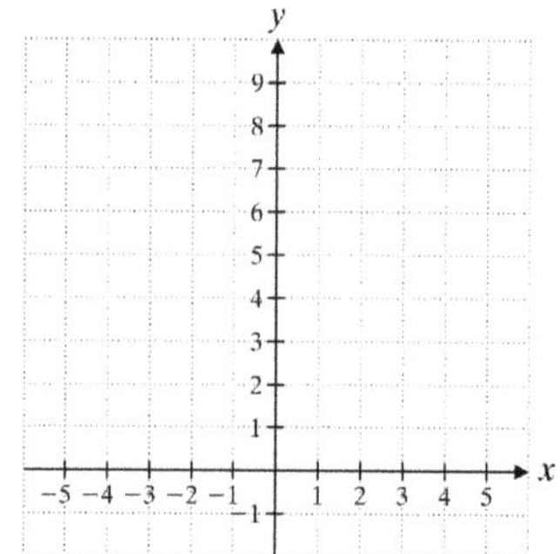

5. Graph $f(x)=e^x$.

The letter e is like the number π. It is an irrational number. If you do not have a scientific calculator, approximate the value for e to use the number in calculations, $e\approx 2.7183$. Otherwise, use a scientific calculator to produce a table of values for this function. Then, draw the graph.

6. Graph $f(x)=e^{1+x}$.

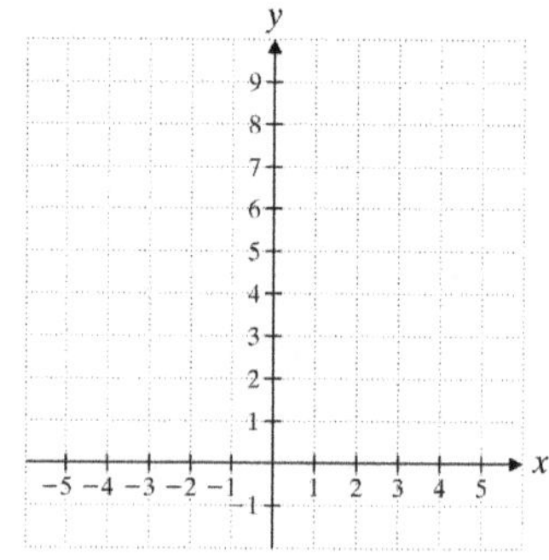

Example	Student Practice
7. Solve. $2^x = \frac{1}{16}$ To use the property of exponential equations, we must have the same base on both sides of the equation. Rewrite 16 as 2^4. $2^x = \frac{1}{16}$ $2^x = \frac{1}{2^4}$ $2^x = 2^{-4}$ Recall the property of exponential equations, which states that if $b^x = b^y$, then $x = y$ for $b > 0$ and $b \neq 1$. Since $2^x = 2^{-4}$, $x = -4$.	**8.** Solve. $3^x = 27$
9. If we invest \$8000 in a fund that pays 15% annual interest compounded monthly, how much will we have in 6 years? In this situation, $P = 8000$, $r = 15\% = 0.15$, $t = 6$, and $n = 12$ because interest is compounded monthly, or 12 times a year. $A = 8000\left(1 + \frac{0.15}{12}\right)^{(12)(6)}$ $= 8000(1 + 0.125)^{72}$ $= 8000(2.445920268)$ $\approx 19{,}567.36$	**10.** If you invest \$5500 in a fund that pays 18% annual interest compounded quarterly, how much will you have in 10 years?

Example	Student Practice
11. The radioactive decay of the element americium 241 can be described by the equation $A = Ce^{-0.001608t}$, where C is the original amount of the element in the sample, A is the amount of the element remaining after t years, and $k = -0.0016008$, the decay constant for americium 241. If 10 milligrams (mg) of americium 241 is sealed in a laboratory container today, how much will theoretically be present in 2000 years? Round your answer to the nearest hundredth.	**12.** The radioactive decay of the element cobalt 60 can be described by the equation $A = Ce^{-0.131527t}$, where C is the original amount of the element in the sample, A is the amount of the element remaining after t years, and $k \approx -0.131527$, the decay constant for cobalt 60. If 5 milligrams (mg) of cobalt 60 is sealed in a laboratory container today, how much will theoretically be present in 10 years? Round your answer to the nearest hundredth.
Substitute $C = 10$ and $t = 2000$ into the equation and simplify using a calculator. $A = Ce^{-0.0016008t} = 10e^{-0.0016008(2000)}$ $\approx 10(0.040697)$ ≈ 0.41 mg	

Extra Practice

1. Graph $f(x) = 3^{-x}$.

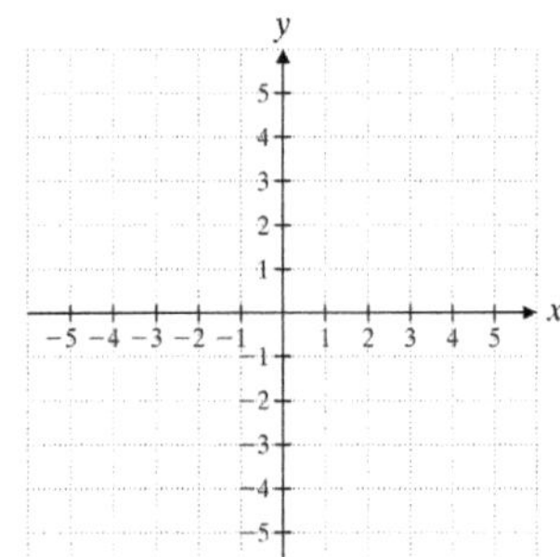

2. Graph $f(x) = 2^{x+4}$.

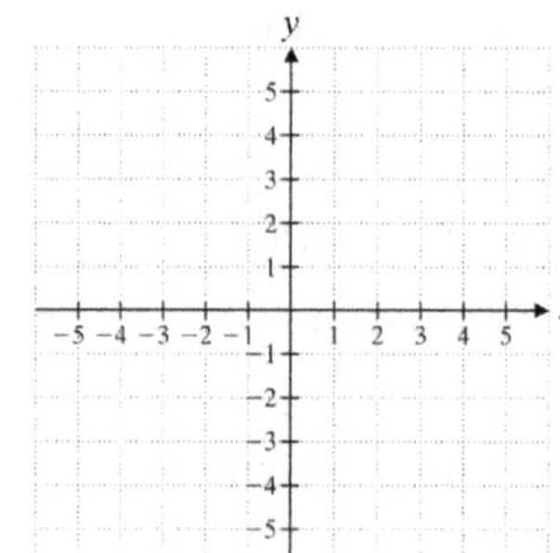

3. Solve for x. $5^{x+3} = 25$

4. Anton is investing \$4000 at an annual rate of 4.6% compounded annually. How much money will Anton have after 5 years? Round your answer to the nearest cent.

Concept Check

Explain how you would solve $4^{-x} = \frac{1}{64}$.

Name: ______________________________ Date: ______________

Instructor: ___________________________ Topic: ______________

Module 19 Logarithmic and Exponential Functions
Topic 2 The Logarithmic Function

Vocabulary

logarithmic function • logarithm • base • power • inverse

1. The logarithmic function $y = \log_b x$ is the __________ of the exponential function $x = b^y$.

2. A(n) __________ is an exponent.

3. In the function $y = \log_b x,\ b$ is called the __________.

Example	**Student Practice**
1. Write in logarithmic form. $81 = 3^4$ Use the fact that $x = b^y$ is equivalent to $\log_b x = y$. Here, $x = 81,\ b = 3,$ and $y = 4.$ So the logarithmic form of $81 = 3^4$ is $4 = \log_3 81.$	**2.** Write in logarithmic form. $\frac{1}{216} = 6^{-3}$
3. Write in exponential form. $-4 = \log_{10}\left(\frac{1}{10{,}000}\right)$ Use the fact that $x = b^y$ is equivalent to $\log_b x = y$. Here, $y = -4,\ b = 10,$ and $x = \frac{1}{10{,}000}$. So the exponential form of $-4 = \log_{10}\left(\frac{1}{10{,}000}\right)$ is $\frac{1}{10{,}000} = 10^{-4}.$	**4.** Write in exponential form. $5 = \log_3 243$

Vocabulary Answers: 1. inverse 2. logarithm 3. base

Example	Student Practice
5. Solve for the variable. **(a)** $\log_5 x = -3$ Convert the logarithmic equation to an equivalent exponential equation and solve for x. $5^{-3} = x$ $\frac{1}{5^3} = x$ $\frac{1}{125} = x$ **(b)** $\log_a 16 = 4$ $a^4 = 16$ $a^4 = 2^4$ $a = 2$	**6.** Solve for the variable **(a)** $\log_2 x = -8$ **(b)** $\log_b 64 = 3$
7. Evaluate. $\log_3 81$ This is asking, "To what power must we raise 3 to get 81?" Write an equivalent exponential expression using x as the unknown power. $\log_3 81 = x$ $81 = 3^x$ Write 81 as 3^4 and solve for x. $3^4 = 3^x$ $x = 4$ Thus, $\log_3 81 = 4$.	**8.** Evaluate. $\log_2 64$

Example	**Student Practice**

9. Graph $y = \log_2 x$ and $y = 2^x$ on the same set of axes.

Make a table of values (ordered pairs) for each equation.

x	y
-1	$\frac{1}{2}$
0	1
1	2
2	4

x	y
$\frac{1}{2}$	-1
1	0
2	1
4	2

Coordinates of ordered pairs are reversed

Graph the ordered pairs.

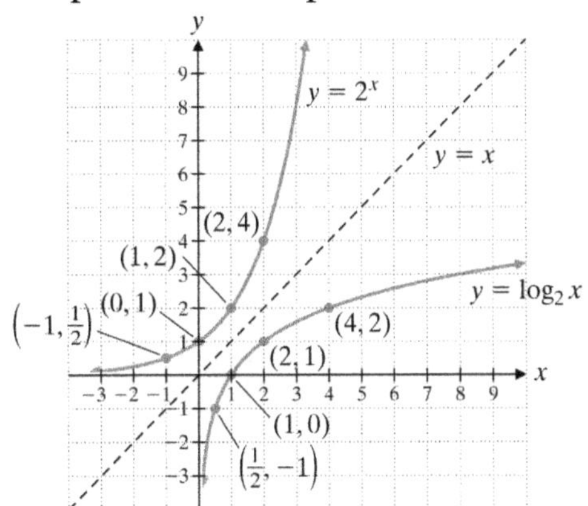

Note that $y = \log_2 x$ is the inverse of $y = 2^x$ because the ordered pairs (x, y) are reversed. The sketch of the two equations shows that they are inverses.

Recall that in function notation, f^{-1} means the inverse function of f. Thus, if we write $f(x) = \log_2 x$, then $f^{-1}(x) = 2^x$.

10. Graph $y = \log_4 x$ and $y = 4^x$ on the same set of axes.

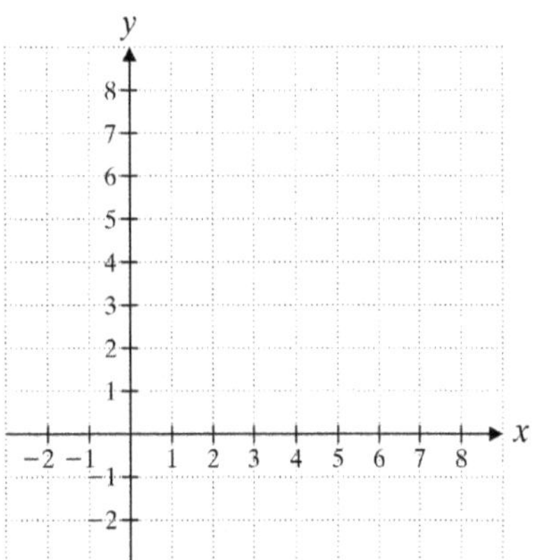

Extra Practice

1. Write in logarithmic form. $0.0001 = 10^{-4}$

2. Write in exponential form. $-6 = \log_e x$

3. Evaluate. $\log_{10} \sqrt{10}$

4. Graph. $\log_2 x = y$

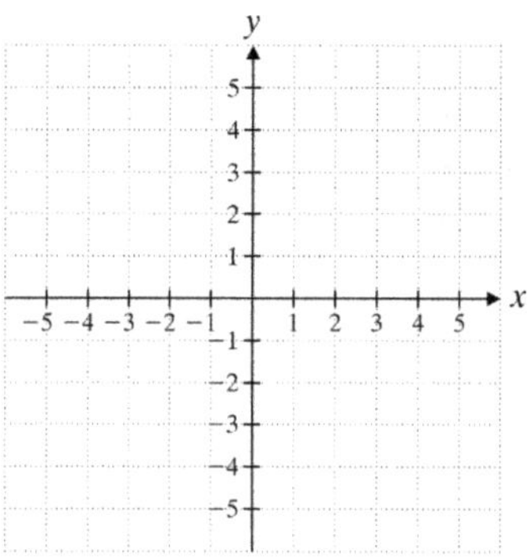

Concept Check

Explain how you would solve for x. $-\frac{1}{2} = \log_e x$

Name: ______________________________ Date: ________________
Instructor: ____________________________ Topic: ________________

Module 19 Logarithmic and Exponential Functions
Topic 3 Properties of Logarithms

Vocabulary
logarithms • Logarithm of a Product Property • Logarithm of a Quotient Property
Logarithm of a Number Raised to a Power Property

1. The ____________ states that for any positive real numbers M and N and any positive base $b \neq 1$, $\log_b MN = \log_b M + \log_b N$.

2. The ____________ is similar to the logarithm of a product property, except that it involves two expression that are divided.

3. ____________ are used to reduce complex expressions to addition and subtraction.

Example	**Student Practice**
1. Write $\log_3 XZ$ as a sum of logarithms. Use the logarithm of a product property, $\log_b MN = \log_b M + \log_b N$, to rewrite the logarithm. $b = 3$, $M = X$, and $N = Z$ $\log_3 XZ = \log_3 X + \log_3 Z$	**2.** Write $\log_5 BC$ as a sum of logarithms.
3. Write $\log_3 16 + \log_3 x + \log_3 y$ as a single logarithm. If we extend our rule, we have $\log_b MNP = \log_b M + \log_b N + \log_b P$. Use this to rewrite the logarithm. $b = 3$, $M = 16$, $M = x$, and $P = y$ $\log_3 16 + \log_3 x + \log_3 y = \log_3 16xy$	**4.** Write $\log_2 x + \log_2 20 + \log_2 z$ as a single logarithm.

Vocabulary Answers: 1. logarithm of a product property 2. logarithm of a quotient property 3. logarithms

Example	Student Practice
5. Write $\log_3\left(\frac{29}{7}\right)$ as the difference of two logarithms. To rewrite the logarithm, use the logarithm of a quotient property, $\log_b\left(\frac{M}{N}\right)=\log_b M-\log_b N$. $\log_3\left(\frac{29}{7}\right)=\log_3 29-\log_3 7$	**6.** Write $\log_{10}\left(\frac{19}{5}\right)$ as the difference of two logarithms.
7. Express $\log_b 36-\log_b 9$ as a single logarithm. Use the reverse of the property $\log_b\left(\frac{M}{N}\right)=\log_b M-\log_b N$ to write the difference as a single logarithm. $\log_b 36-\log_b 9=\log_b\left(\frac{36}{9}\right)=\log_b 4$	**8.** Express $\log_b 6-\log_b 24$ as a single logarithm.
9. Write $\frac{1}{3}\log_b x+2\log_b w-3\log_b z$ as a single logarithm. Use the property $\log_b M^p=p\log_b M$ to eliminate the coefficients of the logarithmic terms. Then, combine the sum of the logarithms and then the difference to obtain a single logarithm. $\frac{1}{3}\log_b x+2\log_b w-3\log_b z$ $=\log_b x^{1/3}+\log_b w^2-\log_b z^3$ $=\log_b x^{1/3}w^2-\log_b z^3$ $=\log_b\left(\frac{x^{1/3}w^2}{z^3}\right)$	**10.** Write $5\log_b y-\frac{1}{4}\log_b w+3\log_b z$ as a single logarithm.

Example	Student Practice
11. Write $\log_b\left(\frac{x^4y^3}{z^2}\right)$ as a sum or difference of logarithms. Use the logarithm of a quotient property to rewrite the logarithm as a difference. $\log_b\left(\frac{x^4y^3}{z^2}\right)=\log_b x^4y^3-\log_b z^2$ Now, use the logarithm of a product property. $\log_b x^4y^3-\log_b z^2$ $=\log_b x^4+\log_b y^3-\log_b z^2$ Finally, use the logarithm of a number raised to a power property to eliminate the exponents and get the following result. $4\log_b x+3\log_b y-2\log_b z$	**12.** Write $\log_b\left(\frac{w^6x^5}{y^3}\right)$ as a sum or difference of logarithms.
13. Evaluate. **(a)** Evaluate $\log_7 7$. Since $\log_b b=1,\ \log_7 7=1$. **(b)** Evaluate $\log_5 1$. Since $\log_b 1=0,\ \log_5 1=0$. **(c)** Find x if $\log_3 x=\log_3 17$. If $\log_b x=\log_b y$, then $x=y$. Thus, since $\log_3 x=\log_3 17,\ x=17$.	**14.** Evaluate. **(a)** Evaluate $\log_{20} 20$. **(b)** Evaluate $\log_7 1$. **(c)** Find x if $\log_5 x=\log_5 12$.

Example	Student Practice
15. Find x if $2\log_7 3 - 4\log_7 2 = \log_7 x$.	**16.** Find x if $2\log_{11} 7 - 3\log_{11} 5 = \log_{11} x$.
Use the logarithm of a number raised to a power property to eliminate the coefficients of the logarithmic terms.	
$\log_7 3^2 - \log_7 2^4 = \log_7 x$ $\log_7 9 - \log_7 16 = \log_7 x$	
Then, use the logarithm of a quotient property and then the property that if $\log_b x = \log_b y$, then $x = y$, to find x.	
$\log_7\left(\frac{9}{16}\right) = \log_7 x$ $\frac{9}{16} = x$	

Extra Practice

1. Write as a sum or difference of logarithms. $\log_7 x^3yz^2$

2. Write as a single logarithm. $\frac{1}{3}\log_a 6 + 2\log_a 6 - 5\log_a x$

3. Use the properties of logarithms to simplify the following. $\frac{1}{5}\log_3 3 - \log_{11} 1$

4. Find x if $\log_4 x + \log_4 2 = 3$.

Concept Check

Explain how you would simplify $\log_{10}(0.001)$.

Name: ______________________________ Date: ________________
Instructor: ___________________________ Topic: ________________

Module 19 Logarithmic and Exponential Functions
Topic 4 Common and Natural Logarithms; Change of Base

Vocabulary
common logarithm • antilogarithm • natural logarithm • change of base formula

1. Base 10 logarithms are called ____________ and are usually written with no subscripts.

2. The ____________ is $\log_b x = \dfrac{\log_a x}{\log_a b}$, where a, b, and $x > 0$, $a \neq 1$, and $b \neq 1$.

3. Logarithms with base e are known as ____________ and are usually written $\ln x$.

Example	Student Practice
1. On a scientific calculator or a graphing calculator, find a decimal approximation for each of the following. **(a)** $\log 7.32$ Enter the number 7.32 on a calculator and then press the log key. $\log 7.32 \approx 0.864511081$ **(b)** $\log 73.2$ $\log 73.2 \approx 1.864511081$	**2.** On a scientific calculator or a graphing calculator, find a decimal approximation for each of the following. **(a)** $\log 12.67$ **(b)** $\log 126.7$
3. Find an approximate value for x if $\log x = 4.326$. Find the antilogarithm. We know that $\log_{10} x = 4.326$ is equivalent to $10^{4.326} = x$. Solve this problem by finding the value of $10^{4.326}$. Using a calculator, we have the following. $x \approx 21{,}183.61135$	**4.** Find an approximate value for x if $\log x = 7.438$.

Vocabulary Answers: 1. common logarithm 2. change of base formula 3. natural logarithms

Example	Student Practice
5. Evaluate $\text{antilog}(-1.6784)$. This is the equivalent to asking what the value is of $10^{-1.6784}$. Be sure to enter the negative sign on your calculator. $10^{-1.6784} \approx 0.020970076$	**6.** Evaluate $\text{antilog}(-2.5041)$.
7. On a scientific calculator, approximate the following values. **(a)** $\ln 7.21$ This is asking for the natural log of 7.21. Use the $\boxed{\ln}$ key on your calculator to solve. $\ln 7.21 \approx 1.975468951$ **(b)** $\ln 72.1$ $\ln 72.1 \approx 4.278054044$	**8.** On a scientific calculator, approximate the following values. **(a)** $\ln 3.25$ **(b)** $\ln 55.93$
9. On a scientific calculator, find an approximate value for x for each equation. **(a)** $\ln x = 2.9836$ If $\ln x = 2.9836$, then $e^{2.9836} = x$. Use the $\boxed{e^x}$ key on a scientific calculator to solve. Thus, $x = e^{2.9836} \approx 19.75882051$. **(b)** $\ln x = -1.5619$ If $\ln x = -1.5619$, then $e^{-1.5619} = x$. Thus, $x = e^{-1.5619} \approx 0.209737192$.	**10.** On a scientific calculator, find an approximate value for x for each equation. **(a)** $\ln x = 6.0123$ **(b)** $\ln x = -0.9774$

Example	Student Practice
11. Evaluate using common logarithms. $\log_3 5.12$ Use the change of base formula. Here $b=3$ and $x=5.12$. $\log_b x = \frac{\log_a x}{\log_a b}$ $\log_3 5.12 = \frac{\log 5.12}{\log 3} \approx 1.486561234$ The answer is approximate with nine decimal places, depending on the calculator you may have more or fewer digits.	**12.** Evaluate using common logarithms. $\log_6 7.981$
13. Obtain an approximate value for $\log_4 0.005739$ using natural logarithms. Use the change of base formula where $a=e,\ b=4$ and $x=0.005739$. $\log_4 0.005739 = \frac{\log_e 0.005739}{\log_e 4}$ $= \frac{\ln 0.005739}{\ln 4}$ ≈ -3.722492455 To check, we want to know the following. $4^{-3.722492455} \stackrel{?}{=} 0.005739$ Using a calculator, this can be verified using the $\boxed{y^x}$ key. The answer checks.	**14.** Obtain an approximate value for $\log_2 0.02546$ using natural logarithms.

Example	Student Practice
15. Using a scientific calculator, graph $y = \log_2 x$.	**16.** Using a scientific calculator, graph $y = \log_4 x$.

Use the change of base formula with common logarithms to find $y = \dfrac{\log x}{\log 2}$.
Find values of y for various values of x and organize them into a data table. Graph the data points.

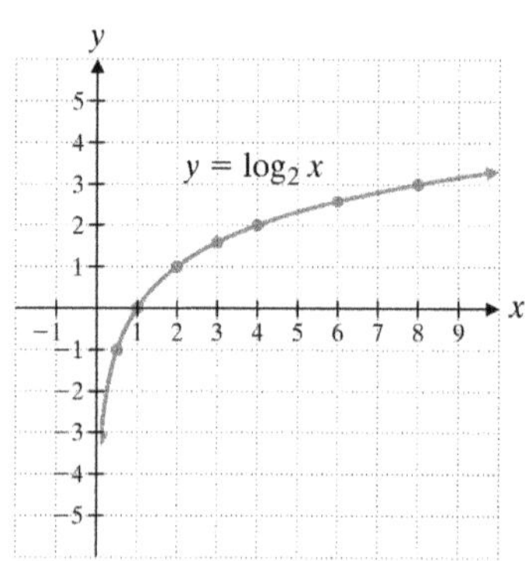

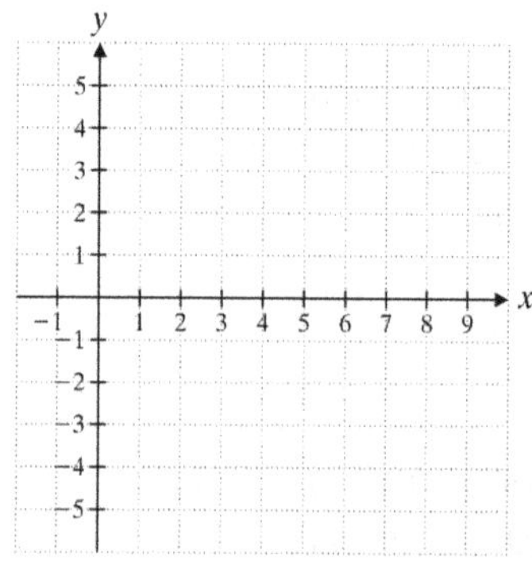

Extra Practice

1. On a scientific calculator or a graphing calculator, find a decimal approximation for $\log 11.2$.

2. On a scientific calculator, find an approximate value for x if $\ln x = 3.2$.

3. Evaluate using common logarithms.
$\log_{12} 0.451$

4. Obtain an approximate value for $\log_4 0.0332$ using natural logarithms.

Concept Check

Explain how you would find x using a scientific calculator if $\ln x = 1.7821$.

Name: ______________________________ Date: ________________
Instructor: ___________________________ Topic: ________________

Module 19 Logarithmic and Exponential Functions
Topic 5 Exponential and Logarithmic Equations

Vocabulary
logarithm • natural logarithm • logarithmic equation • exponential equation

1. If an exponential equation involves e raised to a power, take the ____________ of each side of the equation.

2. It is not possible to take the ____________ of a negative number.

3. To solve a(n) ____________, get one logarithmic term on one side and a numerical value on the other, then convert it to an exponential equation using the definition of a logarithm.

Example	Student Practice
1. Solve. $\log_3(x+6)-\log_3(x-2)=2$	**2.** Solve. $\log_2(x+4)-\log_2(x-2)=2$

Apply property 2, which states that $\log_b\frac{M}{N}=\log_b M-\log_b N$. Then write the equation in exponential form.

$$\log_3(x+6)-\log_3(x-2)=2$$
$$\log_3\left(\frac{x+6}{x-2}\right)=2$$
$$\frac{x+6}{x-2}=3^2$$

Solve the resulting equation.

$$\frac{x+6}{x-2}=9$$
$$x+6=9(x-2)$$
$$x+6=9x-18$$
$$24=8x$$
$$3=x$$

The check is left to the student.

Vocabulary Answers: 1. natural logarithm 2. logarithm 3. logarithmic equation

Example	Student Practice
3. Solve. $\log(x+6)+\log(x+2)=\log(x+20)$ Use property 6, if $b \neq 1$, $x>0$, $y>0$ and $\log_b x=\log_b y$, then $x=y$. $\log(x+6)+\log(x+2)=\log(x+20)$ $\log(x+6)(x+2)=\log(x+20)$ $x^2+8x+12=x+20$ $x^2+7x-8=0$ $(x+8)(x-1)=0$ $x-1=0$ or $x+8=0$ $x=1$ $\quad$ $x=-8$ Check the solutions. $\log(1+6)+\log(1+2)\stackrel{?}{=}\log(1+20)$ $\log(7\cdot 3)\stackrel{?}{=}\log 21$ $\log 21=\log 21$ $\log(-8+6)+\log(-8+2)\stackrel{?}{=}\log(-8+20)$ $\log(-2)+\log(-6)\neq\log(12)$ Discard -8 because it leads to taking the logarithm of a negative number. Thus, the only solution is $x=1$.	**4.** Solve. $\log(x+6)+\log(x-4)=\log(3x-4)$
5. Solve $2^x=7$. Leave your answer in exact form. Take the logarithm of each side. Solve for x. $2^x=7$ $\log 2^x=\log 7$ $x\log 2=\log 7$ $x=\dfrac{\log 7}{\log 2}$	**6.** Solve $5^x=12$. Leave your answer in exact form.

Example	Student Practice
7. Solve $e^{2.5x} = 8.42$. Round your answer to the nearest ten-thousandth.	**8.** Solve $e^{4.7x} = 10.75$. Round your answer to the nearest ten-thousandth.
$\ln e^{2.5x} = \ln 8.42$ $(2.5x)(\ln e) = \ln 8.42$ $2.5x = \ln 8.42$ $x = \dfrac{\ln 8.42}{2.5}$ $x \approx 0.8522$	
9. If P dollars are invested in an account that earns interest at 12% compounded annually, the amount available after t years is $A = P(1+0.12)^t$. How many years will it take for \$300 in this account to grow to \$1500? Round your answer to the nearest whole year.	**10.** If P dollars are invested in an account that earns interest at 15% compounded annually, the amount available after t years is $A = P(1+0.15)^t$. How many years will it take for \$150 in this account to grow to \$1700? Round your answer to the nearest whole year.
Substitute the known values and simplify. $1500 = 300(1+0.12)^t$ $1500 = 300(1.12)^t$ $5 = (1.12)^t$ Now, take the common logarithm of each side and solve for t. $\log 5 = \log(1.12)^t$ $\log 5 = t(\log 1.12)$ $\dfrac{\log 5}{\log 1.12} = t$ $14.20150519 \approx t$ Thus, it would take approximately 14 years.	

Example	Student Practice
11. At the beginning of 2011, the world population was seven billion people and the growth rate was 1.2% per year. If this growth rate continues, how many years will it take for the population to double to fourteen billion? Use the formula $A = A_0e^{rt}$ and write the population in terms of billions. $A = A_0e^{rt}$ $14 = 7e^{(0.012)t}$ $2 = e^{(0.012)t}$ $\ln 2 = \ln e^{(0.012)t}$ $\ln 2 = 0.012t$ $57.76226505 \approx t$ It would take about 58 years.	**12.** At the beginning of 2011, the mosquito population in a certain county was eight million and the growth rate was 1.7% per year. If this growth rate continues, how many years will it take for the population to double to sixteen million?

Extra Practice

1. Solve. $\log 2x + \log 4 = \log(2x + 12)$

2. Solve. $\ln 8 - \ln x = \ln(x - 7)$

3. Solve $6^x = 4^{x+2}$. Round your answer to the nearest thousandth.

4. If P dollars are invested in an account that earns interest at 6% compounded annually, the amount available after t years is $A = P(1 + 0.06)^t$. How many years will it take for \$4000 in this account to grow to \$7000? Round your answer to the nearest whole year.

Concept Check

Explain how you would solve $26 = 52e^{3x}$.

MATH COACH

Mastering the skills you need to do well on the test.

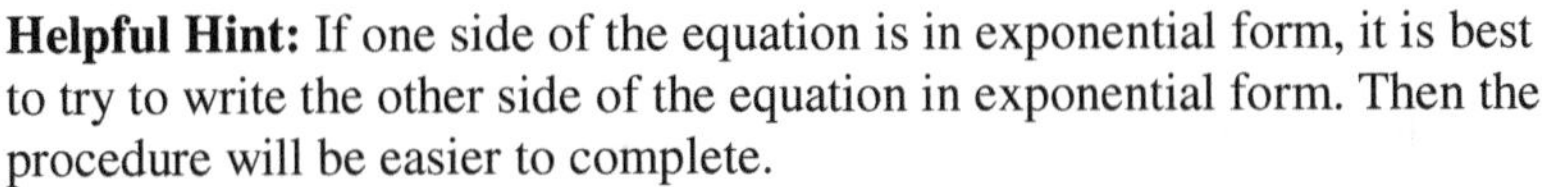

Watch the MATH COACH videos in MyMathLab® or on YouTube while you work the problems below. These helpful hints will help you avoid making common errors on test problems.

Solving an Exponential Equation—Problem 3

Solve. $4^{x+3} = 64$

> **Helpful Hint:** If one side of the equation is in exponential form, it is best to try to write the other side of the equation in exponential form. Then the procedure will be easier to complete.

Did you rewrite the equation as $4^{x+3} = 4^3$?
Yes ____ No ____

If you answered No, remember to write your equation in the form $b^x = b^y$. Note that $64 = 4^3$, and you want the base of the exponent on each side of the equation to be the same number, b, such that $b > 0$ and $b \neq 1$.

Did you use the property of exponential equations to write the equation $x + 3 = 3$?
Yes ____ No ____

If you answered No, remember that if $b^x = b^y$, then $x = y$ for any $b > 0$ and $b \neq 1$. Now solve the equation for x.

If you answered Problem 3 incorrectly, go back and rework the problem using these suggestions.

Using the Properties of Logarithms to Write Sums and Differences of Logarithms as a Single Logarithm—Problem 6 Write as a single logarithm. $2\log_7 x + \log_7 y - \log_7 4$

> **Helpful Hint:** Try your best to memorize the three properties of logarithms in Objectives 11.3.1, 11.3.2, and 11.3.3. They are essential to know when working with logarithmic expressions.

Did you use property 3 to eliminate the coefficient of the first logarithmic term, $2\log_7 x$, and rewrite that term as $\log_7 x^2$?
Yes ____ No ____

If you answered No, review property 3 in Objective 11.3.3 and complete this step again.

Did you use property 1 to combine the sum of the first two logarithmic terms and obtain the expression, $\log_7 x^2 y - \log_7 4$?
Yes ____ No ____

If you answered No, review property 1 in Objective 11.3.1 and complete this step again.

In your last step, you will need to use property 2 in 11.3.2 to combine the difference of two logarithms.

Now go back and rework the problem using these suggestions.

Solving a Logarithmic Equation—Problem 12

Solve the equation and check your solution. $\log_8(x+3)-\log_8 2x=\log_8 4$

Helpful Hint: Use the properties of logarithms to rewrite the equation such that one logarithmic term appears on each side of the equation.

First, did you combine the two logarithms on the left side of the equation and rewrite the equation as

$\log_8\left(\frac{x+3}{2x}\right)=\log_8 4$?

Yes ____ No ____

If you answered No, use property 2 from Objective 11.3.2 to complete this first step again.

Next, did you rewrite the equation as $\frac{x+3}{2x}=4$?

Yes ____ No ____

If you answered No, notice that you now have one logarithmic term on each side of the equation, which is the goal mentioned in the Helpful Hint. You can use property 6 from Objective 11.3.4 to evaluate these two logarithms.

In your final step, solve the equation for x. Check your solution by substituting for x in the original equation, evaluating the logarithms and then simplifying the resulting equation.

If you answered Problem 12 incorrectly, go back and rework the problem using these suggestions.

Solving an Exponential Equation Involving e Raised to a Power—Problem 14

Solve the equation. Leave your answer in exact form. Do not approximate. $e^{5x-3}=57$

Helpful Hint: The first step is to take the natural logarithm of each side of the equation. Then you can simplify the equation further using the properties of logarithms.

Did you take the natural logarithm of each side of the equation to obtain $\ln e^{5x-3}=\ln 57$?

Yes ____ No ____

Next did you rewrite this equation as $(5x-3)(\ln e)=\ln 57$?

Yes ____ No ____

If you answered No to these questions, review property 3 of logarithms from Objective 11.3.3 and complete this step again.

Did you rewrite the equation as $5x-3=\ln 57$?

Yes ____ No ____

If you answered No, remember that $\ln e=\log_e e=1$. This combines the definition of natural logarithms or $\ln e=\log_e e$ and property 4 of logarithms from Objective 11.3.4, which indicates that $\log_e e=1$.

In your final step, solve the equation for x without evaluating $\ln 57$.

Now go back and rework the problem using these suggestions.

Name: ______________________________ Date: ________________
Instructor: ____________________________ Topic: ________________

Module 20 Additional Topics
Topic 1 Quadratic Functions

Vocabulary
vertex • quadratic function

1. A ____________ is a function of the form $f(x) = ax^2 + bx + c$.

2. The ____________ is the lowest point on a parabola opening upward or the highest point on a parabola opening downward.

Example	Student Practice
1. Find the coordinates of the vertex and the intercepts of the quadratic function $f(x) = x^2 - 8x + 15$.	**2.** Find the coordinates of the vertex and the intercepts of the quadratic function $g(x) = x^2 - 7x + 12$.
The vertex occurs at $\frac{-b}{2a} = \frac{-(-8)}{2(1)} = 4.$	
To find the y-coordinate, find $f(4)$.	
$f(4) = 4^2 - 8(4) + 15$ $= 16 - 32 + 15 = -1$	
The vertex is $(4,-1)$. The y-intercept is at $f(0)$.	
$f(0) = 0^2 - 8(0) + 15 = 15$	
The y-intercept is $(0,15)$. The x-intercepts occur when $x^2 - 8x + 15 = 0$. Solve for x.	
$(x-5)(x-3) = 0$ $x - 5 = 0 \quad x - 3 = 0$ $x = 5 \quad\quad x = 3$	
The x-intercepts are $(5,0)$ and $(3,0)$.	

Vocabulary Answers: 1. quadratic function 2. vertex

Example	Student Practice
3. Find the vertex and the intercepts, and then graph the function $f(x)=x^2+2x-4$.	**4.** Find the vertex and the intercepts, and then graph the function $f(x)=x^2-4x+1$.

Since $a>0$, the parabola opens upward. Find the vertex.

$$x=\frac{-b}{2a}=\frac{-2}{2(1)}=\frac{-2}{2}=-1$$

$$f(-1)=(-1)^2+2(-1)-4=-5$$

The vertex is $(-1,-5)$. The y-intercept is at $f(0)$.

$$f(0)=(0)^2+2(0)-4=-4$$

The y-intercept is $(0,-4)$. The x-intercepts occur when $f(x)=0$. We use the quadratic formula.

$$x=\frac{-b\pm\sqrt{b^2-4ac}}{2a}=\frac{-2\pm\sqrt{2^2-4(1)(-4)}}{2(1)}$$
$$=-1\pm\sqrt{5}$$

The x-intercepts are approximately $(-3.2,0)$ and $(1.2,0)$. The vertex is $(-1,-5)$; the y-intercept is $(0,-4)$; and the x-intercepts are approximately $(-3.2,0)$ and $(1.2,0)$. Connect these points by a smooth curve to graph the parabola.

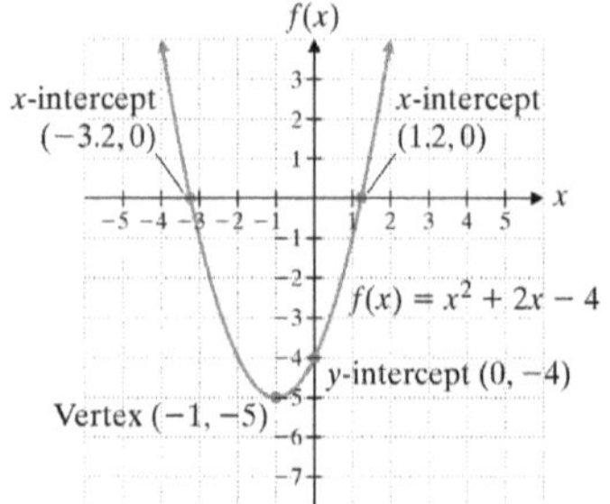

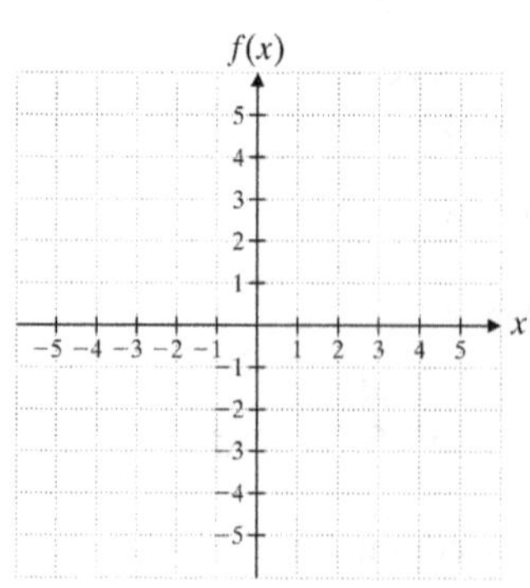

Example	Student Practice
5. Find the vertex and the intercepts, and then graph the function $f(x)=-2x^2+4x-3$.	**6.** Find the vertex and the intercepts, and then graph the function $f(x)=-3x^2-6x-6$.

Since $a<0$, the parabola opens downward. Find the vertex.

$$x=\frac{-4}{2(-2)}=\frac{-4}{-4}=1$$

$$f(1)=-2(1)^2+4(1)-3=-1$$

The vertex is $(1,-1)$. The y-intercept is at $f(0)$. $f(0)=-2(0)^2+4(0)-3=-3$. The y-intercept is $(0,-3)$. The x-intercepts occur when $f(x)=0$. We use the quadratic formula.

$$x=\frac{-4\pm\sqrt{4^2-4(-2)(-3)}}{2(-2)}$$

$$=\frac{-4\pm\sqrt{-8}}{-4}$$

Because $\sqrt{-8}$ yields an imaginary number, there are no x-intercepts for the graph of the function. We will look for three additional points. We try $f(2)$, $f(3)$, and $f(-1)$ and get the points $(2,-3)$, $(3,-9)$, and $(-1,-9)$.

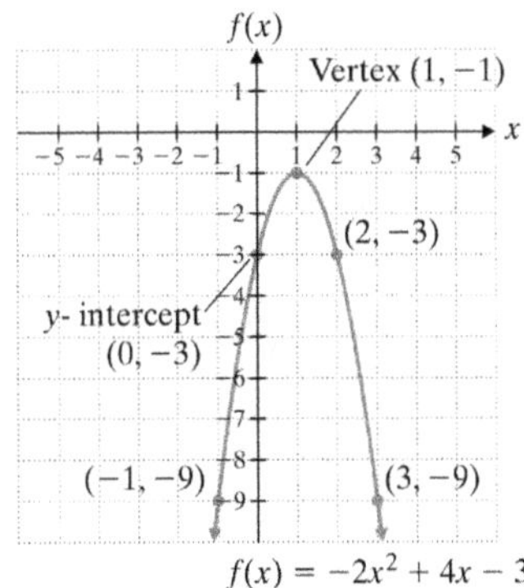

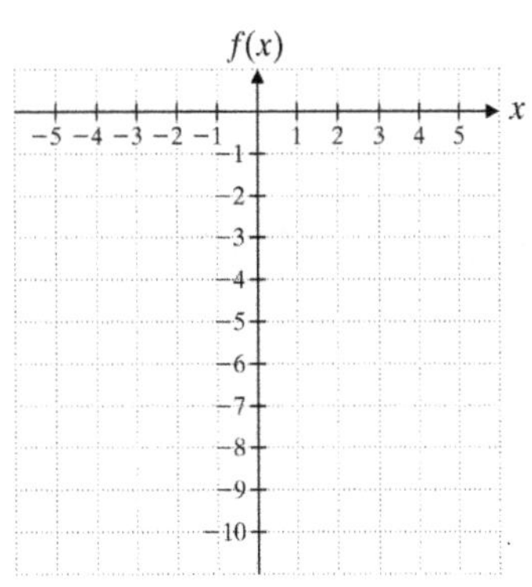

Extra Practice

1. Find the coordinates of the vertex and the intercepts of $g(x) = x^2 + 5x - 6$. When necessary, approximate x- intercepts to the nearest tenth.

2. Find the coordinates of the vertex and the intercepts of $f(x) = 5x^2 + 17x - 12$. When necessary, approximate x- intercepts to the nearest tenth.

3. Find the vertex, the y- intercept, and the x- intercepts (if any exist), and then graph the function $r(x) = 3x^2 - 2x + 1$.

4. Find the vertex, the y- intercept, and the x- intercepts (if any exist), and then graph the function $f(x) = -x^2 + 2x - 1$.

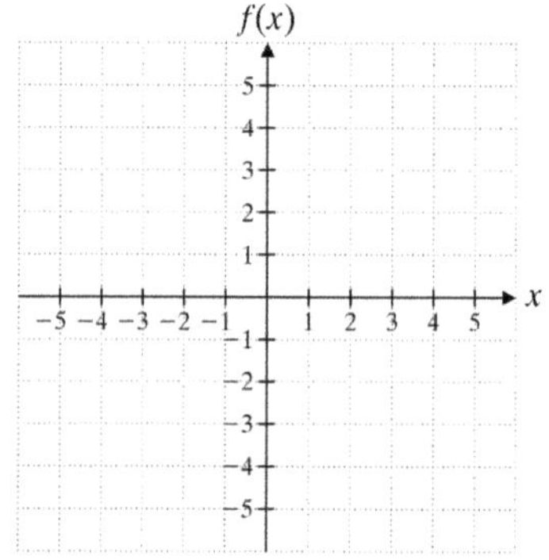

Concept Check

Explain how you would find the vertex of $f(x) = 4x^2 - 9x - 5$.

Name: ______________________________ Date: ________________
Instructor: ____________________________ Topic: ________________

Module 20 Additional Topics
Topic 2 Compound and Quadratic Inequalities

Vocabulary
quadratic inequality • boundary points • compound inequalities
empty set • and • or

1. The notation ∅ represents the ____________.

2. A(n) ____________ has the form $ax^2 + bx + c < 0$ (or replace < by >, ≤, or ≥), where a, b, and c are real numbers and $a \neq 0$.

3. Inequalities that consist of two inequalities connected by the word and or the word or are called ____________.

4. The two points where the expression of a quadratic inequality is equal to zero are called ____________.

Example	Student Practice
1. Graph the values of x where $7 < x$ and $x < 12$. We read the inequality starting with the variable. Thus, we graph all values of x, where x is greater than 7 and where x is less than 12. All such values must be between 7 and 12. Numbers that are greater than 7 and less than 12 can be written as $7 < x < 12$. 5 6 7 8 9 10 11 12 13 14 15	**2.** Graph the values of x where $-2 < x$ and $x < 2$.
3. Graph the region where $x < 3$ or $x > 6$. Read the inequality as "x is less than 3 or x is greater than 6." This includes all values to the left of 3 as well as all values to the right of 6 on a number line. We shade these regions. −3 −2 −1 0 1 2 3 4 5 6 7	**4.** Graph the region where $x < -1$ or $x > 1$.

Vocabulary Answers: 1. empty set 2. quadratic inequality 3. compound inequality 4. boundary points

Example	Student Practice
5. Solve for x and graph the compound solution. $3x+2>14$ or $2x-1<-7$	**6.** Solve for x and graph the compound solution. $4x-6\le -2$ or $5x+2\ge 22$

We solve each inequality separately.

$$3x+2>14 \text{ or } 2x-1<-7$$
$$3x>12 \qquad 2x<-6$$
$$x>4 \qquad x<-3$$

The solution is $x<-3$ or $x>4$.

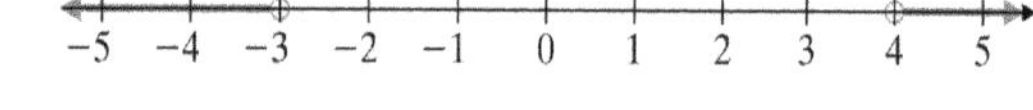

Example	Student Practice
7. Solve and graph $x^2-10x+24>0$.	**8.** Solve and graph $x^2-x-12>0$.

Replace the inequality symbol with an equals sign and solve the resulting equation.

$$x^2-10x+24=0$$
$$(x-4)(x-6)=0$$
$$x-4=0 \text{ or } x-6=0$$
$$x=4 \qquad x=6$$

We use the boundary points to separate the number line into distinct regions.

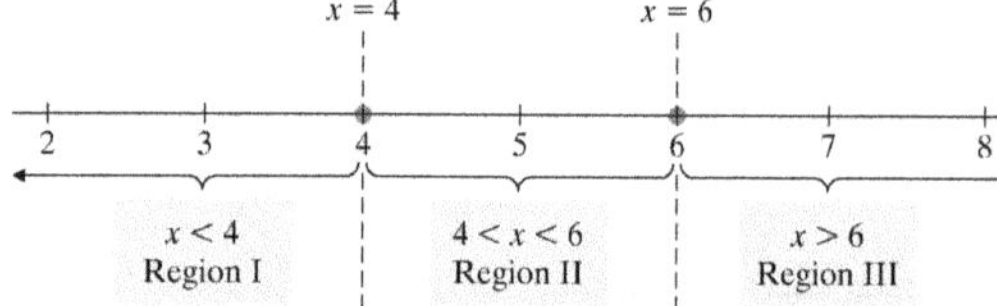

Evaluate the quadratic expression at a test point in each of the regions. Pick the test point $x=1$ and get $15>0$. Pick the test point $x=5$ and get $-1<0$. Pick the test point $x=7$ and get $3>0$. Thus, $x^2-10x+24>0$ when $x<4$ or when $x>6$.

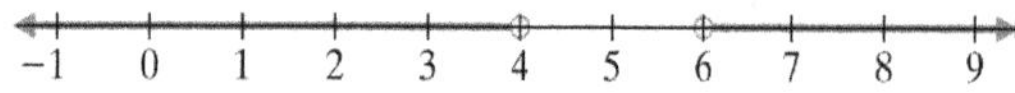

Example

9. Solve and graph $x^2 + 4x > 6$. Round your answer to the nearest tenth.

First we write $x^2 + 4x - 6 > 0$. Because we cannot factor $x^2 + 4x - 6$, we use the quadratic formula to find the boundary points.

$$x = \frac{-4 \pm \sqrt{4^2 - 4(1)(-6)}}{2(1)}$$
$$= \frac{-4 \pm \sqrt{40}}{2}$$
$$= -2 \pm \sqrt{10}$$
$$-2 + \sqrt{10} \approx 1.2 \text{ or } -2 - \sqrt{10} \approx -5.2$$

−5.2 1.2

−7 −6 −5 −4 −3 −2 −1 0 1 2 3

$x < -2 - \sqrt{10}$ Region I

$-2 - \sqrt{10} < x < -2 + \sqrt{10}$ Region II

$x > -2 + \sqrt{10}$ Region III

We will see where $x^2 + 4x - 6 > 0$. Test $x = -6$.

$$(-6)^2 + 4(-6) - 6 = 36 - 24 - 6 = 6 > 0$$

Test $x = 0$.

$$(0)^2 + 4(0) - 9 = 0 + 0 - 6 = -6 < 0$$

Test $x = 2$.

$$(2)^2 + 4(2) - 6 = 4 + 8 - 6 = 6 > 0$$

Our answer is $x < -5.2$ or $x > 1.2$.

−5.2 1.2

−7 −6 −5 −4 −3 −2 −1 0 1 2 3

Student Practice

10. Solve and graph $x^2 + 4x > 1$. Round your answer to the nearest tenth.

Extra Practice

1. Graph the values of x that satisfy the conditions given. $-4 < x \leq \frac{1}{2}$

2. Graph the values of x that satisfy the conditions given. $x \leq 4$ or $x \geq \frac{13}{2}$

3. Solve for x and graph your results. $x - 3 \geq 2$ or $x + 1 \leq 2$

4. Solve and graph. $2x^2 - 3x - 5 \leq 0$

Concept Check

Explain what happens when you solve the inequality $x^2 + 2x + 8 > 0$.

Name: ______________________________ Date: ______________

Instructor: ___________________________ Topic: ______________

Module 20 Additional Topics
Topic 3 Absolute Value Equations and Inequalities

Vocabulary

absolute value • $|a| = |b|$ • $|ax+b| = c$ $|ax+b| < c$ • $|ax+b| > c$

1. The ____________ of a number x can be pictured as the distance between 0 and x on the number line.

2. ____________ is equivalent to $ax+b < -c$ or $ax+b > c$.

Example	Student Practice
1. Solve and check your solutions. $|2x+5| = 11$	**2.** Solve and check your solutions. $|4x+6| = 18$

The solutions of an equation of the form $|ax+b| = c$, where $a \neq 0$ and c is a positive number, are those values that satisfy $ax+b = c$ or $ax+b = -c$. Thus, we have the following:

$$\begin{array}{ccc} 2x+5=11 & \text{or} & 2x+5=-11 \\ 2x=6 & & 2x=-16 \\ x=3 & & x=-8 \end{array}$$

The two solutions are 3 and −8. Check the solutions.

$$\begin{array}{cc} |2x+5| = 11 & |2x+5| = 11 \\ |2(3)+5| \stackrel{?}{=} 11 & |2(-8)+5| \stackrel{?}{=} 11 \\ |6+5| \stackrel{?}{=} 11 & |-16+5| \stackrel{?}{=} 11 \\ |11| \stackrel{?}{=} 11 & |-11| \stackrel{?}{=} 11 \\ 11 = 11 & 11 = 11 \end{array}$$

The solutions check.

Vocabulary Answers: 1. absolute value 2. $|ax+b| > c$

Example	Student Practice
3. Solve $\lvert 3x-1\rvert+2=5$ and check your solutions.	**4.** Solve $\lvert 6x-3\rvert-4=5$ and check your solutions.

First rewrite the equation so that the absolute value expression is alone on one side of the equation.

$$\lvert 3x-1\rvert+2=5$$

$$\lvert 3x-1\rvert+2-2=5-2$$

$$\lvert 3x-1\rvert=3$$

Now solve $\lvert 3x-1\rvert=3$ for x.

$3x-1=3$ or $3x-1=-3$

$x=\frac{4}{3}$ $\quad$ $x=-\frac{2}{3}$

Example	Student Practice
5. Solve and check. $\lvert 3x-4\rvert=\lvert x+6\rvert$	**6.** Solve and check. $\lvert x+4\rvert=\lvert 4x+6\rvert$

The solutions of the given equation must satisfy $3x-4=x+6$ or $3x-4=-(x+6)$. Solve each equation to get $x=5$ and $x=-\frac{1}{2}$.

Check the solutions by substituting them into the original equation.

$x=5$: $\lvert 3(5)-4\rvert\stackrel{?}{=}\lvert 5+6\rvert$

$$\lvert 11\rvert=\lvert 11\rvert$$

$x=-\frac{1}{2}$: $\left\lvert 3\left(-\frac{1}{2}\right)-4\right\rvert\stackrel{?}{=}\left\lvert -\frac{1}{2}+6\right\rvert$

$$\left\lvert -\frac{3}{2}-4\right\rvert\stackrel{?}{=}\left\lvert -\frac{1}{2}+6\right\rvert$$

$$\left\lvert -\frac{3}{2}-\frac{8}{2}\right\rvert\stackrel{?}{=}\left\lvert -\frac{1}{2}+\frac{12}{2}\right\rvert$$

$$\frac{11}{2}=\frac{11}{2}$$

Example	Student Practice
7. Solve and graph the solution. $\|x\| \le 4.5$ The inequality $\|x\| \le 4.5$ means that x is less than or equal to 4.5 units from 0 on a number line. We draw a picture. −5 −4 −3 −2 −1 0 1 2 3 4 5 Thus, the solution is $-4.5 \le x \le 4.5$.	**8.** Solve and graph. $\|x\| \le 3$
9. Solve and graph the solution. $\|x+5\| \le 10$ We want to find the values of x that make $-10 \le x+5 \le 10$ a true statement. We need to solve the compound inequality. Subtract 5 from each part. $-10-5 \le x+5-5 \le 10-5$ $-15 \le x \le 5$ Thus, the solution is $-15 \le x \le 5$. We graph this solution. −25 −20 −15 −10 −5 0 5 10 15 20 25	**10.** Solve and graph the solution. $\|x+2\| \le 3$
11. Solve and graph the solution. $\|x\| \ge 5\frac{1}{4}$ The inequality means that x is more than $5\frac{1}{4}$ units from 0 on a number line. 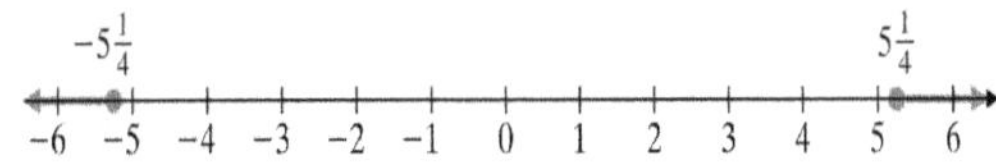 The solution is $x \le -5\frac{1}{4}$ or $x \ge 5\frac{1}{4}$.	**12.** Solve and graph the solution. $\|x\| \ge 4$

Example	Student Practice
13. Solve and graph the solution. $\|-3x+6\|>18$	**14.** Solve and graph the solution. $\|-2x+5\|\geq 3$

Remember to reverse the inequality sign when dividing by a negative number.

$-3x+6>18$ or $-3x+6<-18$

$-3x>12$ $\quad -3x<-24$

$\frac{-3x}{-3}<\frac{12}{-3}$ $\quad \frac{-3x}{-3}>\frac{-24}{-3}$

$x<-4$ $\quad x>8$

−10 −8 −6 −4 −2 0 2 4 6 8 10

Extra Practice

1. Solve. Check your solutions. $|2x+9|=31$

2. Solve. Check your solutions. $|3m+2|-10=-6$

3. Solve and graph the solution. $|x-3|\leq 3$

4. Solve. In a certain company, the measured thickness t of a computer chip must not differ from the standard s by more than 0.05 millimeter. The engineers express this requirement as $|t-s|\leq 0.05$. Find the limits of t if the standard s is 1.33 millimeters.

Concept Check

Explain what happens when you try to solve for x. $|7x+3|<-4$

MATH COACH

Mastering the skills you need to do well on the test.

Watch the MATH COACH videos in MyMathLab® or on YouTube™ while you work the problems below. These helpful hints will help you avoid making common errors on test problems.

Find the Vertex, Intercepts and Graph of a Quadratic Function—Problem 17

Find the vertex and the intercepts of $f(x) = -x^2 - 6x - 5$. Then graph the function.

> **Helpful Hint:** When the function is written in $f(x) = ax^2 + bx + c$ form, we can find the vertex using the vertex formula. We can solve for the intercepts using the substitutions $x = 0$ and $f(x) = 0$ to find the unknown coordinates. And, if $a < 0$, the graph is a parabola opening downward.

Do you see that $a = -1$, $b = -6$, and $c = -5$?
Yes ____ No ____

Did you use the vertex formula to discover that the vertex point has an x-coordinate of -3? Yes ____ No ____

If you answered No to these questions, notice that the function is written in $f(x) = ax^2 + bx + c$ form and review the vertex formula: $x = \frac{-b}{2a}$. Substitute the resulting value for x into the original function to find the y-coordinate of the vertex point.

To find the y-intercept, did you substitute 0 for x into the original function to find the value for y?
Yes ____ No ____

After letting $f(x) = 0$ and substituting the values for a, b, and c into the quadratic formula, did you get the expression $x = \frac{-(-6) \pm \sqrt{(-6)^2 - 4(-1)(-5)}}{2(-1)}$? Yes ____ No ____

If you answered No to these questions, remember that the y-intercept will be an ordered pair in the form $(0, y)$ or in this case, $(0, f(x))$, and the x-intercept will be an ordered pair in the form $(x, 0)$. Be careful when substituting values for a, b, and c into the quadratic formula and remember to evaluate $\sqrt{16}$ as both 4 and -4. Simplify the expression to find the possible x-values.

Since $a < 0$, the parabola will open downward. Plot the vertex, x-intercept, and y-intercept points and connect these points with a curve to find the graph of the function.

If you answered Problem 17 incorrectly, go back and rework the problem using these suggestions.

Worksheet Answers Module 1

Topic 1

Student Practice

2. (a) the hundred thousands place
   (b) the thousands place
4. $5,000,000 + 300,000 + 2000 + 700 + 60 + 9$
6. 3 hundred-dollar bills, 8 ten-dollar bills, and 9 one-dollar bills
8. (a) eight thousand five hundred ninety-two
   (b) five million, two hundred thirty thousand, eighty-nine
10. (a) <
    (b) >
12. (a) $4 < 9$
    (b) $8 > 2$
14. 65,700
16. 5,680,000

Extra Practice

1. eighty thousand, fifty-nine; $80,000 + 50 + 9$
2. >
3. $98 < 114$
4. 2900

Concept Check

Answers may vary. To round 8937 to the nearest hundred, first identify the round-off place digit: $8\underline{9}37$. The digit to the right is less than 5. Do not change the round-off place digit. Replace all digits to the right with zeros. 8900.

Topic 2

Student Practice

2. (a) $6 + 2$
   (b) $x + 10$
4. The sums are shown at the top of the next column. Notice that we only need to learn 3 addition facts because the remaining are either a repeat of these or use the addition property of zero.

   $6 + 0 = 6$
   $5 + 1 = 6$
   $4 + 2 = 6$
   $3 + 3 = 6$
   $2 + 4 = 6$
   $1 + 5 = 6$
   $0 + 6 = 6$
6. $n + 12$
8. 28
10. 1547 people
12. 21 m

Extra Practice

1. $a + (2 + 7)$; $a + 9$
2. 1845
3. 10,951
4. 38 in.

Concept Check

(a) The 1 that is placed above the 9 is in the tens place. Its value is 1×10 or 10.
(b) The 1 that is placed above the 3 is in the hundreds place. Its value is 1×100 or 100.

Topic 3

Student Practice

2. (a) 3
   (b) 0
4. 647
6. $8 - 5$
8. 4
10. 461
12. 102 ft

Extra Practice

1. $400 - 301$
2. 500
3. 41,102
4. $55

Concept Check

Answers may vary. One possible solution follows. We can borrow only from a place value that has a nonzero whole number. When we subtract 35 from 800 we cannot subtract 5 from 0, so we must change 800 to 790 and 10 ones.

Topic 4

Student Practice

2.

```
                   ★★
                   ★★
                  5★★
  2★★★★★           ★★
   ★★★★★           ★★
     5              2
```

Both arrays consist of 10 items.

4. 3 and b are the factors and 24 is the product.
6. $5 \cdot n = 5n$
8. 180
10. $210n$
12. 137,400
14. 340,725
16. $304

Extra Practice

1. Associative property of multiplication
2. $4 \cdot 2$
3. 7,528,014
4. 75 CDs

Concept Check

Answers may vary. One possible solution follows. Before we multiply 546 by 2000, we first separate the 2000 into its nonzero digits (2) and trailing zeros (000). We then multiply the nonzero digits by 546, or 2×546. To this product we then add to the right side the number of trailing zeros.

Topic 5

Student Practice

2. $180 \div 9$
4. $4 \div 48$
6. 4
8. (a) 1
   (b) Undefined
   (c) 0
10. 315 R 30
12. 3 candy bars

Extra Practice

1. Undefined
2. 98 R 5
3. $52 \div x$ (Note: Choice of variable may vary.)
4. 12 plants

Concept Check

The following division problem has been partially solved.

```
     2
13)2645
   26
    04
```

The next step is to ask how many times the divisor (13) will go into the dividend (04). Because it doesn't, the next numeral in the quotient is 0.

```
     20
13)2645
   26
    04
```

Next we multiply the 0 in the quotient by the divisor (13) and subtract the product from the dividend.

```
     20
13)2645
   26
    04
    00
     4
```

Topic 6

Student Practice

2. (a) 7^8
   (b) $9^3 a^4$
4. (a) $y \cdot y \cdot y \cdot y \cdot y \cdot y \cdot y$
   (b) $9 \cdot 9 \cdot 9 \cdot 9 \cdot 9 \cdot 9$
6. (a) 1024

(b) 1
(c) 1000

8. 1,000,000,000
10. 625
12. (a) 12^2
(b) 5^7
14. 43
16. 4

Extra Practice

1. a^5
2. y^6
3. 32
4. 24

Concept Check

Answers may vary. One possible solution follows.
First, evaluate terms with exponents other than 1.
$50+3\times5^2\div25=50+3\times25\div25$
Next, evaluate multiplication operations.
$50+3\times25\div25=50+75\div25$
Next, evaluate division operations.
$50+75\div25=50+3$
Lastly, we evaluate addition operations.
$50+3=53$

Topic 7

Student Practice

2. (a) $8\cdot a+11$
(b) $4(n+6)$
4. 4
6. 6
8. $32+4y$
10. $9y+41$

Extra Practice

1. $8(x-1)$
2. 5
3. 144
4. $3x+y+10$

Concept Check

Answers may vary. One possible solution is to first simplify $5(x+1)$ using the distributive property.
$5(x+1)=5x+5$
Next, evaluate the simplified expression for $x=2$, by substituting 2 for *x*.
$5x+5=5(2)+5=10+5=15$

Topic 8

Student Practice

2. $6b$
4. $11x+15y$
6. $(x+3y)+(6x+y)+(5x+8y)$
$=12x+12y$
8. No
10. $n=5$
12. $a+9=13;\ a=4$
14. $3x=21;\ x=7$

Extra Practice

1. $18ab+11$
2. The sum of a number and eight is twelve.
3. $a=9$
4. $3+x=12$

Concept Check

Answers may vary. One possible solution follows.
The first step in both processes is to combine like terms.
$3x+x+2x=3x+1x+2x$
$=(3+1+2)x=6x$
At this point the process for (a) is complete. Because (b) has an equals sign, it can be solved by isolating the *x*. To isolate, and therefore solve for *x*, divide both sides of the equation by the coefficient of *x*.
$6x=12$
$\frac{x}{6}=\frac{12}{6}$
$x=2$

Topic 9

Student Practice

2. $22,000
4. 8250 frequent-flyer points
6. Assistant office manager

Extra Practice

1. Bedspread $200; Digital Camera $300; Fountain $100; Stained Glass $500
2. $1100
3. $595
4. 90 pages; 180 pages

Concept Check

Answers may vary. One possible solution is to list all deposits to her vacation account over the six months.

Description of deposit	deposit amount
Initial balance	$200
Monthly deposit $\times$ months $\$100 \times 6 = \$600$	$600
Tax return deposit $\div$ 2 $\$900 \div 2 = \$450$	$450
Sum in vacation account	$1250

Sahara will have $1250 after six months in her vacation account. She will not have enough money to take a $1500 vacation.

Worksheet Answers Module 2

Topic 1

Student Practice

2. −5 −4 −3 −2 −1 0 1 2 3 4 5
4. (a) <
   (b) <
6. (a) +\$50
   (b) −30 ft
8. −5 −4 −3 −2 −1 0 1 2 3 4 5
10. 2
12. >
14. −4
16. (a) Albany
    (b) Albany, Boston, and Buffalo

Extra Practice

1. −5 −4 −3 −2 −1 0 1 2 3 4 5
2. 21, 21
3. 42
4. −4 has the greater opposite. The opposites are 4 and −5, and 4 is greater. (Explanations may vary.)

Concept Check

Answers may vary. Possible solution: The numbers are arranged left to right on the number line. They are arranged from smallest to largest.

$-10,\ -6,\ -4,\ -1,\ 0$

Topic 2

Student Practice

2. (a)

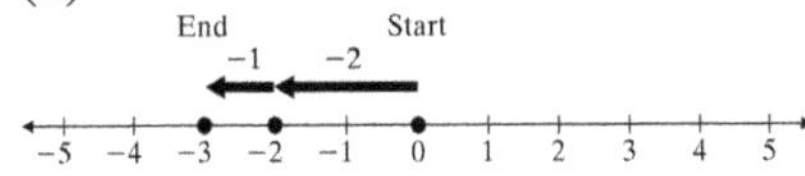

   (b) $-2+(-1)$
   (c) −3
4. −11
6. −7°F
8. 2
10. 18
12. −3
14. 1
16. \$20,000

Extra Practice

1. 42
2. −1
3. −6
4. 175 feet

Concept Check

Answers may vary. Possible solution: Since the values given for x, y, and z are all negative, their sum is negative. However, it should be obvious that the absolute value of this sum is less than 132. So when the sum of x, y, and z is added to 132, we keep the sign of 132, which is positive.

Topic 3

Student Practice

2. (a) −\$25
   (b) −4
4. (a) $50+(-25)=25$
   (b) $8+(-5)=3$
6. (a) −5
   (b) −6
8. (a) −2
   (b) −18
   (c) 10
10. −13
12. 7
14. 3720 ft

Extra Practice

1. 8
2. −3
3. −14
4. 5:00 P.M.

Concept Check

Answers may vary. The problem was not completed correctly. First, subtraction is written as addition of the opposite.

$-6-(-3)+(-7)=-6+3+(-7)$

Next, the like signs are combined.

$-6+3+(-7)=-13+3$

Next, unlike signs are combined.

$-13+3=-10$

Topic 4

Student Practice

2. -32
4. (a) 15
   (b) -15
6. 240
8. 4
10. (a) -16
    (b) 16
12. (a) -4
    (b) 4
    (c) -4
14. 60
16. (a) -8
    (b) -1

Extra Practice

1. -28
2. 9
3. -1
4. -8

Concept Check

Answers may vary. One possible solution is it is true because an even number of negative signs occurs in the expression, the product of which must be positive.

Topic 5

Student Practice

2. 0
4. 8
6. -1
8. 6

Extra Practice

1. -27
2. -9
3. 10
4. $570+2(125)+4(75)$; \$1120

Concept Check

Answers may vary. One possible solution is to first evaluate terms with exponents.

$3^2+5(2-4)=9+5(2-4)$

Next, evaluate operations inside parentheses.

$9+5(2-4)=9+5(-2)$

Next, evaluate multiplication operations.

$9+5(-2)=9-10$

Lastly, combine like terms.

$9-10=-1$

Topic 6

Student Practice

2. $-5y+5x$
4. $4m+2n$
6. (a) 5
   (b) $5x$
8. $-3m-2n-5mn$
10. -2
12. $-7x+14$
14. $s=-194$; 194 feet per second, downward

Extra Practice

1. $109-3a-6ab+5b$
2. -96
3. $7a+42$
4. $13a+1$

Concept Check

Answers may vary. The possible alternatives to writing $-3b+7$ are $7-3b$ or $7+(-3b)$.

Worksheet Answers Module 3

Topic 1

Student Practice

2. 5
4. $x = -26$
6. $-9 = y$
8. (a) $\angle a = \angle b - 40°$
   (b) $100° = \angle b$
10. $x = 120°$, $\angle b = 124°$

Extra Practice

1. -124
2. $n = 38$
3. $x = 12$
4. $\angle a = 77°$

Concept Check

Answers may vary. It is not correct. He should have added 9 to each side.

Topic 2

Student Practice

2. -17
4. $x = -19$
6. $y = 10$
8. $N = 7Q$
10. (a) $C = 8B$
    (b) $B = 16$
12. 17 shares

Extra Practice

1. $M = 7E$
2. $x = 3$
3. $-6 = x$
4. 23 feet

Concept Check

Answers may vary. First multiply $3(2x)$ to get $6x$. Then combine the like terms, $4x$ and $6x$, to get $10x$. Finally divide each side by 10 and simplify to get $x = -2$.

Topic 3

Student Practice

2. 56 yards
4. 3.75 ft
6. 54 ft^2
8. 81 ft^2
10. $b = 23$ ft
12. $W = 7$ m

Extra Practice

1. $P = 60$ yd; $A = 225$ yd^2
2. 285 $in.^2$
3. 80 ft^3
4. 16 ft^3

Concept Check

Answers may vary.

(a) To determine the amount of sand needed, find the volume.
(b) To find the volume of a box, use the formula $V = LWH$.
(c) Since the length is double the height, the length is 10 inches. The volume is given in cubic feet, and the length and height are given in inches. First change the units on these measurements so that all are in terms of the same unit. Then write an equation using the formula $V = LWH$, where V, L, and H are replaced with the values from the previous step, and solve for W.

Topic 4

Student Practice

2. $7 \cdot 7 \cdot 7 \cdot 7 \cdot 7 \cdot 7 \cdot 7 \cdot 7$, 7^8
4. (a) 3^7
   (b) $3^2 \cdot 2^5$
6. $168a^8b^5$
8. $-56a^3b^6$
10. (a) Trinomial
    (b) Monomial
    (c) Binomial
12. $5x^8 + 6x^7 - 24x^4$
14. $A = 5x^9 - 4x^5$

Extra Practice

1. 10^{18}
2. $-18a^6 + 42a^5$
3. $-4x^4 - x^3y + 20x^3 + y$
4. $V = 60n^2 - 60n$

Concept Check

Answers may vary.

(a) Multiply both terms in parentheses by $2x^4$ to obtain $2x^4 \cdot x^2 + 2x^4 \cdot y$. Then add exponents on the powers of x being multiplied in the first term: $2x^6 + 2x^4y$.

(b) Add the exponents on the powers of x being multiplied: $-3x^6$.

(c) To complete the problem, add the simplified forms obtained in parts (a) and (b): $\left(2x^6 + 2x^4y\right) - 3x^6$. Since $2x^6$ and $-3x^6$ are like terms, complete the simplification by combining these terms: $-x^6 + 2x^4y$.

Worksheet Answers Module 4

Topic 1

Student Practice

2. (a) 3 and 5
   (b) 2, 3, and 5
4. neither: 0; prime: 5, 11, 19, 29, 37, 41; composite: 9, 18, 34, 60
6. $2\cdot2\cdot2\cdot2\cdot2\cdot2$ or 2^6
8. $2\cdot2\cdot2\cdot2\cdot5$ or $2^4\cdot5$
10. $2^4\cdot3\cdot7$
12. $2^3\cdot3^2$

Extra Practice

1. 3
2. composite
3. $2\cdot7^2$
4. $2^5\cdot3^2\cdot5^2$

Concept Check

(a) Answers may vary. One possible solution is by noticing that the number ends in an even number.
(b) If any whole number ends with an even number, the number has a factor of 2.

Topic 2

Student Practice

2. (a) $\frac{5}{6}$
   (b) $\frac{2}{9}$
   (c) 1
4. (a) 1
   (b) 0
6. $\frac{5}{37}$
8. (a) Improper fraction
   (b) Improper fraction
   (c) Mixed number
10. $7\frac{1}{4}$
12. $\frac{49}{5}$

Extra Practice

1. $\frac{3}{8}$
2. $\frac{13}{25}$
3. $10\frac{3}{8}$
4. $\frac{151}{20}$

Concept Check

Answers may vary. One possible solution is to multiply the whole number portion by the denominator. This product is then added to the numerator.

Topic 3

Student Practice

2. $\frac{28x}{32x}$
4. $\frac{35x}{49x}$
6. $-\frac{9}{4}$
8. $\frac{a}{3}$
10. (a) $\frac{7}{40}$
    (b) $\frac{59}{80}$

Extra Practice

1. $\frac{40a}{64a}$
2. $\frac{6x}{45x}$
3. $\frac{5x}{8}$
4. $\frac{3}{4}$

Concept Check

Answers may vary. One quick way to determine that the fraction may be reduced is to see that both the numerator and the denominator are divisible by three. The common factor of three will cancel.

Topic 4

Student Practice

2. (a) 5^4

   (b) $\frac{1}{a^5}$

4. $\frac{1}{2a^3}$

6. 5^{18}

8. (a) 5^8

   (b) 1

   (c) $5^7 y^{21}$

10. $\frac{x^4}{625}$

Extra Practice

1. $\frac{2b^2}{3a^2}$

2. $\frac{5y^8}{7z^4}$

3. $\frac{2^2a^4b^6}{c^8}$

4. $2^3 3^2 y^{12} z^4$

Concept Check

Answers may vary. One possible approach is to simplify the rational expression in the parentheses first. The result is the quantity $2x^2$, raised to the third power. Distribution of the exponents to the quantity $2x^2$ yields $8x^6$.

Topic 5

Student Practice

2. (a) $\frac{5}{7}$

   (b) $\frac{4}{9}$

4. 19 miles per gallon

6. 46 tornados per year

8. (a) 15

   (b) 20

   (c) 16

10. (a) \$7 per towel; \$9 per towel

    (b) The package of 8 towels

Extra Practice

1. $\frac{11}{2}$

2. $\frac{47}{23}$

3. $\frac{5}{7}$

4. pork

Concept Check

Answers may vary. One possible solution is to write a ratio.

$$\frac{8 \text{ sales people}}{160 \text{ customers}} = \frac{1 \text{ sales person}}{20 \text{ customers}}$$

The answer is that there are 0.05 sales people per customer, or 1 sales person per 20 customers.

Topic 6

Student Practice

2. $\frac{5 \text{ hours}}{150 \text{ miles}} = \frac{7 \text{ hours}}{210 \text{ miles}}$

4. $\frac{5}{17} \neq \frac{19}{24}$

6. This is a proportion.

8. $n = 8$

10. 56 feet

12. \$3200

Extra Practice

1. $\frac{4}{36} \neq \frac{14}{117}$

2. $x = 169$

3. 70 minutes

4. 15 books

Concept Check

Answers may vary. One possible method is to write a proportion. Let $x =$ amount to Justin.

$$\frac{5}{7} = \frac{x}{840}$$
$$5(840) = 7x$$
$$4200 = 7x$$
$$600 = x$$

Justin will receive \$600 when Sara receives \$840.

Worksheet Answers Module 5

Topic 1

Student Practice

2. $\frac{2}{5}$
4. $\frac{3}{10}$
6. $4x^7$
8. 25 ft
10. (a) -3
    (b) $\frac{1}{x}$
12. $-\frac{119}{195}$
14. $-\frac{5x^3}{18}$
16. $6\frac{1}{4}$ feet

Extra Practice

1. $-\frac{3}{19}$
2. $\frac{9}{35x}$
3. $-\frac{3}{5}$
4. $12x^7$

Concept Check

Answers may vary. One possible solution is to first write both terms with a denominator.

$$\frac{-16x^2}{3} \div \frac{8x}{1}$$

Next rewrite the expression as a multiplication problem by multiplying the first term by the reciprocal of the second term.

$$\frac{-16x^2}{3} \cdot \frac{1}{8x}$$

Multiply, and simplify.

$$-\frac{2x}{3}$$

Topic 2

Student Practice

2. (a) $6y$, $12y$, $18y$, $24y$, $30y$, $36y$, $42y$
   $10y$, $20y$, $30y$, $40y$, $50y$, $60y$, $70y$
   (b) $30y$
4. 18
6. 2856
8. $20x^3$
10. 2 P.M.

Extra Practice

1. 180
2. 42
3. $560x^3$
4. 12:06 A.M.

Concept Check

Answers may vary. No, the LCM of $63x^2$ and $75x^3$ has two factors of 3 since $63x^2$ has two factors of 3 and three factors of x since $75x^3$ has three factors of x.

Topic 3

Student Practice

2. $\frac{13}{17}$
4. $-\frac{7}{5}$
6. (a) $\frac{7}{5y}$
   (b) $\frac{x-3}{8}$
8. 100
10. (a) 30
    (b) 5
12. $\frac{8}{45}$
14. $\frac{7x}{25}$

16. $\frac{8}{45}$

Extra Practice

1. $\frac{2}{15}$
2. $\frac{23}{50}$
3. $\frac{x}{6}$
4. $\frac{23}{40}$

Concept Check

(a) $20 = 2 \cdot 2 \cdot 5$

$6 = 2 \cdot 3$

$\text{LCD} = 2 \cdot 2 \cdot 3 \cdot 5 = 2^2 \cdot 3 \cdot 5 = 60$

(b) Answers may vary. One possible solution is to multiply each term by a rational number, the numerator and denominator of which are the quotient of the LCD divided by the denominator of each respective term.

Topic 4

Student Practice

2. $6\frac{6}{7}$
4. $12\frac{13}{20}$
6. $8\frac{19}{36}$
8. $1\frac{4}{5}$
10. $3\frac{1}{4}$
12. $\frac{156}{7}$ or $22\frac{2}{7}$
14. $-\frac{29}{24}$ or $-1\frac{5}{24}$
16. 8 times

Extra Practice

1. $14\frac{9}{14}$
2. $8\frac{13}{16}$
3. $\frac{481}{40}$ or $12\frac{1}{40}$
4. $-\frac{34}{3}$ or $-11\frac{1}{3}$

Concept Check

Answers may vary. One possible solution is to first express both terms as improper fractions. Multiply the numerators, multiply the denominators, then simplify.

Topic 5

Student Practice

2. $\frac{431}{875}$
4. $\frac{3}{128}$
6. $3x$
8. $\frac{45}{116}$
10. 203 inches of yarn

Extra Practice

1. $\frac{1}{125}$
2. $\frac{7}{100}$
3. $6x$
4. $\frac{1}{2}$

Concept Check

Answers may vary. One possible solution is to first multiply both the numerator and the denominator by 2, eliminating the denominator.

$$\frac{1+2\times 3}{\frac{1}{2}} = 2(1+2\times 3)$$

Next, perform the multiplication operation inside the parentheses, then add that product to 1. Multiply that sum by the 2 outside the parentheses. The expression is completely simplified.

Topic 6

Student Practice

2. (a) $27\frac{1}{2}$ feet by $23\frac{1}{2}$ feet

   (b) $357

4. 46 pieces of wood

Extra Practice

1. $4\frac{5}{7}$
2. 9 pounds of meat; $16\frac{1}{5}$ pounds of potato salad; $13\frac{1}{2}$ pounds of fruit
3. 5 bundles of plastic tubing
4. $106\frac{2}{3}$ parts cement

Concept Check

(a) Subtract
(b) Multiply
(c) Divide

Topic 7

Student Practice

2. $y = 60$
4. $y = 6$
6. $y = 14$

Extra Practice

1. $x = 176$
2. $x = 135$
3. $x = -112$
4. $x = 35$

Concept Check

Answers may vary. One possible solution is that Amy intended to eliminate the denominator. Amy failed, however, to include the sign of the denominator, and this error yielded an answer with the right absolute value, but the wrong sign.

Worksheet Answers Module 6

Topic 1

Student Practice

2. (a) Seven hundred thirty-four thousandths
   (b) Six and twenty-three hundredths
4. Three hundred forty-three and $\frac{51}{100}$
6. $1\frac{4928}{10,000}$
8. 1.023
10. >
12. 34.055
14. 151.00

Extra Practice

1. Three and forty-eight hundredths
2. 19.243
3. <
4. 3.142

Concept Check

Answers may vary. One possible solution follows:
There are four places after the decimal so there are four zeros in the denominator.

$$8.6711 = 8\frac{6711}{10,000}$$

Topic 2

Student Practice

2. 32.07
4. 27.86
6. 7.59
8. $9.7b - 8.2a$
10. −1.32
12. $52
14. (a) 5.4 rebounds
    (b) 114.6 points

Extra Practice

1. 28.374
2. −6.4
3. $2.63x + 3.4y$
4. 12.26

Concept Check

Answers may vary. One possible solution follows:
We replace the variable with 0.866.
$x - 3.1 = 0.866 - 3.1$
We keep the sign of the larger absolute value and subtract.

$$\begin{array}{r} -3.100 \\ \underline{0.866} \\ -2.234 \end{array}$$

Topic 3

Student Practice

2. 94.798
4. −39.65
6. 425,600
8. 0.45
10. −0.050
12. 4.55
14. 0.8125
16. $3.58\overline{3}$

Extra Practice

1. −45.69
2. 2368.1
3. 3.25
4. $11.2\overline{6}$

Concept Check

Answers may vary. One possible solution follows:
Check Marc's answer.

$$0.097 \times 0.5 \stackrel{?}{=} 0.485$$

$$\begin{array}{r} 0.097 \\ \underline{\times \quad 0.5} \\ 0.0485 \end{array}$$

$$0.0485 \stackrel{?}{=} 0.485$$

No, the answers are not the same.
$0.0485 \neq 0.485$

Topic 4

Student Practice

2. (a) 7500
   (b) 750
4. (a) 320
   (b) 640
   (c) 352
6. $2
8. (a) $230
   (b) $70
   (c) $160

Extra Practice

1. 120
2. 240
3. 16,800
4. $14,000

Concept Check

Answers may vary. Two possible solutions follow:

Find 35% of 200.

1. 35% of 200
$$= 3\times(10\% \text{ of } 200)+5\times(1\% \text{ of } 200)$$
$$= 3\times 20+5\times 2$$
$$= 60+10$$
$$= 70$$
2. $$35\% \text{ of } 200 = 35\times(1\% \text{ of } 200)$$
$$= 35\times 2$$
$$= 70$$

Topic 5

Student Practice

2. 3%
4. 134%
6. 0.5%
8. (a) 0.015
   (b) 0.8%
10.

Decimal Form	Percent Form
0.423	42.3%
0.644	64.4%
0.003	0.3%
90.1	9010%

12. (a) 53.6%
    (b) $\frac{87}{250}$
14. $\frac{1}{250}$

Extra Practice

1. 2.45
2. 7%
3. 533.33%
4. $\frac{1}{800}$

Concept Check

Answers may vary. One possible solution follows:

Change 0.43% to a decimal. Move the decimal point two places to the left.

$0.43\% = 0.0043$

Then change it to a fraction. There are 4 decimal places, so there are 4 zeros in the denominator.

$$0.43\% = 0.0043 = \frac{43}{10{,}000}$$

Topic 6

Student Practice

2. (a) $n = 30\%\times 30;\ n = 9$
   (b) $9 = 30\%\times n;\ n = 30$
   (c) $9 = n\%\times 30;\ n = 30$
4. $80 = n\%\times 50;\ n = 160$
6. 41.76
8. 81.25%
10. 16.77%
12. $184

Extra Practice

1. $n = 83\%\times 155;\ n = 128.65$
2. $240 = 80\%\times n;\ n = 300$
3. $280 = n\%\times 70;\ n = 400$
4. $5.21

Concept Check

Answers may vary. One possible solution follows:

0.8% of windows are defective. There

are 375 windows in a shipment. How many windows in the shipment are defective?
Let $x =$ the number of defective windows in the shipment.
$x = 0.8\%$ of 375 windows
Write the equation.
$x = 0.8\% \times 375 = 0.008 \times 375 = 3$
3 of the windows in the shipment should be defective.

Topic 7

Student Practice

2. (a) 20
   (b) 34
   (c) p
4. (a) $b = 42$; $a = 8.4$
   (b) b is unknown; $a = 35$
6. (a) $p = 23$; $b = 524$; a is unknown
   (b) p is unknown; $b = 4$; $a = 37$

8 270
10. 300
12. 125%
14. 8%

Extra Practice

1. $p = 88$; $b = 198$; a is unknown
2. 0.19
3. 800
4. 40%

Concept Check

Answers may vary. One possible solution follows:
In the percent proportion, what can you say about the *percent number* if the value of the *amount* is larger than the *base*?

$$\frac{\text{amount}}{\text{base}} = \frac{\text{percent number}}{100}$$

If a is larger than b, we can write $a = b(1+n)$ where n is a positive number. Substitute $b(1+n)$ for a.

$$\frac{b(1+n)}{b} = \frac{p}{100}$$

$$1+n = \frac{p}{100}$$

$$100(1+n) = p$$

Since n is a positive number, $(1+n)$ is larger than 1, and p is larger than 100.

Topic 8

Student Practice

2. $19,180
4. 35%
6. $38,480
8 1001
10. (a) $120
    (b) $1620
12. $130

Extra Practice

1. $1750
2. $3500
3. $45,892.50
4. $300

Concept Check

Answers may vary. One possible solution follows:

Use the simple interest formula.

$I = P \times R \times T$

$P = \$6500,\ \ R = 9\%$ per year,

$T = 4 \text{ months} = \dfrac{4}{12}$ year

$$\begin{aligned} I &= P \times R \times T \\ &= \$6500 \times 9\% \times \frac{4}{12} \\ &= 6500 \times 0.09 \times \frac{1}{3} \\ &= 195 \end{aligned}$$

The interest is $195.

Worksheet Answers Module 7

Topic 1

Student Practice

2. 6 hours
4. 8 pounds
6. \$37.50
8. (a) 500 cm
   (b) 400 mg
10. (a) 0.005 L
    (b) 0.00049 km
12. \$0.12 per mL

Extra Practice

1. 1 mile
2. 23.5 hours
3. 3200 meters
4. $0.01\text{ L} = 0.00001\text{ kL} = 10\text{ mL}$

Concept Check

Answers may vary. Multiply 240 ounces by the unit fraction $\frac{1\text{ pound}}{16\text{ ounces}}$.

Topic 2

Student Practice

2. (a) 2.745 m
   (b) 16.082 L
   (c) 5.808 gal
   (d) 141.75 g
4. 6.37 ft
6. 53.32 mi/hr
8. 0.8668 in.
10. 85°C
12. 39.4°F

Extra Practice

1. 5.15 km
2. 54.96 L
3. 27.9 mi/hr
4. 25°C

Concept Check

Answers may vary. Multiply $\frac{50\text{ km}}{\text{hr}}$ by the unit fraction $\frac{0.62\text{ mi}}{1\text{ km}}$.

Topic 3

Student Practice

2. (a) $\angle Y$
   (b) $\angle ZYX$, $\angle Y$, $\angle XYZ$
4. 66°
6. 70°
8. $\angle b = 115°$, $\angle c = 65°$, $\angle d = 115°$, $\angle e = 65°$

Extra Practice

1. 112°
2. 180°
3. 90°
4. 37°

Concept Check

Answers may vary. $\angle a$ and $\angle d$ are supplementary. Subtract the measure of $\angle a$ from 180° to find the measure of $\angle d$.

10.4

Student Practice

2. (a) Not a perfect square
   (b) Perfect square
   (c) Perfect square
4. $n = \sqrt{144}$
6. 2
8. $\frac{11}{13}$
10. 6.557
12. 22 ft^2
14. 21.4 ft

Extra Practice

1. 10
2. 11
3. 5.4 miles
4. 27.713

Concept Check

Answers may vary. The square of the hypotenuse of a right triangle is equal to the sum of the squares of the legs. The Pythagorean Theorem can be used to find the length of the third side of a right triangle given the lengths of two of the sides.

Topic 4

Student Practice

2. $C \approx 26.062$ m
4. (a) 759.88 in.
   (b) 14
6. $209.33

Extra Practice

1. 53.4 in.
2. 659.4 in.
3. 9.6 in.^{2}
4. 18.84 ft

Concept Check

Answers may vary. First find the radius of the semicircle using $r = \frac{d}{2}$. Then find the area of a full circle with radius r using $A = \pi r^2$. Finally, divide this area by 2 to find the area of the semicircle.

Topic 5

Student Practice

2. 3791.14 cm^3
4. 3052.08 in.^{3}
6. (a) 9498.5 cm^3
   (b) 5887.5 cm^3
8. 1526.0 m^3
10. 216 m^3

Extra Practice

1. 2.6 ft^3
2. 1766.3 cm^3
3. 711.9 in.^{3}
4. 512 m^3

Concept Check

Answers may vary. Since 5 inches is one-half of 10 inches, divide the volume of the can, 1130.40 in.^{3}, by 2.

Topic 6

Student Practice

2. $n \approx 11.2$
4. 132 yd
6. 27 ft
8. 11.34 m

Extra Practice

1. $n = 21$
2. $n = 3.2$ ft
3. 10 ft
4. 360 ft

Concept Check

Answers may vary. Multiply the length of the tree's shadow earlier in the day, 3 feet, by $\frac{1}{2}$ to find the length of its shadow that afternoon.

Worksheet Answers Module 8

Topic 1

Student Practice

2. $+a^2b, +5b^2, -6a, -3b$
4. $a^2 - 8a - 1$
6. $-3x - 7y + 4z$
8. $-3x^2 + 10x - 11$
10. $-13a^2 - 53a + 24$

Extra Practice

1. $+5a^4, -3a^3, +2a^2, -7a, +1$
2. $7y^2 - 8y - 1$
3. $4a^4 + 6a^2 - 9$
4. $5x - 6$

Concept Check

Answers may vary. One possible error was not multiplying the negative one coefficient through the entire second expression.

Topic 2

Student Practice

2. $-20a^2 + 40ab - 10a$
4. $-20z^7 + 40z^6 - 15z^5$
6. $8a^3 + 10a^2 + 17a + 42$
8. $x^2 + 9x + 20$
10. $y^2 + 6y + 5$
12. $x^2 - 9x + 14$
14. $7x^2 + 17x - 12$

Extra Practice

1. $2x^3 + 5x^2 - x - 1$
2. $x^2 + 2x - 35$
3. $2x^2 + 9x - 81$
4. $-2x^3 - 5x^2 - x - 12$

Concept Check

1. $(x+1)(x+2) = x^2 + 2x + x + 2$
$= x^2 + 3x + 2$
2. $(x-1)(x-2) = x^2 - 2x - x + 2$
$= x^2 - 3x + 2$

a. Answers may vary. One possible solution is that in number 1 all products resulting in a first order variable have a positive coefficient, whereas in number 2 the opposite is true.

b. Answers may vary. One possible solution is that all products resulting in a zero order variable are either products of two positive coefficients or two negative coefficients, both resulting in a positive coefficient.

Topic 3

Student Practice

2. (a) $x - 5$
(b) $3x$
(c) $\frac{x}{5}$ or $\frac{1}{5}x$
(d) $x + 10$
4. a. $3x + 8$
b. $3(x+8)$
c. $\frac{1}{4}(x+6)$
6. c = Chuck's car's mileage;
$c + 12{,}000$ = Tom's car's mileage
8. w = the width; $3w - 7$ = the length

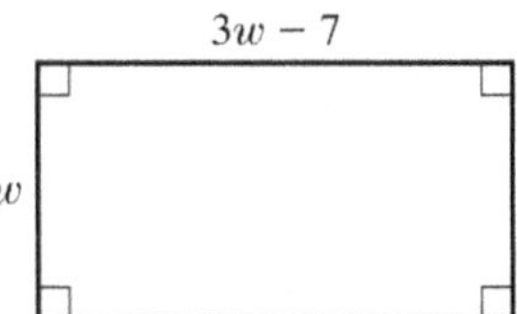

10. s = the second angle; $5s$ = the first angle; $s + 40$ = the third angle

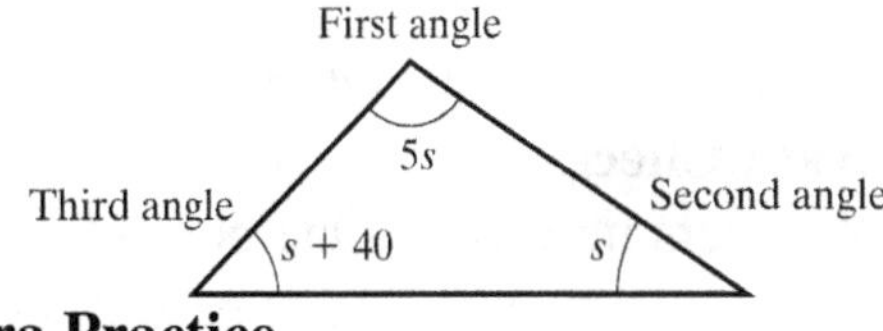

Extra Practice

1. $x + 12$
2. $\frac{1}{4}x - 1$

3. $x =$ value of Allison's car; $x + \$2300 =$ value of Alicia's car
4. $s =$ number of Scott's comic books; $s+17 =$ number of Patricia's comic books; $3s =$ number of Walter's comic books; $4s-5 =$ number of Adrienne's comic books

Concept Check

Answers may vary. Possible solution: "one-third of the sum" means you multiply $\frac{1}{3}$ times the sum. The sum will be the quantity in the parentheses because the $\frac{1}{3}$ is multiplied by the whole sum; $\frac{1}{3}(x+7)$.

Topic 4

Student Practice

2. 5
4. 5
6. (a) $2x^2$
   (b) y^2z
8. $5(4x+7)$
10. $6(2x+4y-5)$
12. $7x^2y(2+5x^2y^2)$

Extra Practice

1. x^3y^5
2. $3(x-3)$
3. $6(5x-2y+3)$
4. Not factorable; GCF is 1.

Concept Check

(a) Answers may vary. One possible solution is that xy is not part of the GCF because y is not a factor of $16x$.

(b) $12xy = 12\cdot x\cdot y = 3\cdot 4\cdot x\cdot y = 2\cdot 6\cdot x\cdot y$

$16x = 16\cdot x = 4\cdot 4\cdot x$

$\text{GCF} = 4x$

(c) $12xy+16x = 4x\cdot 3y + 4x\cdot 4$
$= 4x(3y+4)$

Worksheet Answers Module 9

Topic 1

Student Practice

2. $x = 34$

4. No. $x = 36$
6. $x = -\frac{5}{21}$
8. $x = 21$
10. $x = 7$
12 $x = -6$

Extra Practice

1. $x = 3.53$
2. Yes.
3. $x = -25$
4. Yes

Concept Check

Answers may vary. Possible solution: Substitute the value 3.8 for x in the equation. Simplify. If the resultant equation is true, $x = 3.8$ is the solution. If the resultant equation is not true, $x = 3.8$ is not the solution.

Topic 2

Student Practice

2. $x = 2$
4. $x = -\frac{2}{3}$
6. $x = \frac{17}{4}$
8. $x = 1$
10 $z = 20$

Extra Practice

1. $x = 5$
2. $x = 12$
3. $x = \frac{1}{4}$
4. $z = 2$

Concept Check

Answers may vary. Possible solution: Use the distributive property to remove parentheses. Combine like terms on the left side of the equation. Move the variable terms to the left side of the equation and the constants to the right side of the equation. Simplify.

Topic 3

Student Practice

2. $x = 3$
4. $x = -15$
6. $x = 112$
8. $x = \frac{1}{3}$

Extra Practice

1. $p = \frac{3}{2}$
2. $x = -2$
3. $x = -5$
4. $x = -5$

Concept Check

Answers may vary. Possible solution: Multiply both sides of the equation by the LCD, 12. Add or subtract terms on both sides of the equation to get all terms containing x on one side of the equation. Add or subtract a constant value to both sides of the equation to get all terms not containing x on the other side of the equation. Divide both sides by the coefficient of x and simplify the solution if necessary. Finally, check the solution.

Topic 4

Student Practice

2. $r = 424$ miles per hour
4. $r = \frac{C}{2\pi}$
6. y = x/4 - 3
8. $x = \frac{M - 8y}{4}$

Extra Practice

1. $s = \frac{P}{4}$
2. $x = \frac{y - b}{m}$
3. (a) $W = \frac{V}{LH}$
   (b) $W = 7$ ft
4. (a) $y = -\frac{3}{7}x + \frac{13}{7}$
   (b) $y = 1$

Concept Check

Answers may vary. Possible solution: Add 9 to both sides of the equation. Then multiply both sides of the equation by the LCD, 8. Finally, divide both sides of the equation by 3.

Topic 5

Student Practice

2. 85
4. 65
6. 32°
8. 12 hours for John; 13 hours for Yuri

Extra Practice

1. 546
2. 15
3. 18 energy bars
4. 300 minutes

Concept Check

Answers may vary. Possible solution: Since we want to know how many pairs of socks, let $x =$ the number of pairs of socks. Set up an equation to represent the total amount spent.

$2(\$23) + \$0.75x = \$60.25$

Topic 6

Student Practice

2. 16 hours
4. $800
6. $2000 invested at 6%; $1500 invested at 5%
8. 21 nickels; 17 dimes

Extra Practice

1. $16,800
2. $975
3. $54,000
4. 129 bills

Concept Check

Answers may vary. Possible solution: Since each amount is given in terms of quarters, let $x =$ the number of quarters. Then $2x =$ the number of dimes, and $x + 1 =$ the number of nickels. Now set up an equation and solve, given the value of each coin and the total.

$0.25x + 0.10(2x) + 0.05(x + 1) = 2.55$

Topic 7

Student Practice

2. (a) $>$
   (b) $>$
   (c) $<$
   (d) $>$
   (e) $<$
4. (a) x is less than 3. 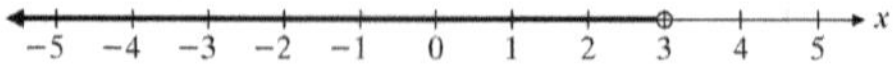
   (b) five halves is greater than or equal to x. 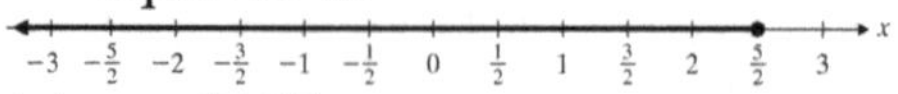
6. (a) $c < \$125$
   (b) $p \leq 35$

8. $x > \frac{3}{2}$

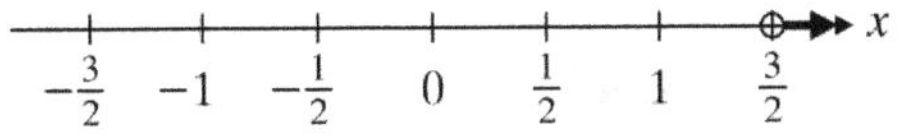

10. $x > -\frac{1}{3}$

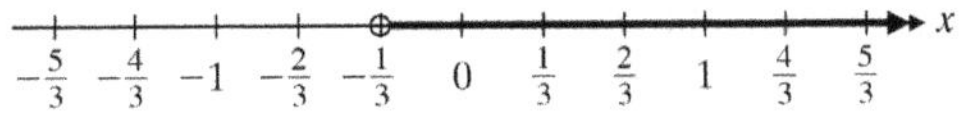

12. $x > -\frac{23}{72}$

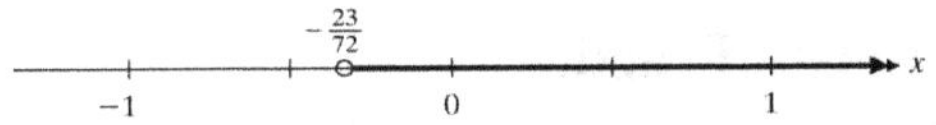

Extra Practice

1. $<$
2. 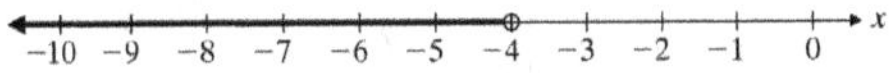
3. $w \geq 1{,}000{,}000$
4. $x < 3$

Concept Check

Answers may vary. Possible solution: $12 < x$ is the same as $x > 12$, but written in different ways. The graphs will be exactly the same.

Worksheet Answers Module 10

Topic 1

Student Practice

2.

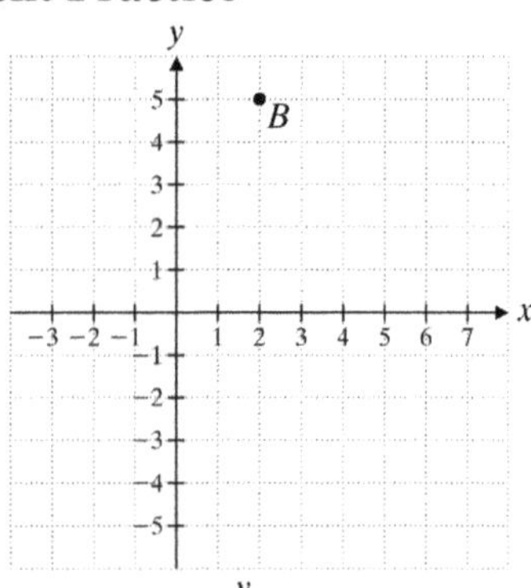

4.

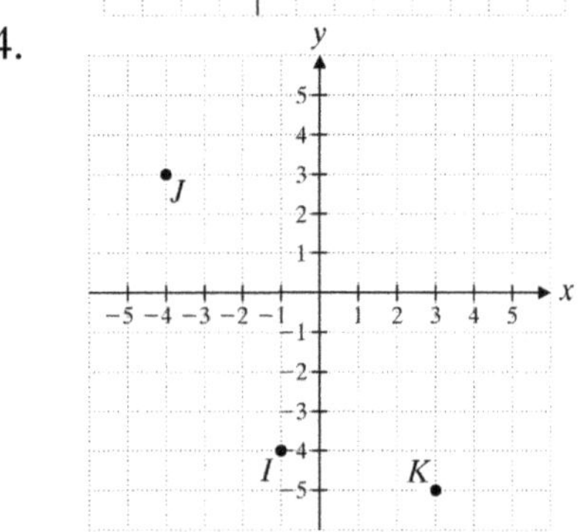

6. $(-3,6)$

8. No

10. $(-2,3)$

Extra Practice

1.

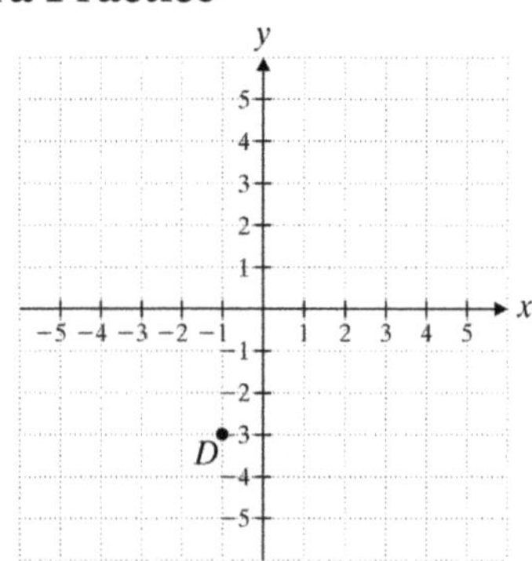

2. $(3,5)$

3. (a) $(-1,5)$

   (b) $(3,-3)$

4. (a) $(-1,-8)$

   (b) $(4,2)$

Concept Check

Answers may vary. Possible solution: Isolate x on the left side of the equation. Substitute the given value for y and solve for x.

Topic 2

Student Practice

2.

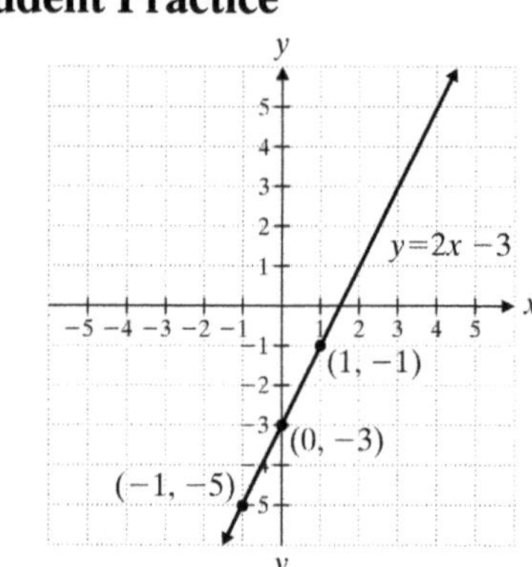

4.

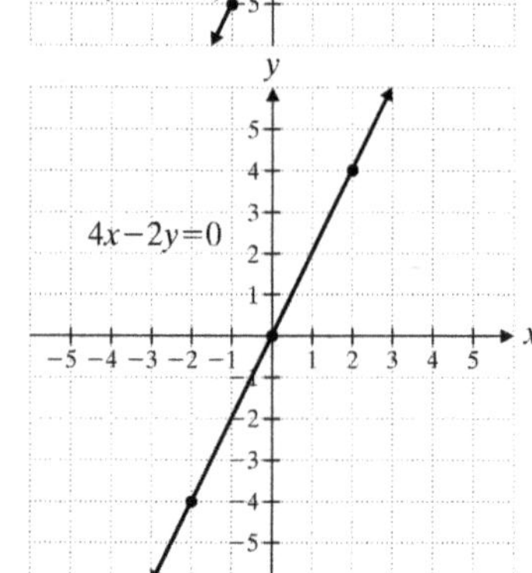

6. (a) x-intercept: $(-1,0)$

   y-intercept: $(0,-2)$

   (b)

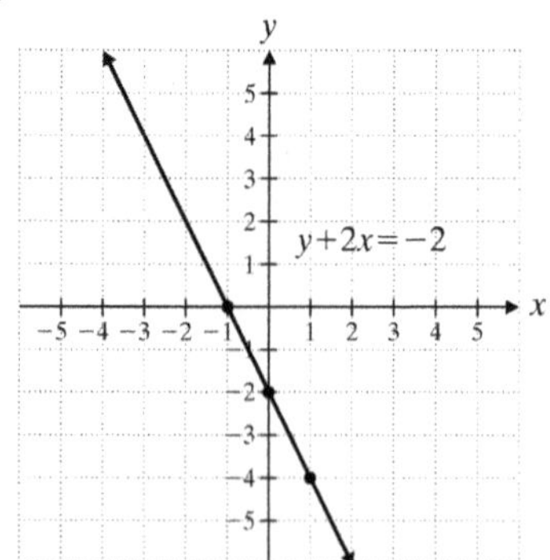

8.

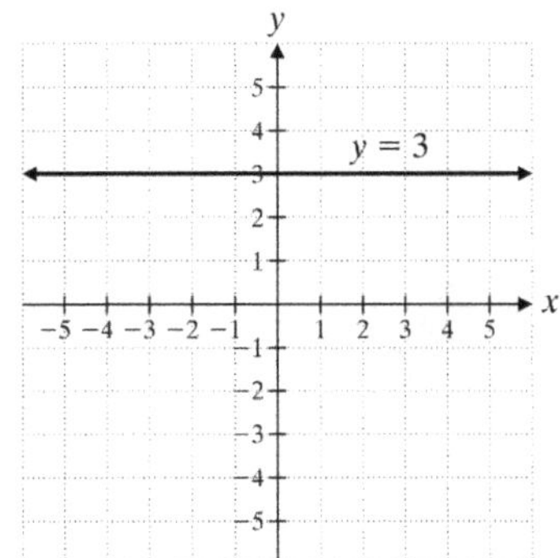

Extra Practice

1.

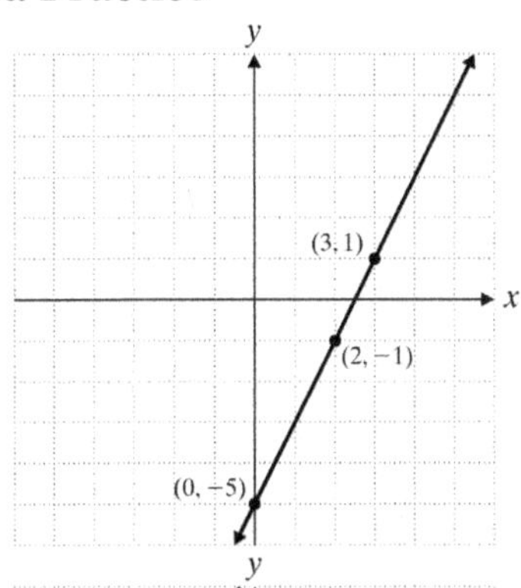

2.

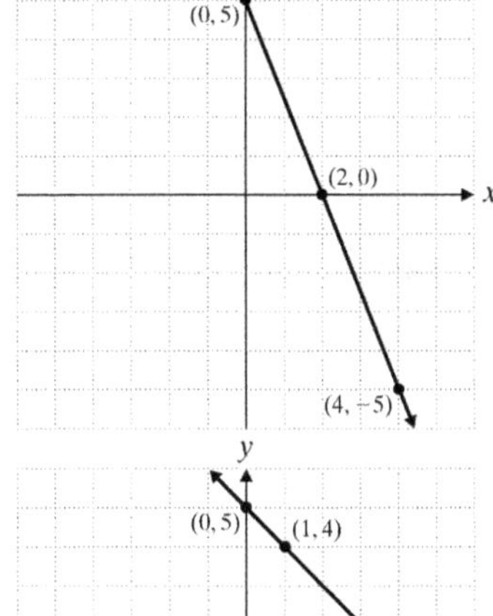

3.

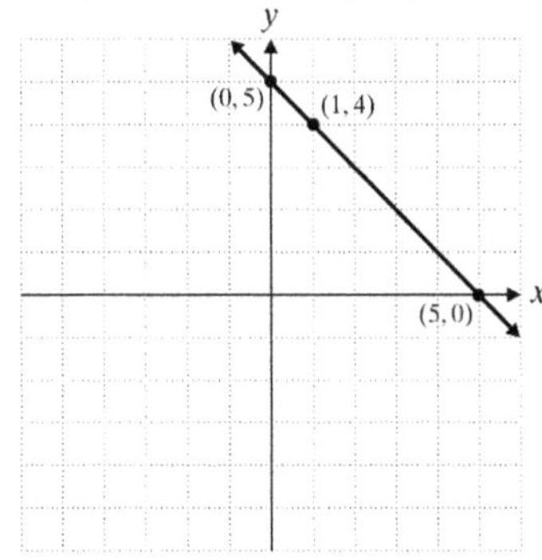

4.

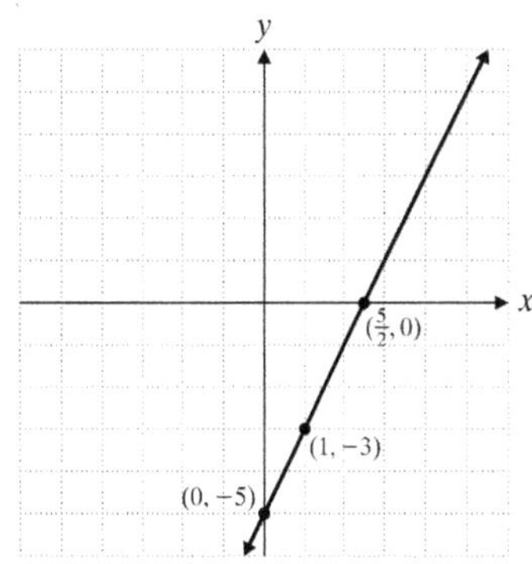

Concept Check

Answers may vary. Possible solution: The most important ordered pair is $(0,0)$, since it is both the x- and y- intercept of the equation.

Topic 3

Student Practice

2. $m=\frac{2}{3}$

4. (a) No slope
   (b) $m=0$

6. $m=-3$, y-intercept: $(0,4)$

8. (a) $y=\frac{3}{4}x-7$
   (b) $3x-4y=28$

10.

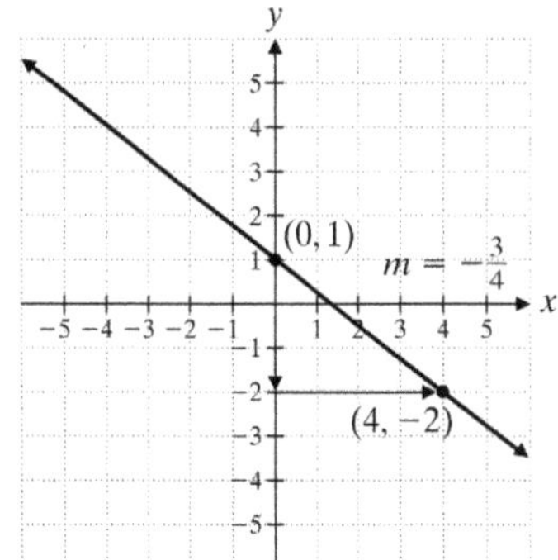

Extra Practice

1. $-\frac{1}{3}$

2. $m=5$, y-intercept: $(0,0)$

3. $y=-4x+\frac{4}{5}$

4. $y=-2$

Concept Check

Answers may vary. Possible solution: Slope measures the vertical change per one unit of horizontal change.

Topic 4

Student Practice

2. $y = -\frac{1}{2}x - 4$

4. $y = -\frac{3}{5}x + 7$

6. $y = -\frac{5}{2}x + 5$

8. (a) $\frac{5}{7}$

   (b) $-\frac{7}{5}$

Extra Practice

1. $y = -\frac{1}{2}x + 2.5$

2. $y = 5x - 11$

3. $y = \frac{7}{6}x - 1$

4. $y = -5x + 8$

Concept Check

Answers may vary. Possible solution: Zero slope indicates a horizontal line. Since the line is horizontal, and it passes through $(-2, -3)$, the equation must be $y = -3$.

Topic 5

Student Practice

2.

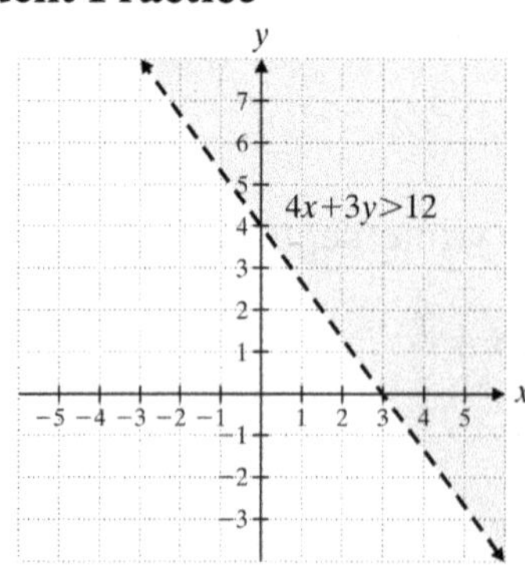

4.

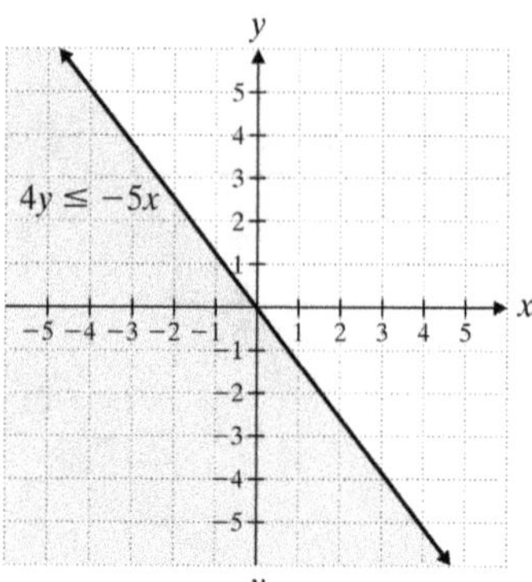

6.

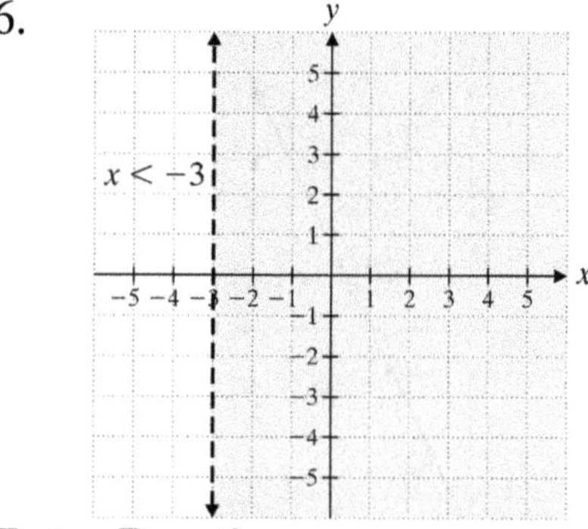

Extra Practice

1.

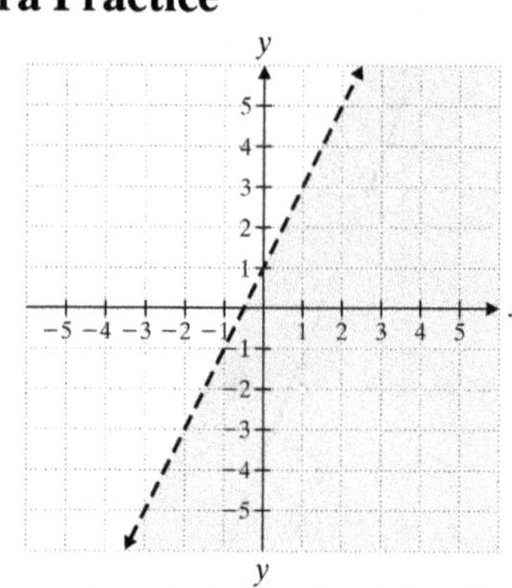

2.

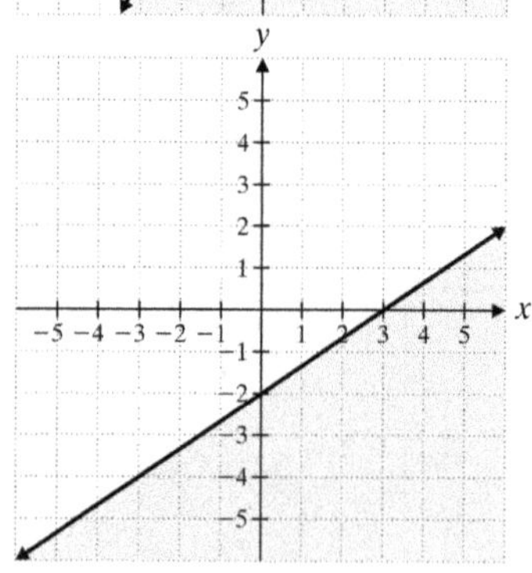

3.

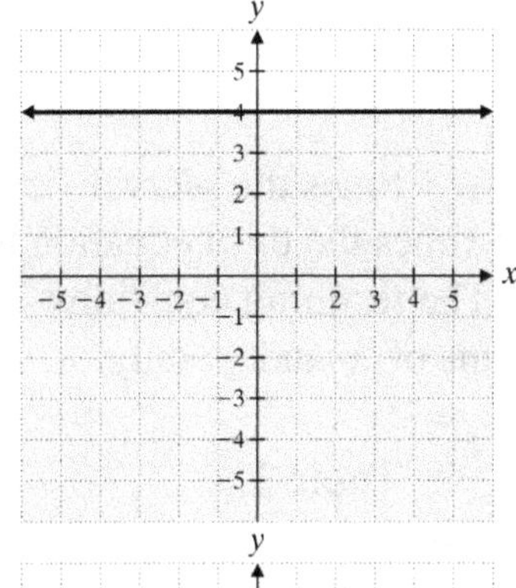

4.

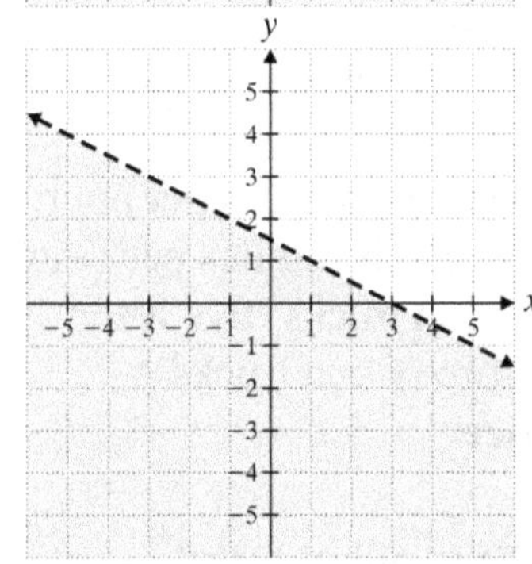

Concept Check

Answers may vary. Possible solution: The inequality is first graphed without shading. The test point coordinates are then substituted into the inequality. If the result of the substitution results in a true statement, the area where the test point lies is shaded. If false, the opposite area is shaded.

Topic 6

Student Practice

2. The domain is $\{-3,-2,4,7\}$.

   The range is $\{1,6\}$.

4. (a) Not a Function
   (b) Function

6.

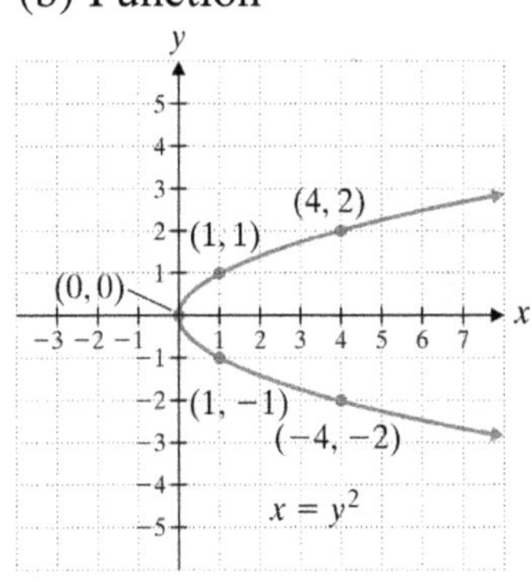

8. (a) Not a function
   (b) Function

10. (a) 49
    (b) 19
    (c) 7

Extra Practice

1. The domain is $\{2.5,3.5,5.5,8.5\}$.

   The range is $\{-6,-2,0,3\}$.

   Function

2. (a) 1
   (b) 28
   (c) 10

3. Not a function

4.

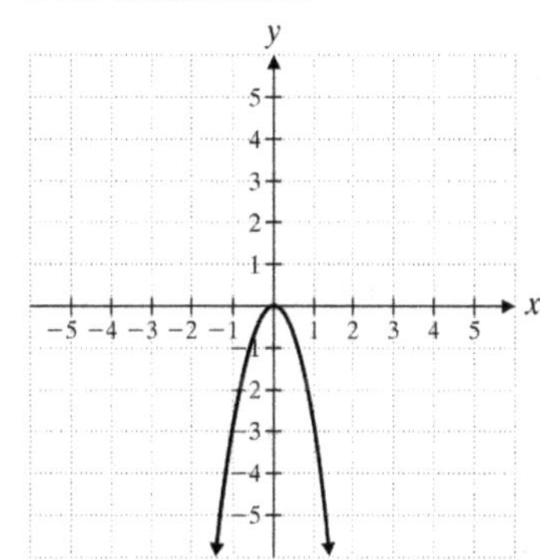

Concept Check

Answers may vary. Possible solution: Duplicate elements are recorded only once in the domain and in the range.

Worksheet Answers Module 11

Topic 1

Student Practice

2. $(3,2)$ is a solution
4. $(4,-2)$
6. $(10,3)$
8. $(4,5)$
10. No solution; inconsistent system
12. Infinite number of solutions; dependent system
14. (a) $(650,1585)$
    (b) $(4,-2)$

Extra Practice

1. $(-1,2)$
2. No solution; inconsistent system
3. $(0,4)$
4. Infinite number of solutions; dependent system

Concept Check

Answers may vary. Possible solution: When the addition method is used, the result of the addition, $0=0$, is an identity. The system is dependent and has an infinite number of solutions.

Topic 2

Student Practice

2. $(2,8,-7)$ is a solution.
4. $(4,8,-2)$
6. $(-1,2,-5)$

Extra Practice

1. $(-2,-2,5)$
2. $(-1,-1,1)$
3. $(6,6,1)$
4. No solution; inconsistent system

Concept Check

Answers may vary. Possible solution: Add 5 times the first equation to 2 times the second equation to eliminate z. Then, add -3 times the second equation to 5 times the third equation to eliminate z. The resulting equations will be in terms of x and y only.

Topic 3

Student Practice

2. 250 advance tickets; 125 door tickets
4. Boat speed is 15 miles per hour; current speed is 5 miles per hour
6. Eight 12-ton trucks; five 8-ton trucks; three 6-ton trucks

Extra Practice

1. 71, 31
2. 185 acres of soybeans, 315 acres of corn
3. 5 binders, 7 pens, and 3 erasers
4. 122 adults, 80 students, 48 senior citizens

Concept Check

Answers may vary. Possible solution: The equations would be set up the same except the right sides would be set to 1500 rather than 1200.

Topic 4

Student Practice

2.

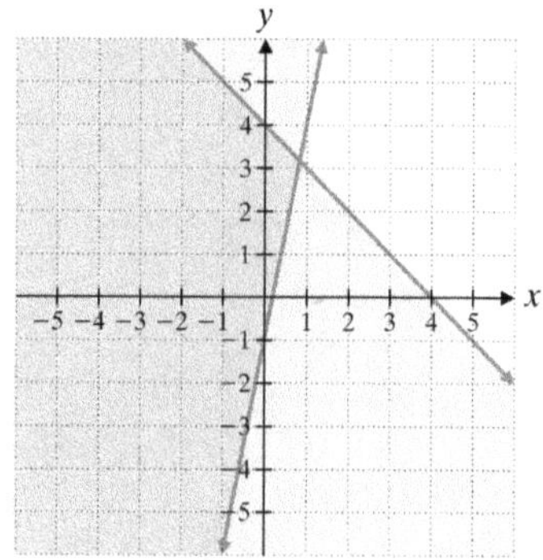

4.

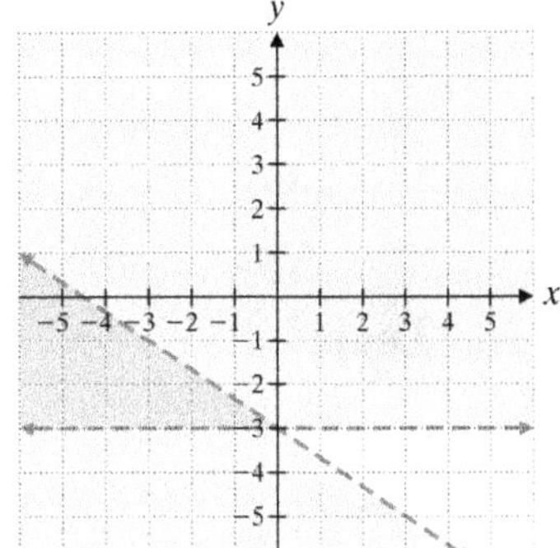

6.

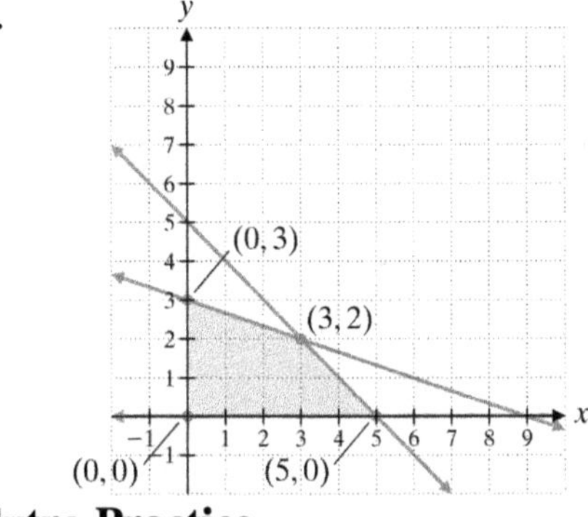

Extra Practice

1.

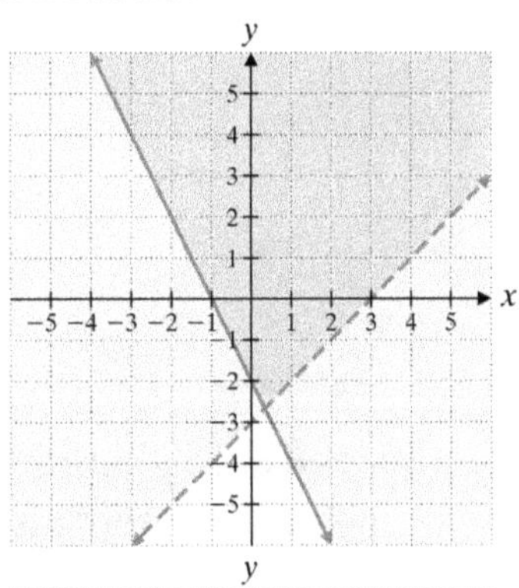

2.

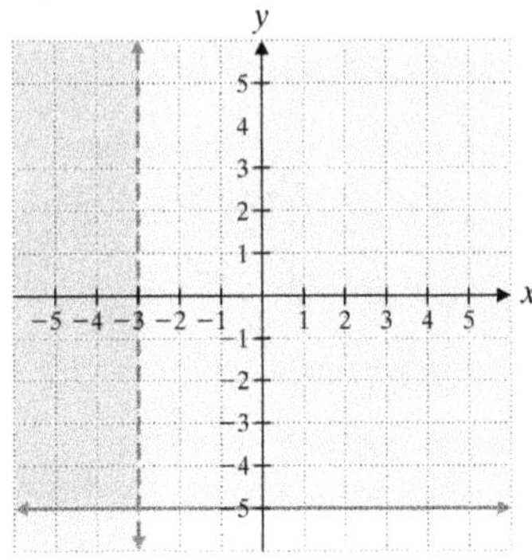

3.

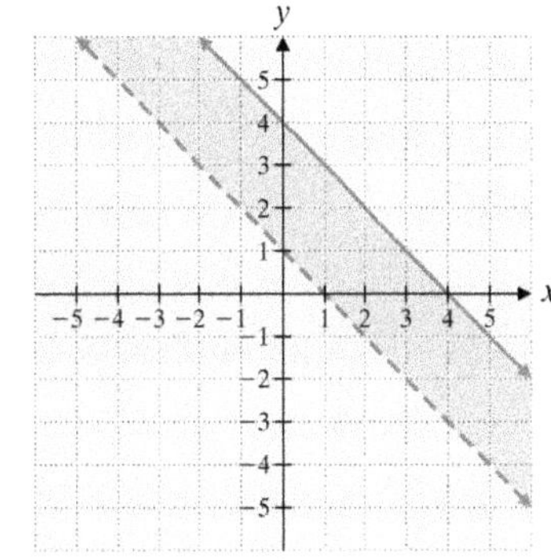

4.

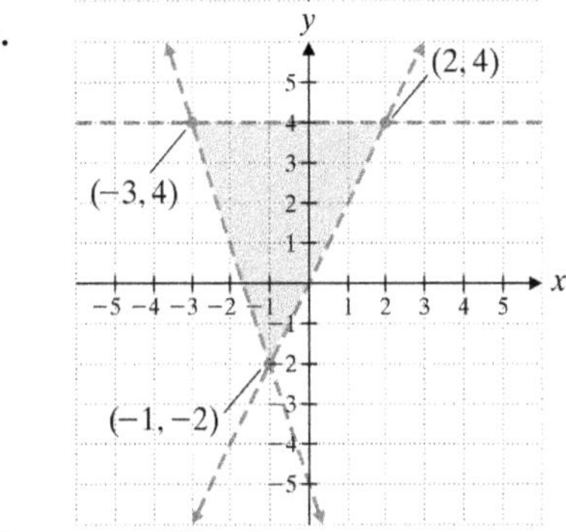

Concept Check

Answers may vary. Possible solution: $y > x + 2$ is graphed using a dashed line and shaded above the line. $x < 3$ is graphed using a dashed line and shaded to the left of the line. The region that satisfies both inequalities is the overlap in the shaded regions.

Worksheet Answers Module 12

Topic 1

Student Practice

2. (a) z^5
   (b) 2^8
4. $-8x^6y^3$
6. x^4
8. $\dfrac{1}{n^4}$
10. $\dfrac{x^2}{2y^5}$
12. (a) y^{32}
    (b) -1
14. $\dfrac{y^3}{z^6}$
16. $-\dfrac{x^{15}}{64y^6}$

Extra Practice

1. $24x^6y^5$
2. 9
3. $\dfrac{1}{b^4}$
4. $\dfrac{9^2}{14^2x^3}$, or $\dfrac{81}{196x^6}$

Concept Check

Answers may vary. Possible solution:
In the numerator and the denominator, raise each factor inside the parentheses to the power.

$$\frac{4^2\left(x^3\right)^2}{2^3\left(x^4\right)^3}$$

Evaluate the constants, and multiply exponents on the variable expressions.

$$\frac{16x^6}{8x^{12}}$$

Divide the numbers and subtract exponents on the variable expressions. Because the larger exponent is in the denominator, the variable expression will be in the denominator, $\dfrac{2}{x^6}$.

Topic 2

Student Practice

2. $\dfrac{1}{2401}$
4. $\dfrac{y^8}{16x^4z^{12}}$
6. 2.564×10^3
8. (a) 1.3×10^{-3}
   (b) 1.0×10^{-6}
10. (a) 0.000528
    (b) 3,221,400
12. 3.86×10^{11} hr

Extra Practice

1. $\dfrac{1}{343}$
2. $\dfrac{b^8c^{12}}{81a^{12}}$
3. 6340
4. 2.0×10^{-1}

Concept Check

Answers may vary. Possible solution:
Raise each factor inside the parentheses to the power.

$$4^{-3}\left(x^{-3}\right)^{-3}\left(y^4\right)^{-3}$$

Multiply exponents on the variable expressions.

$$4^{-3}x^9y^{-12}$$

Rewrite as a fraction.

$$\frac{x^9}{4^3y^{12}}$$

Evaluate 4^3.

$$\frac{x^9}{64y^{12}}$$

Topic 3

Student Practice

2. (a) Degree 7; monomial
   (b) Degree 2; trinomial
   (c) Degree 7; binomial
4. $4x^2-8x+5$
6. $\frac{7}{4}x^2-\frac{8}{5}x+\frac{5}{8}$
8. $-3x^2-10x+5$
10. 26.4 miles per gallon

Extra Practice

1. Degree 6; trinomial
2. $10.8x-14$
3. $5r^4-r^2+8$
4. 9.01 miles per gallon

Concept Check

Answers may vary. Possible solution: To determine the degree of the polynomial, first determine the degree of each term by finding the sum of the exponents on the variables in each term. The degree of the first term, $2xy^2$, is $1+2=3$, and the degree of the second term, $-5x^3y^4$, is $3+4=7$. The degree of the polynomial is the greater of these, which is 7. To determine whether the polynomial is a monomial, a binomial, or a trinomial, we must count the number of terms in the polynomial. The polynomial has two terms, $2xy^2$ and $-5x^3y^4$, so it is a binomial.

Topic 4

Student Practice

2. $-5y^3+20y^2$
4. $2x^4y-14x^3y-20x^2y$
6. $15x^2+4x-3$
8. $7x^2+14xz-3xy-6yz$
10. $16x^2+24xy+9y^2$
12. $20x^4-20x^2y^4-15y^8$
14. $8x^2-2$

Extra Practice

1. $-10x^4+3x^2$
2. $x^2-14x+33$
3. $15x^2-11xy-56y^2$
4. $8x^4-2x^2y^3-15y^6$

Concept Check

Answers may vary. Possible solution: First write the square of the binomial as the product of the binomial and itself.

$(7x-3)(7x-3)$

Then use FOIL and collect like terms.

$49x^2-21x-21x+9=49x^2-42x+9$

Topic 5

Student Practice

2. $64y^2-16$
4. $4x^2-81y^2$
6. $49x^2+14xy+y^2$
8. $2x^4+2x^3-29x^2+9x+30$
10. $x^4-7x^3+6x^2-22x+40$
12. $2x^3+6x^2-8x-24$

Extra Practice

1. x^2-100
2. $25a^2-9b^2$
3. $24x^3+2x^2+7x+3$
4. $x^3-4x^2-7x+10$

Concept Check

Answers may vary. Possible solution: To use the formula $(a+b)^2=a^2+2ab+b^2$ to multiply $(6x-9y)^2$, first identify *a* and *b*: $a=6x$ and $b=-9y$. Then substitute these values for *a* and *b* in the formula and simplify.

Math displayed on next page.

$(6x-9y)^2$

$=(6x)^2+2(6x)(-9y)+(-9y)^2$

$=36x^2-108xy+81y^2$

Topic 6

Student Practice

2. $3x^3+x-4$
4. $2x^2+2x+1+\dfrac{5}{x+1}$
6. $4x^2-2x-3+\dfrac{8}{2x+1}$
8. $4x^2+4x+6+\dfrac{24}{2x-3}$

Extra Practice

1. $6a^5-2a^3+4a-1$
2. $4y^2-6y+9$
3. $3x^2-2x+5$
4. $y^2+y-1-\dfrac{5}{y-1}$

Concept Check

Answers may vary. Possible solution: Multiply the quotient and the divisor. Then add the remainder. You should get the original dividend.

$(x-2)(x^2+2x+8)+13$

$=x^3+2x^2+8x-2x^2-4x-16+13$

$=x^3+4x-3$

Yes, the answer checks.

Worksheet Answers Module 13

Topic 1

Student Practice

2. (a) $5(y+3z)$

   (b) $y(12-5z)$

4. $11y(4y^2+5y-x)$

6. (a) $7(3m^2-4n^2)$

   (b) $m^2(mn^2+9n^2+3m^2)$

8. $9xy^2(3y-4x-x^2y)$

10. $11m^3n^2(m+1)$

12. $(3x+4y)(5y^2-1)$

14. $\pi(m^2+4n^2+9z^2)$

Extra Practice

1. $3x(x^2+4x-7)$
2. $6x(10xy+3y-4)$
3. $(x+3y)(8a-b)$
4. $(5a-1)(4x-3)$

Concept Check

Answers may vary. Possible solution: Determine that the largest integer that will divide into the coefficient of all terms is 36. Determine that the variables common to all terms are a^2 and b^2.

Write the above common factors as the first part of the answer (the first factor).

Remove common factors, and what remains is the second part of the answer (the second factor).

Topic 2

Student Practice

2. $(2z+5)(z-4)$
4. $(3x+1)(4x+5)$
6. $(y+3z)(7+x)$
8. $(z+y)(n+6)$
10. $(2x+3)(x-3)$
12. $(5z-1)(n-m)$
14. $(5x-3w)(7y-3z)$

Extra Practice

1. $(x-2)(x+1)$
2. $(x+5)(x-2)$
3. $(3x-2)(2x+5)$
4. $(3a+2b)(4x+5y)$

Concept Check

Answers may vary. Possible solution: Start by grouping terms $10ax$ with $5ab$ and $2bx$ with b^2.

$(2x+b)$ can be factored out of both groups leaving the second factor to be $(5a+b)$.

Topic 3

Student Practice

2. $(x+3)(x+6)$
4. $(x+11)(x+3)$
6. $(x-4)(x-7)$
8. $(x-6)(x+4)$
10. $(z-2)(z+14)$
12. $(x^2+2)(x^2-5)$
14. $4(x-7)(x+3)$
16. $(x+10)(x-7)$

Extra Practice

1. $(x+2)(x+4)$
2. $(a-3)(a-4)$
3. $(x-4)(x+3)$
4. $2(x+2)(x+7)$

Concept Check

Answers may vary. Possible solution: The first step is to factor out the greatest common factor of 4, leaving $4(x^2 - x - 30)$. Next write the expression in factored form using variables m and n, $4(x+m)(x+n)$. Next determine that the product of m and n is -30, and the sum is -1. m and n may equal 5 and -6. Substitute the values of m and n, then check.

Topic 4

Student Practice

2. $(2x+5)(x+4)$
4. $(3x-1)(x-1)$
6. $(x-3)(4x+7)$
8. $(x+1)(2x+3)$
10. $(x-3)(5x+2)$
12. $2(4x-1)(x+3)$
14. $5(2x-3)(3x-2)$

Extra Practice

1. $(2x-1)(2x-5)$
2. $(7x+3)(x-1)$
3. $(5x-2)(x+9)$
4. $3(5y-3)(y+4)$

Concept Check

Answers may vary. Possible solution: First step is to factor out common coefficients and variables.

$2x(5x^2 + 9xy - 2y^2)$

Next step is to factor the inside expression using grouping. The grouping number is -10.

$2x\left[5x^2 + 10xy - xy - 2y^2\right]$

$= 2x\left[5x(x+2y) - y(x+2y)\right]$

$= 2x(2x-y)(x+2y)$

Lastly, check the solution by multiplying the factors.

Topic 5

Student Practice

2. $(6x+1)(6x-1)$
4. $(8x+3)(8x-3)$
6. $(4x^2+1)(2x+1)(2x-1)$
8. $(x+5)^2$
10. (a) $(8x+5y)^2$
    (b) $(7x^2-1)^2$
12. $(4x+1)(9x+4)$
14. $3(x+5)(x-5)$
16. $3(4x-3)^2$

Extra Practice

1. $(5a-6b)(5a+6b)$
2. $(6a+5b)^2$
3. $(x^2-10)(x^2+10)$
4. $(3x^2-5)^2$

Concept Check

Answers may vary. Possible solution: First factor out the common factor of 2.

$2(12x^2 + 60x + 75)$

Next use grouping to factor the inside expression. The grouping number is 900.

$2(12x^2 + 30x + 30x + 75)$

$= 2\left[6x(2x+5) + 15(2x+5)\right]$

$= 2(6x+15)(2x+5)$

Lastly, check by multiplying.

Topic 6

Student Practice

2. (a) $y(3y+1)^2$
   (b) $-4x(x-8)(x+1)$
4. $(b-3)(x-4)(x+4)$

6. (a) prime
   (b) prime

Extra Practice

1. prime
2. $7x(3-x)(3+x)$
3. $-x(x+3)(x-15)$
4. $(x+3)(x+2)$

Concept Check

Answers may vary. Possible solution: The first step is to group terms and factor out common factors.

$2x(x+3w)-5(x+3w)$

Next, factor out $(x+3w)$.

$(x+3w)(2x-5)$

Finally, check by multiplying.

Topic 7

Student Practice

2. $\frac{1}{3}$ and -3
4. 0 and $\frac{4}{5}$
6. $x=7,\ x=-9$
8. width $=12$ in., length $=5$ in.
10. 3 seconds

Extra Practice

1. $3,\ -\frac{1}{2}$
2. $-3, 3$
3. $7, -5$
4. $3, 4$

Concept Check

Answers may vary. Possible solution: Let $x=$width of rectangle, then length$=(2x+3)$.

$A=(\text{width})(\text{length})$

$65=x(2x+3)$

$65=2x^2+3x$

$0=2x^2+3x-65$

$0=(2x+13)(x-5)$

Set each factor equal to 0 and solve for x.

$2x+13=0 \qquad x-5=0$

$2x=-13 \qquad x=5$

$x=-\frac{13}{2}$

x cannot be negative in this case, because it describes a length, So $x=5$.

length$=2(5)+3=13$ feet

width$=5$ feet

Worksheet Answers Module 14

Topic 1

Student Practice

2. $\frac{8}{11}$
4. $\frac{2}{5}$
6. $\frac{x+7}{x+4}$
8. $\frac{x+2}{x-6}$
10. $-\frac{4}{7}$
12. $-\frac{3x+4}{5+2x}$
14. $\frac{5x+3y}{9x+7y}$
16. $\frac{5a+4b}{2a+b}$

Extra Practice

1. $\frac{3}{x}$
2. $\frac{3x-4}{3x+4}$
3. $-\frac{x+9}{3(x+2)}$
4. $\frac{5x-2y}{3x+y}$

Concept Check

Answers may vary. Possible solution: Completely factoring both numerator and denominator is the only way to see what factors are shared, and may consequently be eliminated. In this case, it can be seen that $(x-y)$ is a common factor.

Topic 2

Student Practice

2. $\frac{3x+7}{2x-3}$
4. $\frac{x}{(x+5)(3x+4)}$ or $\frac{x}{3x^2+19x+20}$
6. $\frac{4x-3}{2x+1}$
8. $-\frac{1}{(x-5)(x-2)}$

Extra Practice

1. $\frac{x}{x+1}$
2. $\frac{x}{5x+7}$
3. $\frac{1}{(x-2)(x+5)}$
4. $\frac{5(x-1)}{3x+4}$

Concept Check

Answers may vary. Possible solution: The first step is to change the operation from division to multiplication, by changing the operator and inverting the second fraction. Secondly, all terms must be factored completely. Next, common factors in the numerators and denominators may be canceled. Lastly the multiplication operation is performed.

Topic 3

Student Practice

2. $\frac{8}{3x+4}$
4. $\frac{2x+11}{(3x-2)(x+9)}$
6. $28(4x+5)$
8. $72a^2b^4c^3$
10. $\frac{2xz+4}{xyz}$

12. $\frac{11x+22}{4x^2-25}$ or $\frac{11(x+2)}{(2x+5)(2x-5)}$

14. $\frac{18x+10y}{9x^2-y^2}$ or $\frac{2(9x+5y)}{(3x+y)(3x-y)}$

16. $\frac{2x+17}{6x+21}$ or $\frac{2x+17}{3(2x+7)}$

Extra Practice

1. $(2x-3)(x-4)(x+5)$
2. $\frac{3x+5}{x+1}$
3. $\frac{2x+5y}{x^2y-y^3}$ or $\frac{2x+5y}{y(x+y)(x-y)}$
4. $\frac{-x-5}{x^3-7x-6}$ or $-\frac{x+5}{(x-3)(x+1)(x+2)}$

Concept Check

Answers may vary. Possible solution: First factor each denominator completely. The LCD will be the product containing each different factor. If a factor occurs more than once in any one denominator, the LCD will contain that factor repeated the greatest number of times that it occurs in any one denominator.

Topic 4

Student Practice

2. $\frac{4m}{n^2(3m+4)}$

4. $\frac{15(m+n)}{mn(3x-5y)}$

6. $\frac{x-4}{5(x+2)(x+6)}$

8. $\frac{7x-3y}{7}$

10. $\frac{16xy-7y^2}{20xy^2-36}$

12. $\frac{7x-3y}{7}$

Extra Practice

1. $\frac{5a+4b}{2}$
2. $\frac{2}{xy}$
3. $\frac{2x}{x-35}$
4. $\frac{-ab(32x-9y)}{12xy(7b+5a)}$

Concept Check

Answers may vary. Possible solution: The first step is to find the LCD for the fractions in the numerator.

$x-3=(x-3)$

$2x-6=2\cdot(x-3)$

$\text{LCD}=2(x-3)$

Next, multiply the first fraction in the numerator by the 2 to obtain common denominators in the numerator. Lastly, add the two fractions in the numerator.

Topic 5

Student Practice

2. $x=-44$
4. $x=-6$
6. $x=\frac{7}{3}$
8. no solution

Extra Practice

1. $x=32$
2. no solution
3. $a=60$
4. $x=-10$

Concept Check

Answers may vary. Possible solution: To find the LCD first factor each denominator, then multiply one instance of each factor.

The math is displayed on the following page.

$x^2 - 9 = (x-3)(x+3)$

$3x - 9 = 3(x-3)$

$2x + 6 = 2(x+3)$

$2x^2 - 18 = 2(x-3)(x+3)$

$\text{LCD} = 2 \cdot 3 \cdot (x-3)(x+3)$

Topic 6

Student Practice

2. 594 miles
4. $17\frac{11}{15}$ centimeters
6. Train A traveled 120 kilometers per hour. Train B traveled 105 kilometers per hour.
8. 2 hours and 24 minutes

Extra Practice

1. $x = 38$
2. $133\frac{1}{3}$ miles
3. 16 feet
4. 3 hours and 22 minutes

Concept Check

Answers may vary. Possible solution: One of the fractions needs to be inverted in order for the equation to be an accurate statement.

Worksheet Answers Module 15

Topic 1

Student Practice

2. $\frac{27}{64}a^3b^{-18}$
4. (a) $x^{12/5}$
   (b) $x^{3/4}$
   (c) $6^{9/13}$
6. (a) $-12x^{9/10}$
   (b) $5x^{7/8}y^{-2/3}$
8. $-8x^{7/6}+24x^{1/2}$
10. (a) 64
    (b) 16
12. $\frac{5+x}{x^{3/4}}$
14. $6z\left(3z^{2/3}-4z^{1/3}\right)$

Extra Practice

1. $x^{16/5}$
2. $ab^{4/3}$
3. $\frac{1+3y}{y^{2/3}}$
4. $5x\left(2x^{3/4}+5x^{1/8}\right)$

Concept Check

Answers may vary. Possible solution: Change the exponents to have equal denominators, then add the numerators over the common denominator. This is the combined, simplified exponent for x.

Topic 2

Student Practice

2. (a) 4
   (b) 1
   (c) 3.3
4. The domain is all real numbers x, where $x \geq 9$.
6. 7
8. (a) $z^{3/5}$
   (b) $a^{8/3}$
10. (a) $\sqrt[3]{a^5b^5}$ or $\sqrt[3]{(ab)^5}$
    (b) $\sqrt[5]{5x}$
12. (a) $\frac{1}{4}$
    (b) Not a real number
14. (a) $5|a|$
    (b) $4z^2$
    (c) $3x^2|y|$

Extra Practice

1. 7
2. $(4m-3n)^{5/8}$
3. $\frac{1}{\left(\sqrt[7]{4}\right)^4}$ or $\frac{1}{\sqrt[7]{256}}$
4. $7|a^3|b^8$

Concept Check

Answers may vary. Possible solution: Factor the coefficient completely to identify the fourth root. Remove the fourth root from under the radical. Divide the exponents of the variables by 4 to remove the variables from under the radical.

Topic 3

Student Practice

2. $5\sqrt{2}$
4. $6\sqrt{3}$
6. $-3\sqrt[3]{6}$
8. (a) $4a^2b^2\sqrt{2a}$
   (b) $4yz\sqrt[3]{3x^2yz^2}$
10. $4\sqrt{3a}$
12. $4\sqrt{5}$
14. $5\sqrt{a}-3\sqrt{2a}$
16. $14x^2z\sqrt[3]{5z}$

Extra Practice

1. $5x\sqrt{x}$
2. $5b^4\sqrt[4]{ab^3}$
3. $10\sqrt{7}-4\sqrt{5}$
4. $17x\sqrt{3}$

Concept Check

Answers may vary. Possible solution: Completely factor the coefficient of the radicand to identify the fourth root. Divide the exponents of the variables by 4 to remove the variables from under the radical. The results are moved outside the radical. The remainders stay under the radical.

Topic 4

Student Practice

2. $-12\sqrt{21z}$
4. $39-10\sqrt{15}$
6. $12+18\sqrt{2}+8\sqrt{3}+12\sqrt{6}$
8. $11-2\sqrt{22a}+2a$
10. (a) $\frac{2}{3}$
    (b) $6ab^2\sqrt{a}$
12. $\frac{5\sqrt{2z}}{8z}$
14. $\frac{3\sqrt[3]{a^2}}{a}$
16. $\frac{31+7\sqrt{35}}{13}$

Extra Practice

1. $15a\sqrt{ab}$
2. $265+30\sqrt{70}$
3. $\frac{3\sqrt{2x}}{5y^2}$
4. $x\sqrt{5}+x\sqrt{3}$

Concept Check

Answers may vary. Possible solution: To rationalize the denominator, the numerator and the denominator must be multiplied by the conjugate of the denominator. In this case, the conjugate of the denominator is $3\sqrt{2}+2\sqrt{3}$. The product will not contain a radical in the denominator.

Topic 5

Student Practice

2. $x=6$
4. $x=-\frac{1}{3}$ or $x=3$
6. $x=5$
8. $y=3$

Extra Practice

1. $x=3$
2. No solution
3. $x=0$ or $x=4$
4. $x=5$

Concept Check

Answers may vary. Possible solution: Substitute the found values for x back into the original equation to test for validity.

Topic 6

Student Practice

2. (a) $i\sqrt{41}$
   (b) $5i$
4. -18
6. $x=-3,\ y=4\sqrt{2}$
8. $-3+2i$
10. $8-27i$
12. $-24-12i$
14. (a) -1
    (b) 1
16. $\frac{9-38i}{25}$ or $\frac{9}{25}-\frac{38}{25}i$

Extra Practice

1. $20-3i\sqrt{2}$
2. $-1+i$
3. $4+7i$
4. $\frac{30+12i}{29}$

Concept Check

Answers may vary. Possible solution: Multiply by the FOIL method. Replace i^2 with -1. Combine like terms.

Topic 7

Student Practice

2. 144 feet
4. $y = \frac{25}{6}$
6. 19 lumens
8. $y = \frac{175}{6}$

Extra Practice

1. 40 inches
2. $y \approx 3.9$
3. 80 mph
4. $y \approx 4.2$

Concept Check

Answers may vary. Possible solution: First write the equation as $y = \sqrt{x}k$. Substitute the known values for x and y and solve for k.

Worksheet Answers Module 16

Topic 1

Student Practice

2. 5 or −5
4. $2\sqrt{6}$ or $-2\sqrt{6}$
6. 4 or −4
8. $5i$ or $-5i$
10. $\dfrac{-8+\sqrt{6}}{3}$ or $\dfrac{-8-\sqrt{6}}{3}$
12. $-2+\sqrt{3}$ or $-2-\sqrt{3}$
14. $\dfrac{2+\sqrt{19}}{5}$ or $\dfrac{2-\sqrt{19}}{5}$

Extra Practice

1. $8i$ or $-8i$
2. $\dfrac{-1+\sqrt{15}}{3}$ or $\dfrac{-1-\sqrt{15}}{3}$
3. −5 or −7
4. $\dfrac{1+i\sqrt{47}}{8}$ or $\dfrac{1-i\sqrt{47}}{8}$

Concept Check

Answers may vary. Possible solution: Divide the coefficient of x (1) by 2 to get $\frac{1}{2}$. Since $\left(\frac{1}{2}\right)^2=\frac{1}{4}$, add $\frac{1}{4}$ to both sides of the equation.

Topic 2

Student Practice

2. $-6\pm\sqrt{29}$
4. $\pm\sqrt{15}$
6. $x\approx 20$ or $x\approx 10$
8. $-1\pm\sqrt{11}$
10. $\dfrac{2\pm i\sqrt{14}}{6}$
12. Complex roots
14. $x^2-7x+12=0$
16. $x^2+45=0$

Extra Practice

1. $\dfrac{-6\pm\sqrt{15}}{3}$
2. $\dfrac{-7\pm\sqrt{89}}{2}$
3. One rational solution
4. $2x^2+7x+3$

Concept Check

Answers may vary. Possible solution: Subtract 3 from both sides to put the equation in standard form. Identify the values of a, b, and c, and find the value of the discriminant, b^2-4ac. If the discriminant is a perfect square, there are two rational solutions; if it is a positive number that is not a perfect square, there are two irrational solutions; if it is zero, there is one rational solution; if it is negative, there are two nonreal complex solutions.

Topic 3

Student Practice

2. $x=\pm 4,\ x=\pm 5i$
4. $x=3,\ x=\dfrac{\sqrt[3]{18}}{3}$
6. $x=64,\ x=1$
8. $x=625$
10. $x=\dfrac{1}{3},\ x=-\dfrac{1}{4}$

Extra Practice

1. $x=\pm i\sqrt{7},\ x=\pm i\sqrt{3}$
2. $x=0,\ x=-3$
3. $x=-1,\ x=-32$
4. $x=\dfrac{1}{7},\ x=-\dfrac{1}{8}$

Concept Check

Answers may vary. Possible solution: Substitute y for x^4, y^2 for x^8 which yields $y^2-6y=0$. Next factor out the

y from the left side of the equation yielding $y(y-6)=0$. Set each term equal to zero and solve for y yielding $y=0$ and $y=6$. Substitute x^4 for y and solve for x yielding $x=0$ and $x=\pm\sqrt[4]{6}$.

Topic 4

Student Practice

2. $$r=\sqrt[3]{\frac{3V}{4\pi}}$$
4. $x=2y$ or $x=-5y$
6. $$x=\frac{3y\pm 3\sqrt{y^2-4z}}{4}$$
8. (a) $c=\sqrt{a^2+b^2}$
   (b) 25
10. 10 mi, 24 mi, and 26 mi
12. width $=7$ yd; length $=12$ yd

Extra Practice

1. $$t=\sqrt{\frac{2s}{g}}$$
2. $$x=\frac{7\pm\sqrt{79-12ay-24y}}{2a+4}$$
3. $a=\sqrt{10};\ b=3\sqrt{10}$
4. 25 mph, then 45 mph

Concept Check

Answers may vary. Possible solution:
Set one leg's length $= x$, then the other leg's length $= 3x$.
Use $c^2=a^2+b^2$ with $c=12$, $a=x$, and $b=3x$. Solve for x to find one leg length then multiply the found value of x by 3 to find the other leg's length.

Worksheet Answers Module 17

Topic 1

Student Practice

2. $\sqrt{29}$

4. Center at $(1,4)$; radius is $r=4$

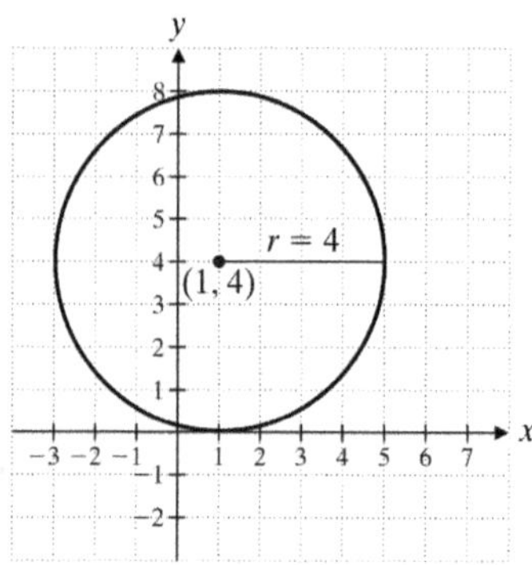

6. $(x-14)^2+(y+5)^2=7$

8. $(x+2)^2+(y-1)^2=9$; Center at $(-2,1)$; radius is $r=3$

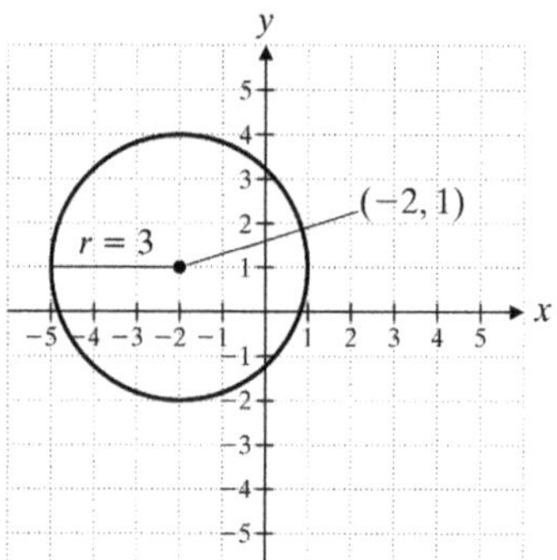

Extra Practice

1. $2\sqrt{5}$

2. $x^2+\left(y-\frac{6}{5}\right)^2=13$

3. Center at $(0,0)$; radius at $r=6$

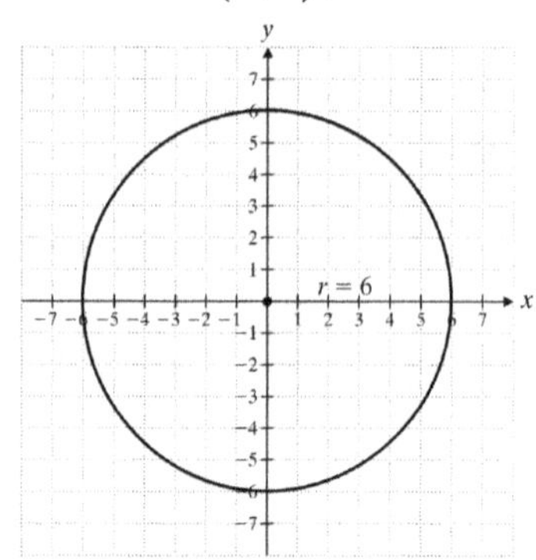

4. $(x+5)^2+(y-3)^2=5$; center at $(-5,\ 3)$; radius at $r=\sqrt{5}$

Concept Check

Answers may vary. Possible solution: Using the distance formula, let $(x_1,y_1)=(-6,8)$, $(x_2,y_2)=(x,12)$, and $d=4$. Then, solve for the unknown variable x.

Topic 2

Student Practice

2. Vertex at $(-1,0)$; axis of symmetry is $x=-1$

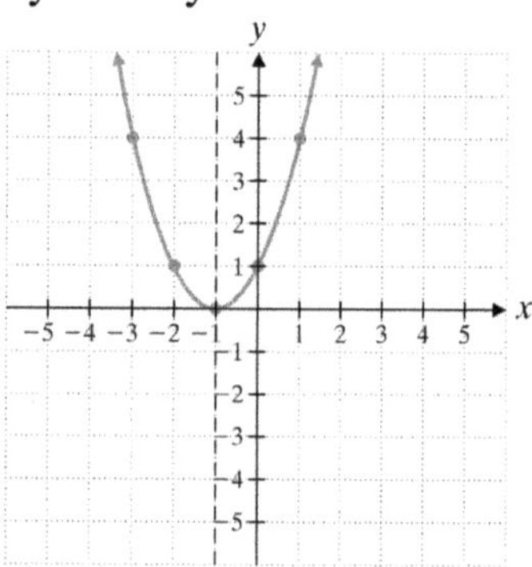

4.

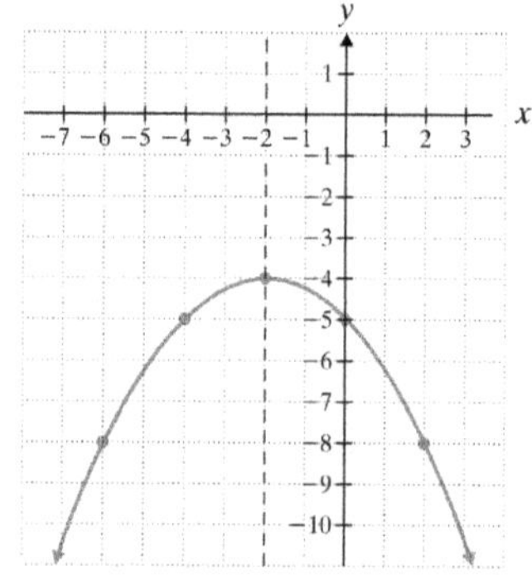

6.

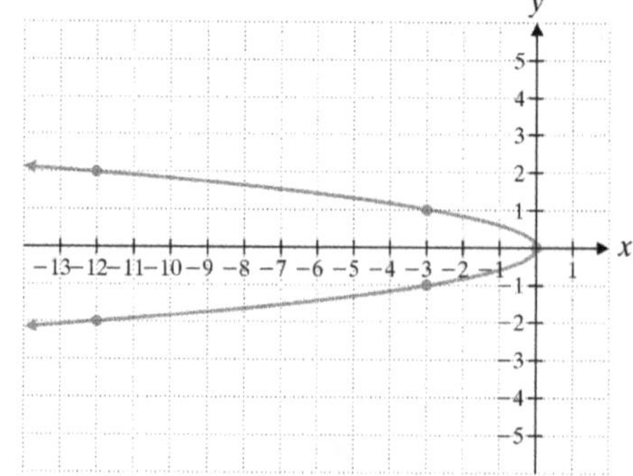

8. $x=(y+3)^2-2$

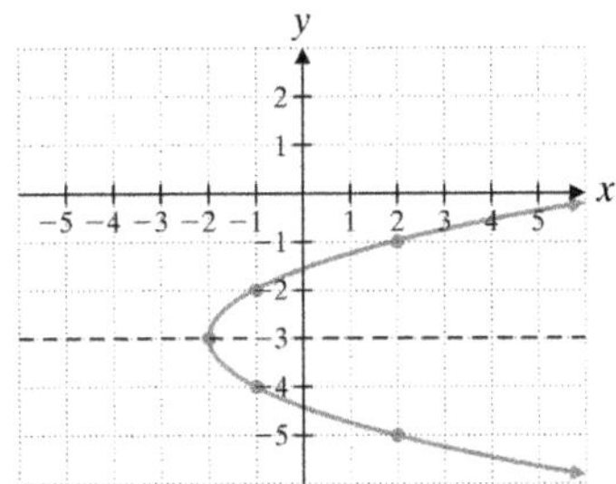

10. $y=-2(x+2)^2+4$

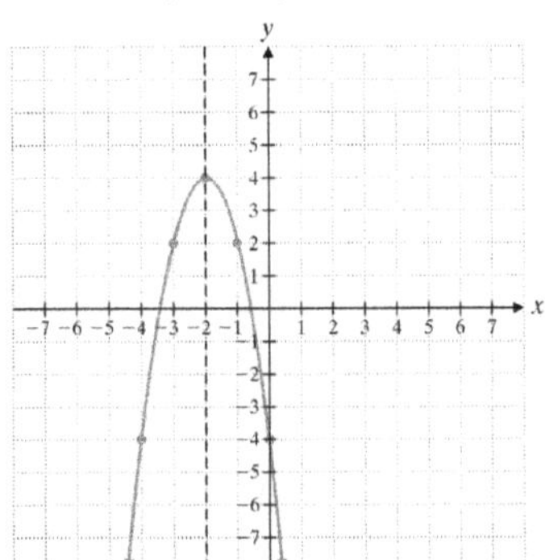

Extra Practice

1. Vertex at $(-4,-3)$; y-intercept at $(0,29)$

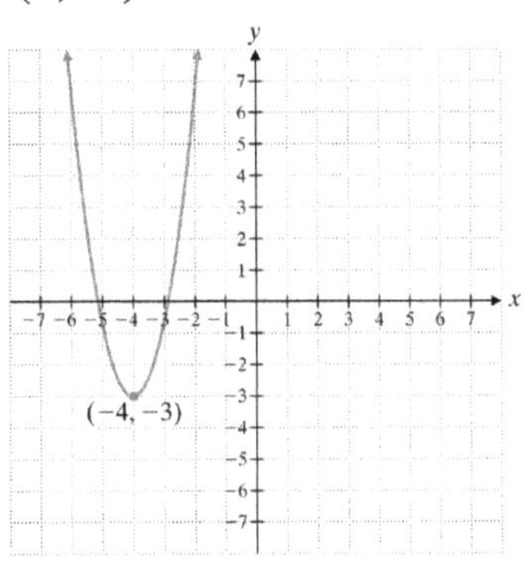

2. Vertex at $\left(\frac{3}{2},3\right)$; y-intercept at $(0,-1)$

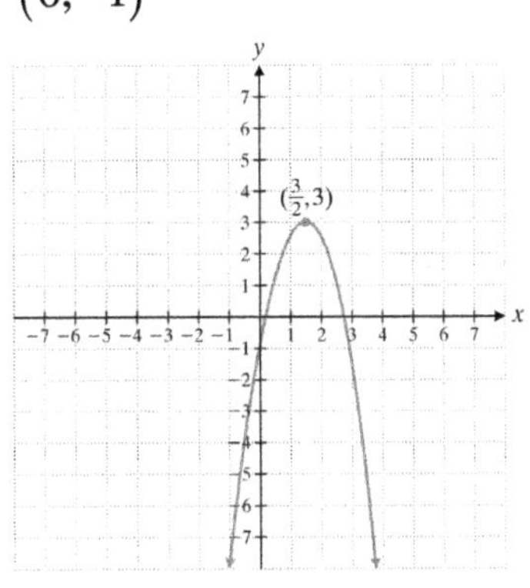

3. Vertex at $(-3,-1)$; x-intercept at $(0,0)$

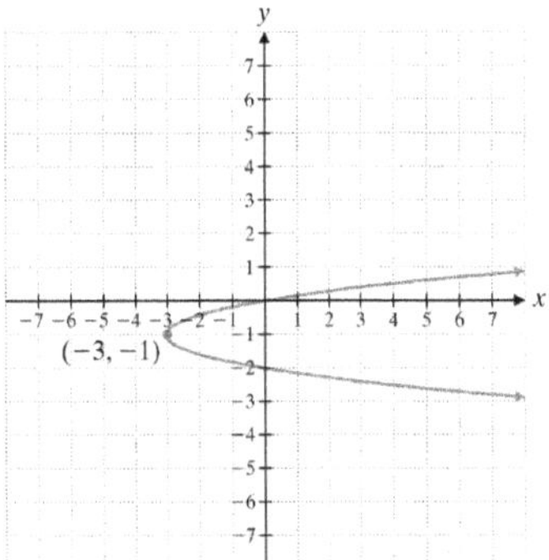

4. $x=-3(y-2)^2+18$
   (a) Horizontal
   (b) Opens left
   (c) $(18,2)$

Concept Check

Answers may vary. Possible solution:
$y=ax^2$, vertical, opens up
$y=-ax^2$, vertical, opens down
$x=ay^2$, horizontal, opens right
$x=-ay^2$, horizontal, opens left
Using these rules you can tell which way the parabolas will open.

Topic 3

Student Practice

2.

4. 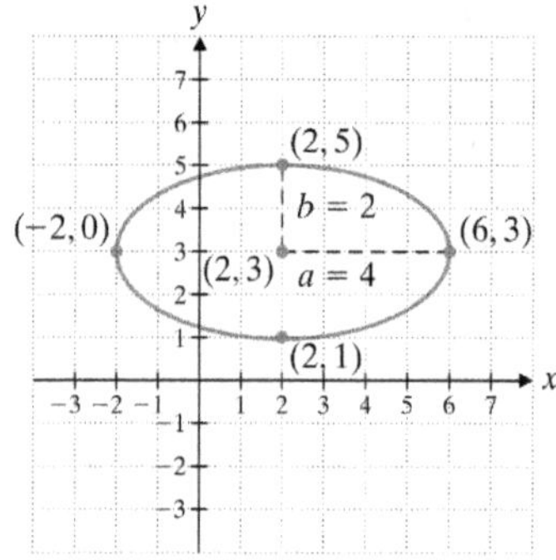

Extra Practice

1. 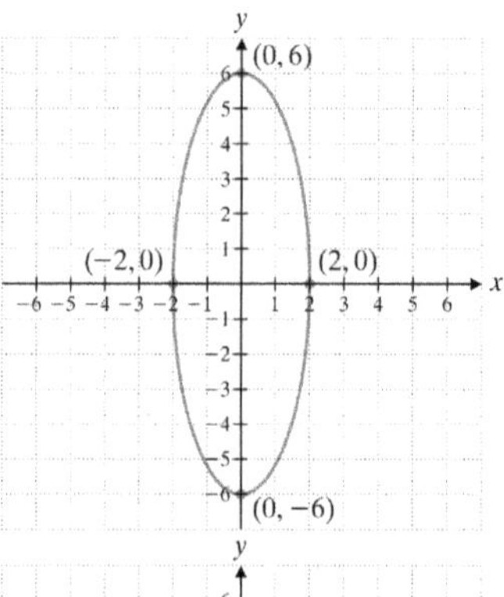

2. 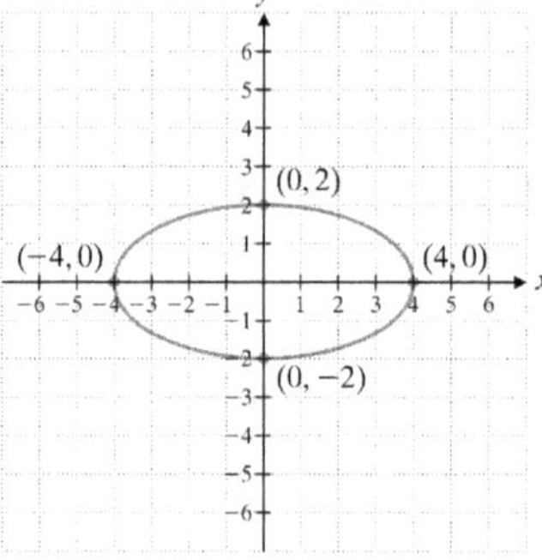

3. 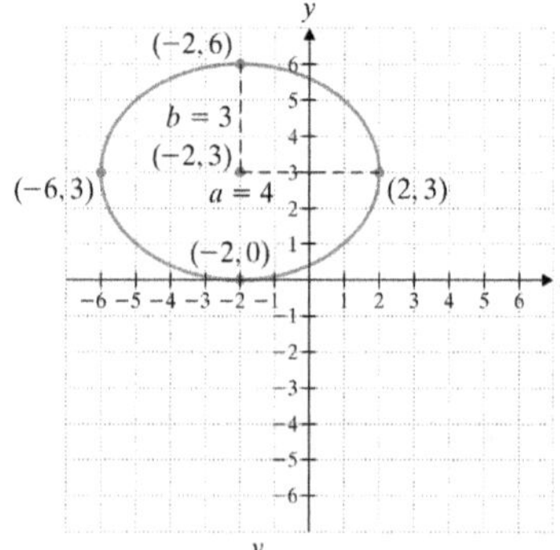

4. 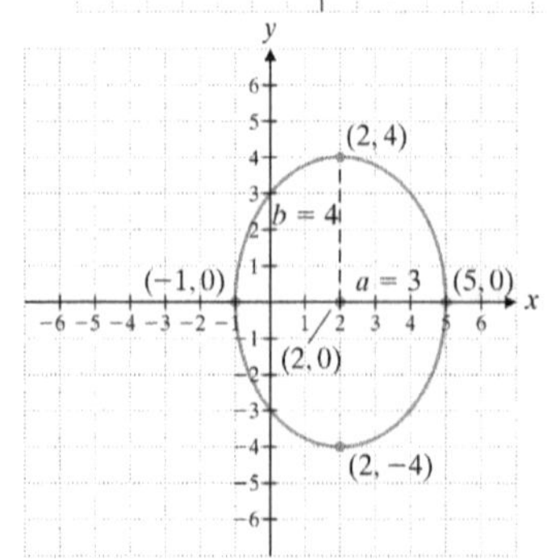

Concept Check

Answers may vary. Possible solution: Add 36 to both sides of the equation, then divide both sides of the equation by 36 to put the equation in standard form. The center of the ellipse is at $(0,0)$ so y-intercepts are $\left(0,\pm\frac{b}{2}\right)$ and x-intercepts are $\left(\pm\frac{a}{2},0\right)$.

Topic 4

Student Practice

2. 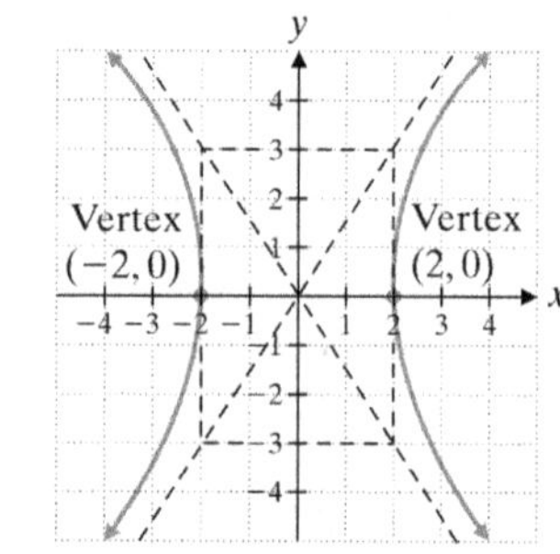

4. 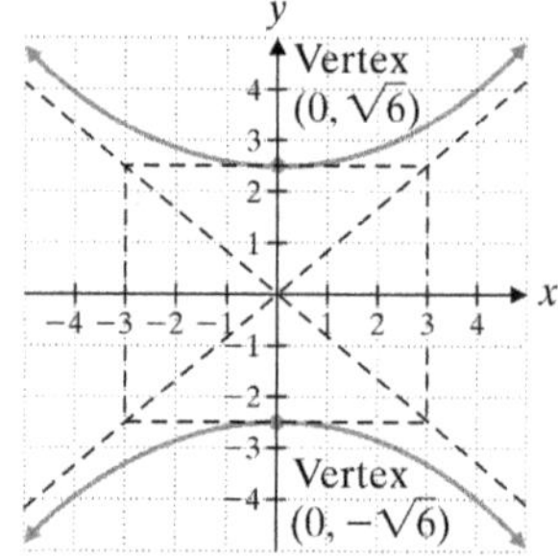

6. 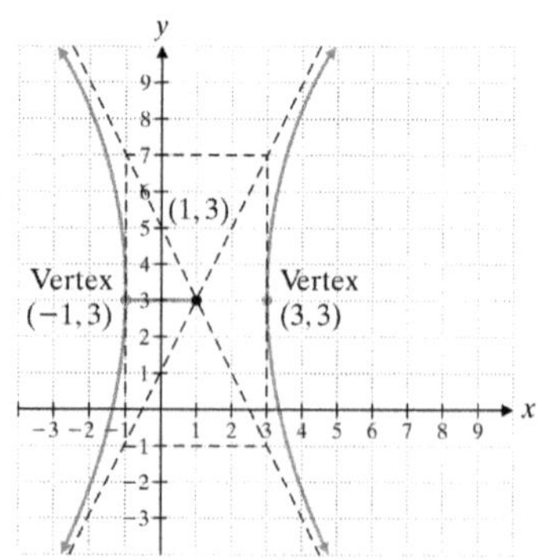

Extra Practice

1.

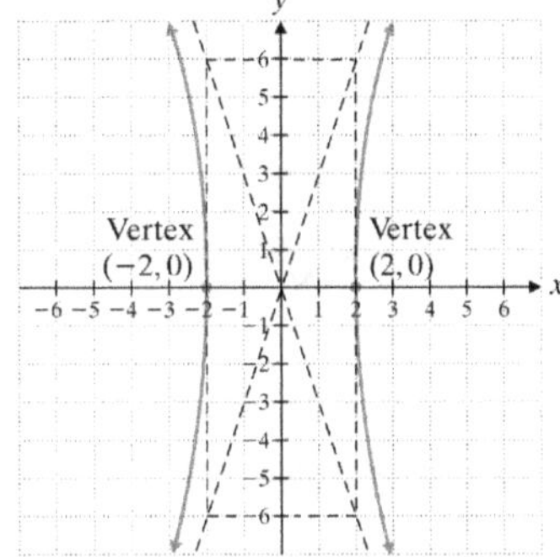

2.

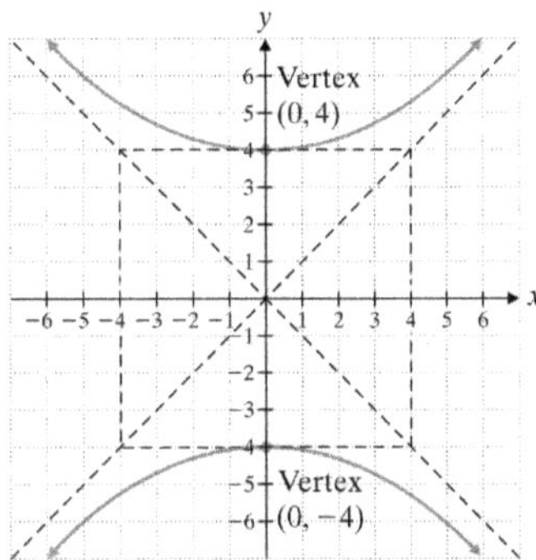

3.

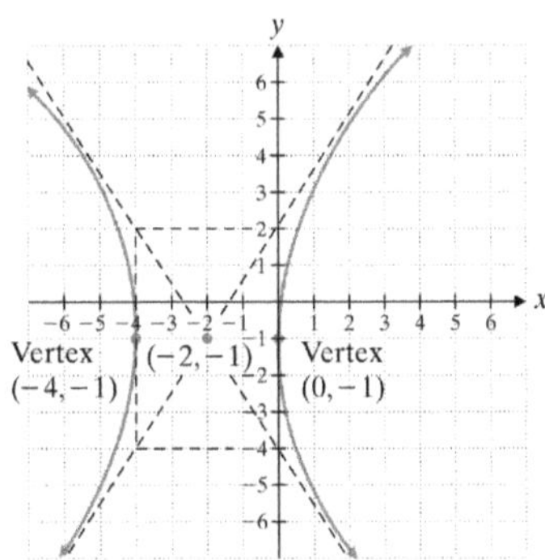

4.

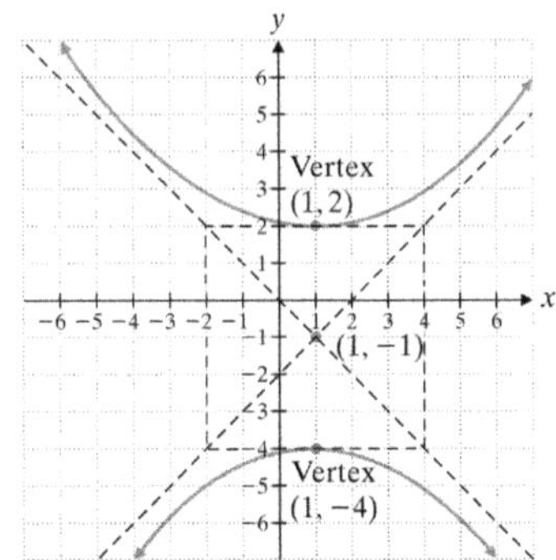

Concept Check

Answers may vary. Possible solution: Divide both sides of the equation by 196 in order to put the equation in standard form. This equation describes a horizontal hyperbola, the asymptote of which is $y = \frac{b}{a}x.$ Substitution yields $y = \frac{7}{2}x.$

Topic 5

Student Practice

2. $(0,5),\ (-7,12)$

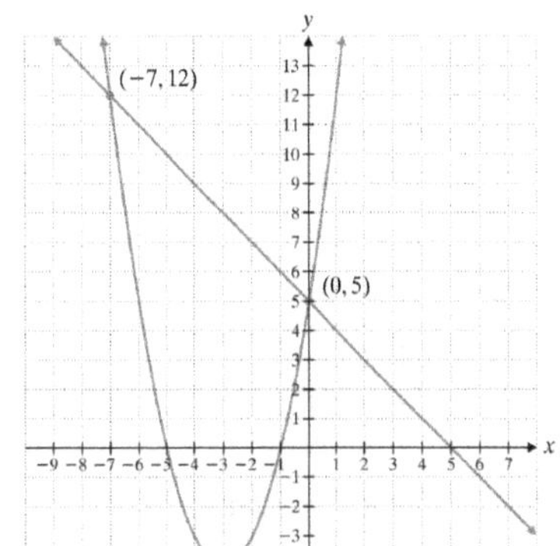

4. $\left(\frac{\sqrt{2}}{2}, -\sqrt{2}\right), \left(-\frac{\sqrt{2}}{2}, \sqrt{2}\right)$

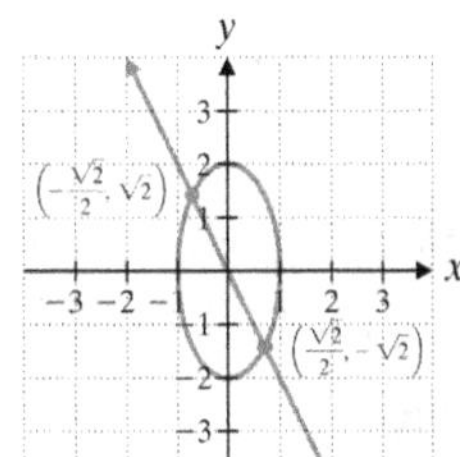

6. $\left(\frac{\sqrt{10}}{10}, \frac{\sqrt{10}}{10}\right), \left(\frac{\sqrt{10}}{10}, -\frac{\sqrt{10}}{10}\right),$ $\left(-\frac{\sqrt{10}}{10}, \frac{\sqrt{10}}{10}\right), \left(-\frac{\sqrt{10}}{10}, -\frac{\sqrt{10}}{0}\right)$

Extra Practice

1. $(3,5),\ (-2,0)$

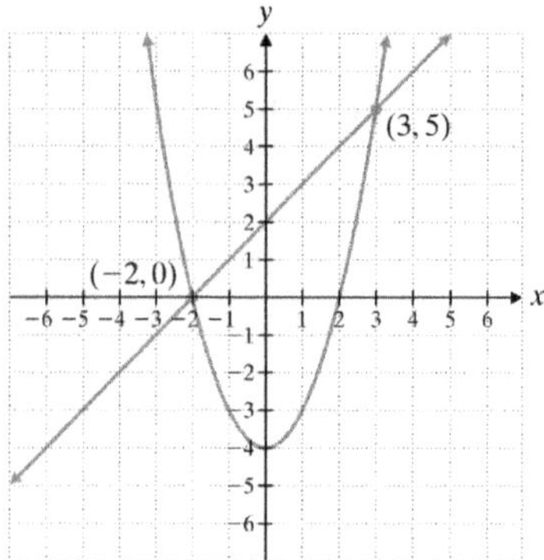

2. $\left(\frac{29}{4}, -\frac{3}{4}\right)$

3. $(3,4)$, $(3,-4)$, $(-3,4)$, $(-3,-4)$

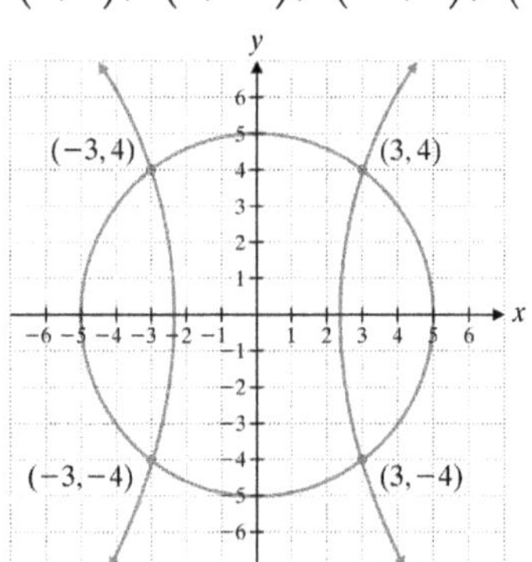

4. $(3,2)$, $(3,-2)$, $(-3,2)$, $(-3,-2)$

Concept Check

Answers may vary. Possible solution: Use the substitution method. Start by labeling the equations.

$$y^2 + 2x^2 = 18 \quad (1)$$
$$xy = 4 \quad (2)$$

Because (2) is a linear equation, and because the y^2 term of (1) has a coefficient of 1, choose to solve (2) for the variable y and substitute the found value for y into (1).

$$y = \frac{4}{x} \ (2)$$

$$\left(\frac{4}{x}\right)^2 + 2x^2 = 18 \ (2) \text{ and } (1)$$

Solve the resulting equation, in terms of x only, for x.

$$x = \pm 1, \ \pm 2\sqrt{2}$$

Solve (2) for y, and substitute each of the four found values of x to find corresponding y values.

$(1,4)$, $(-1,-4)$, $\left(2\sqrt{2},\sqrt{2}\right)$, $\left(-2\sqrt{2},-\sqrt{2}\right)$

Check for extraneous answers.

Worksheet Answers Module 18

Topic 1

Student Practice

2. (a) $6b-7$
   (b) $6b+29$
   (c) $6b+22$
4. $\dfrac{-10}{(a+8)(a+6)}$
6. 5
8. (a) 452.16
   (b) $S(e)$
   $=452.16+150.72e+12.56e^2$
   (c) 498.51 cm^2, the surface area is to large by approximately 46.35 cm^2.

Extra Practice

1. $5a^2+2a+6$
2. $2\sqrt{a^2+2}$
3. $\dfrac{5}{2}$
4. $8x+4h$

Concept Check

Answers may vary. Possible solution: For the function $k(x)=\sqrt{3x+1}$ evaluated at $k(2a-1)$, substitute $2a-1$ for x in the function and solve.
$k(2a-1)=\sqrt{3(2a-1)+1}=\sqrt{6a-2}$

Topic 2

Student Practice

2. (a) A function
   (b) Not a function
4.

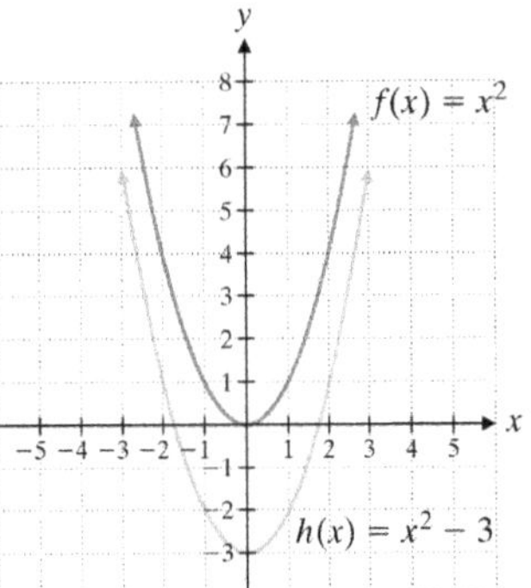

6.

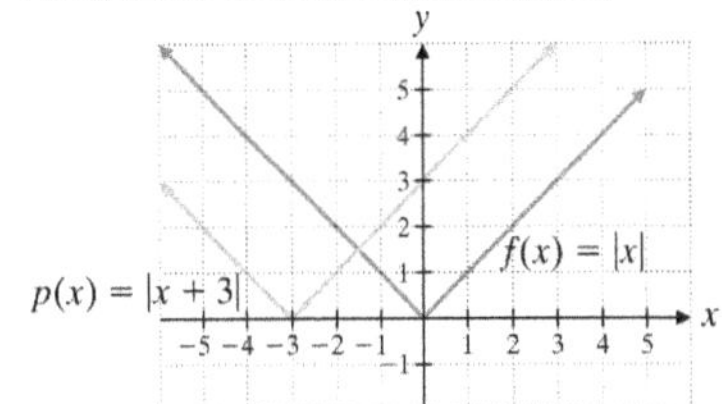

8.

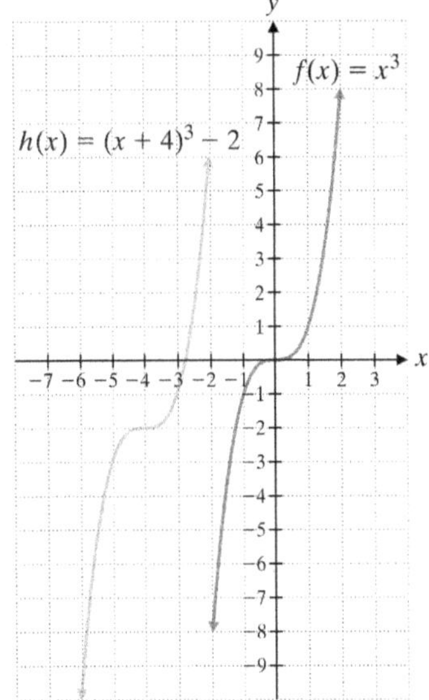

Extra Practice

1. Not a function
2.

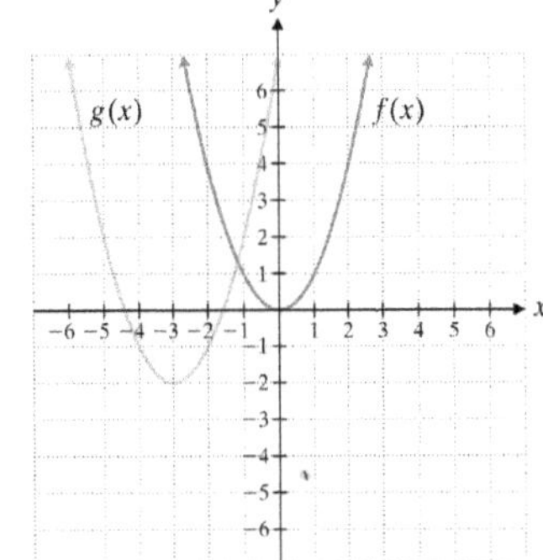

3.

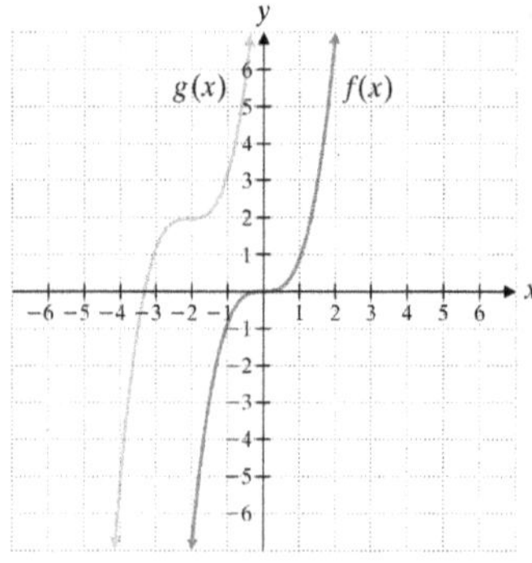

4.

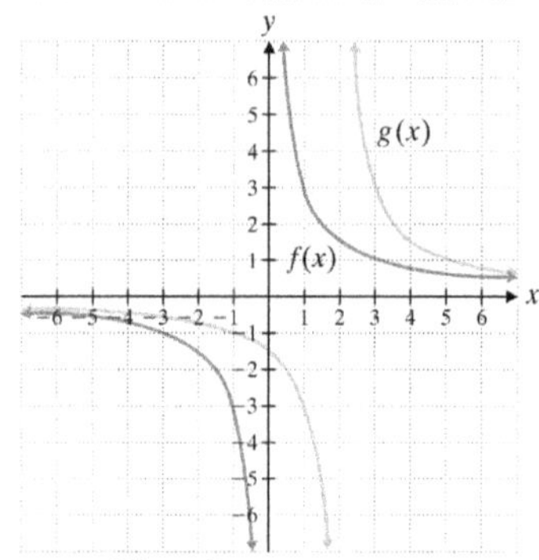

Concept Check

Answers may vary. Possible solution: If a vertical line passes through more than one point of the graph of a relation, the relation is not a function.

Topic 3

Student Practice

2. (a) $2x^2+12x-6$
   (b) 74
4. (a) $4x^3-15x^2+41x-24$
   (b) -198
6. (a) $\dfrac{2x+3}{x-4}$, where $x \neq 4$
   (b) $\dfrac{1}{2x+3}$, where $x \neq -\dfrac{3}{2}$
8. $15x-17$
10. (a) $\sqrt{4x+5}$
    (b) $4\sqrt{x+6}-1$
12. (a) $\dfrac{4}{2x-5}$
    (b) -1

Extra Practice

1. (a) $1.6x^3+4.6x^2-2.7x+7.6$
   (b) $1.6x^3-4.6x^2-2.7x-7.6$
   (c) 84.1
   (d) -33.4
2. (a) $x^3-12x^2+48x-64$
   (b) $x-4$
   (c) -1
   (d) -7
3. $1-2x$
4. $\left|-4x-\dfrac{5}{3}\right|$

Concept Check

Answers may vary. Possible solution: Evaluate both functions for -4, then subtract the results of $g(-4)$ from the results of $f(-4)$.

Topic 4

Student Practice

2. (a) Not one-to-one
   (b) One-to-one
4. (a) Not one-to-one
   (b) One-to-one
6. $Q^{-1}=\{(4,2),(2,6),(7,9),(3,11)\}$.
8. $f^{-1}(x)=\dfrac{x-8}{5}$
10. $f^{-1}(x)=\dfrac{x-75}{3}$
12.

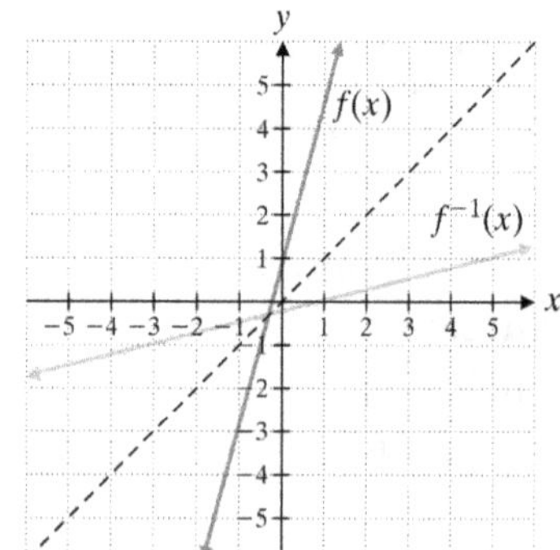

Extra Practice

1. One-to-One
2. $f^{-1}(x)=x+4$
3. $f^{-1}(x)=\dfrac{5}{3x}+\dfrac{4}{3}$ or $\dfrac{5+4x}{3x}$
4. $g^{-1}=-\dfrac{x+4}{3}$

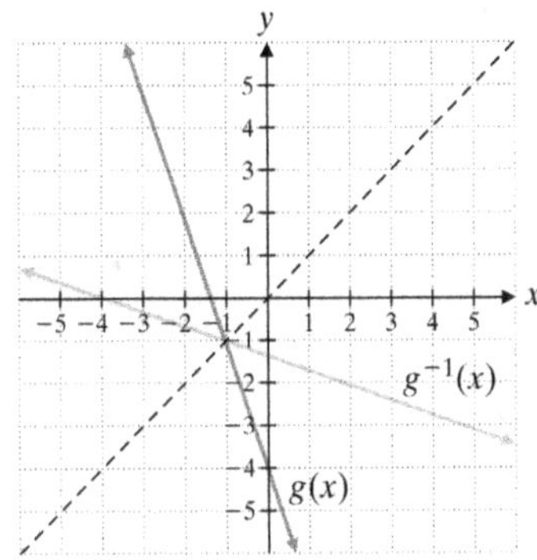

Concept Check

Answers may vary. Possible solution:
To find the inverse of the function
$f(x)=\dfrac{x-5}{3}$, substitute y for $f(x)$.

$$y=\frac{x-5}{3}$$

Interchange x and y.

$$x=\frac{y-5}{3}$$

Solve for y in terms of x.

$$x=\frac{y-5}{3}$$

$$3x=y-5$$

$$y=3x+5$$

Replace y with $f^{-1}(x)$.

$$f^{-1}(x)=3x+5$$

Worksheet Answers Module 19

Topic 1

Student Practice

2.

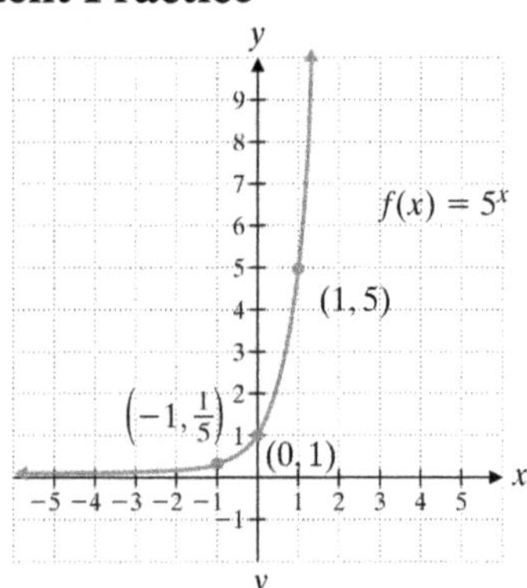

4.

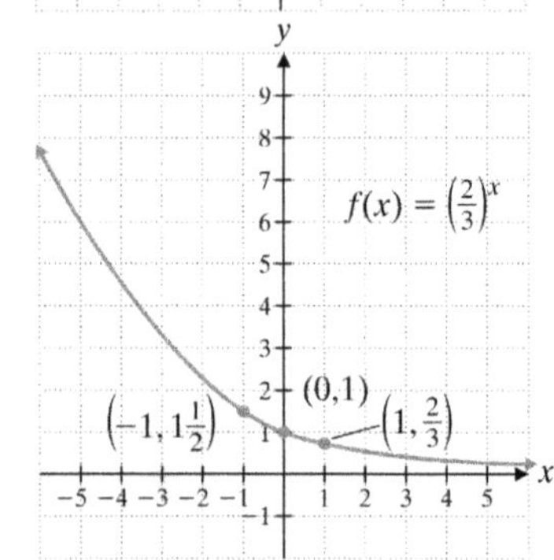

6.

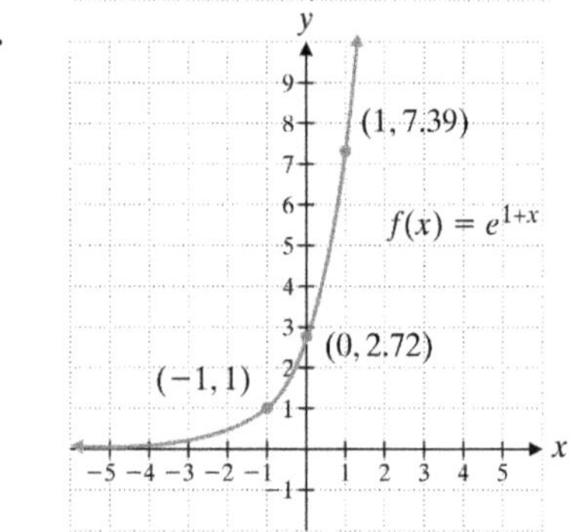

8. $x = 3$
10. $31,990.00
12. 1.34 mg

Extra Practice

1.

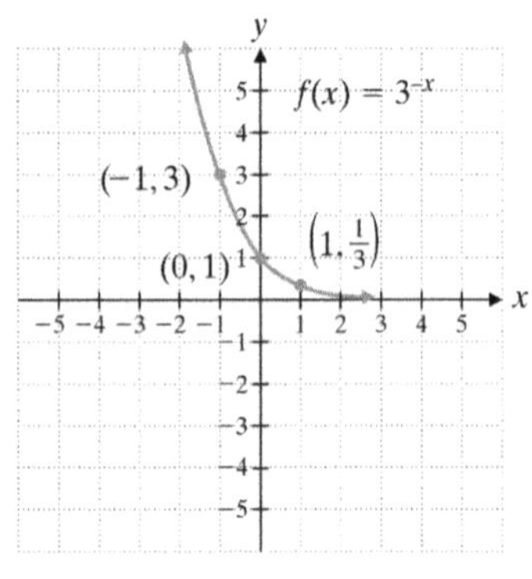

2.

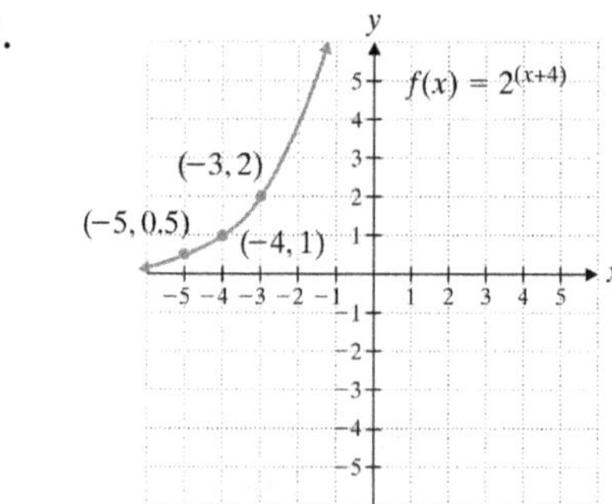

3. $x = -1$
4. $5008.62

Concept Check

Answers may vary. Possible solution:

To solve $4^{-x} = \frac{1}{64}$ for x, remember that $64 = 4^{-3}$. Replace the right side of the equation with 4^{-3}, $4^{-x} = 4^{-3}$. This yields the same base on both sides of the equation, and allows the use of the property of exponential functions to simplify, which states that the exponents may be isolated and compared. Thus, $-x = -3$ or $x = 3$.

Topic 2

Student Practice

2. $-3 = \log_6 \frac{1}{216}$
4. $243 = 3^5$
6. (a) $x = \frac{1}{256}$
   (b) $b = 4$
8. 6

10. 

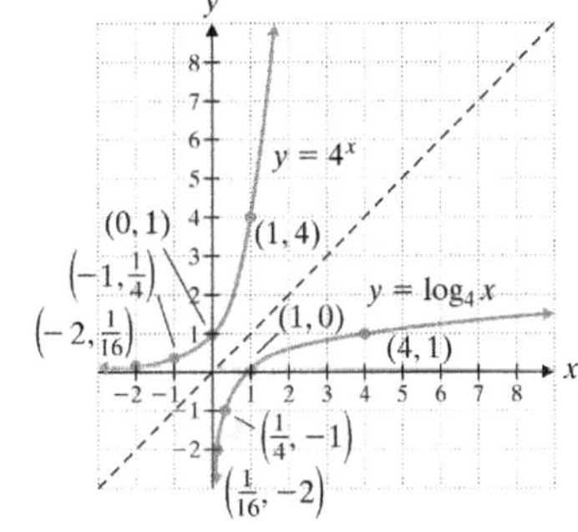

Extra Practice

1. $\log_{10} 0.0001 = -4$
2. $e^{-6} = x$
3. $\frac{1}{2}$
4. 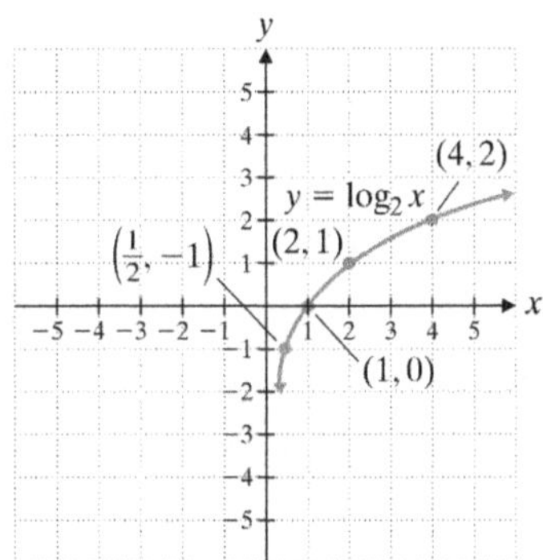

Concept Check

Answers may vary. Possible solution: To solve $-\frac{1}{2} = \log_e x$ for x, write an equivalent exponential equation, $x = e^{-1/2}$.

Topic 3

Student Practice

2. $\log_5 B + \log_5 C$
4. $\log_2 20xz$
6. $\log_{10} 19 - \log_{10} 5$
8. $\log_b \left(\frac{1}{4}\right)$
10. $\log_b \left(\frac{y^5 z^3}{w^{1/4}}\right)$
12. $6\log_b w + 5\log_b x - \log_b y$
14. (a) 1
    (b) 0
    (c) $x = 12$
16. $x = \frac{49}{125}$

Extra Practice

1. $3\log_7 x + \log_7 y + 2\log_7 z$
2. $\log_a \left(\frac{36(6)^{1/3}}{x^5}\right)$
3. $\frac{1}{5}$
4. $x = 32$

Concept Check

Answers may vary. Possible solution: To simplify $\log_{10}(0.001)$, start by setting the expression equal to x, $\log_{10}(0.001) = x$. Then, write 0.001 as a power of 10, $0.001 = 10^{-3}$. Now, rewrite the logarithm, $\log_{10} 10^{-3} = x$. Finally, use the logarithm of a number raised to a power property to write $-3\log_{10} 10 = x$, and the property $\log_b b = 1$ to get $-3(1) = x$, or $-3 = x$.

Topic 4

Student Practice

2. (a) 1.102776615
   (b) 2.102776615
4. $x \approx 27{,}415{,}741.72$
6. 0.003132564
8. (a) 1.178654996
   (b) 4.024100909
10. (a) 408.4216105
    (b) 0.376288177
12. 1.159231332
14. −5.295623771

16.

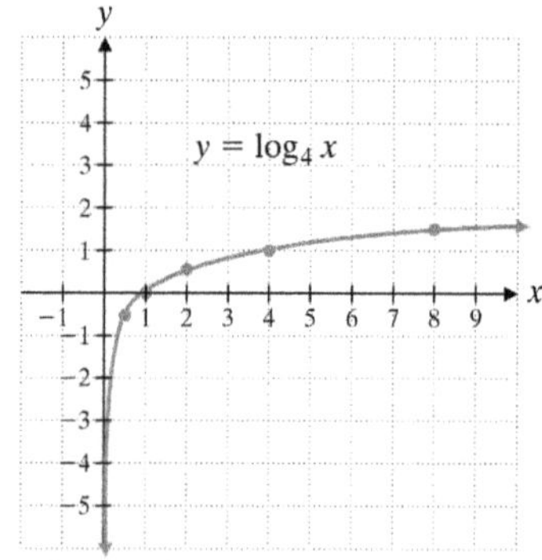

Extra Practice

1. 1.049218023
2. 24.53253019
3. −0.320449841
4. −2.456336474

Concept Check

Answers may vary. Possible solution: To solve $\ln x = 1.7821$ for x using a scientific calculator, first recall that $\ln x = 1.7821$ is equivalent to $e^{1.7821} = x$. Then, use the calculator to evaluate $e^{1.7821}$, resulting in $x \approx 5.942322202$.

Topic 5

Student Practice

2. $x = 4$
4. $x = 5$
6. $x = \dfrac{\log 12}{\log 5}$
8. $x \approx 0.5053$
10. Approximately 17 years
12. Approximately 41 years

Extra Practice

1. $x = 2$
2. $x = 8$
3. $x \approx 6.838$
4. Approximately 10 years

Concept Check

Answers may vary. Possible solution: To solve $26 = 52e^{3x}$, start by dividing both sides by 52.

$$\frac{1}{2} = e^{3x}$$

Then take the natural logarithm of both sides.

$$\ln\left(\frac{1}{2}\right) = 3x$$

Isolate the x.

$$\frac{\ln\left(\frac{1}{2}\right)}{3} = x$$

Approximate the solution with a calculator.

$x \approx -.0231$

Worksheet Answers Module 20

Topic 1

Student Practice

2. vertex $= (3.5, -0.25)$;

   y- intercept $= (0, 12)$;

   x- intercepts $= (4, 0), (3, 0)$

4. vertex $= (2, -3)$;

   y- intercept $= (0, 1)$;

   x- intercepts $\approx (3.7, 0), (0.3, 0)$

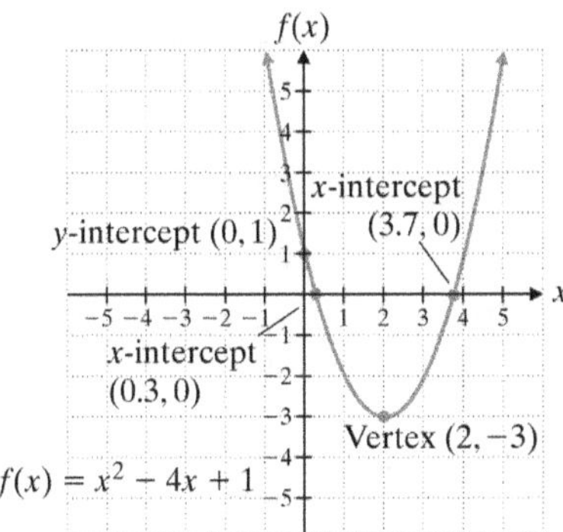

6. vertex $= (-1, -3)$;

   y- intercept $= (0, -6)$;

   No x- intercepts.

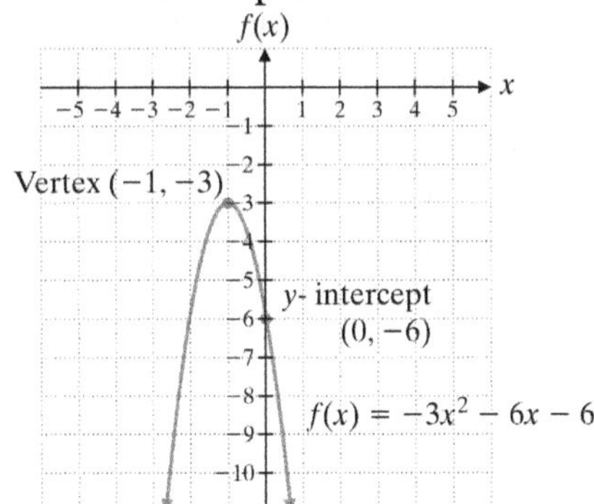

Extra Practice

1. vertex $= (-2.5, -12.25)$;

   y- intercept $= (0, -6)$;

   x- intercepts $= (1, 0), (-6, 0)$

2. vertex $= (-1.7, -26.45)$;

   y- intercept $= (0, -12)$;

   x- intercepts $= (0.6, 0), (-4, 0)$

3. vertex $= \left(\frac{1}{3}, \frac{2}{3}\right)$ or $\approx (0.3, 0.7)$

   y- intercept $= (0, 1)$;

   No x- intercepts

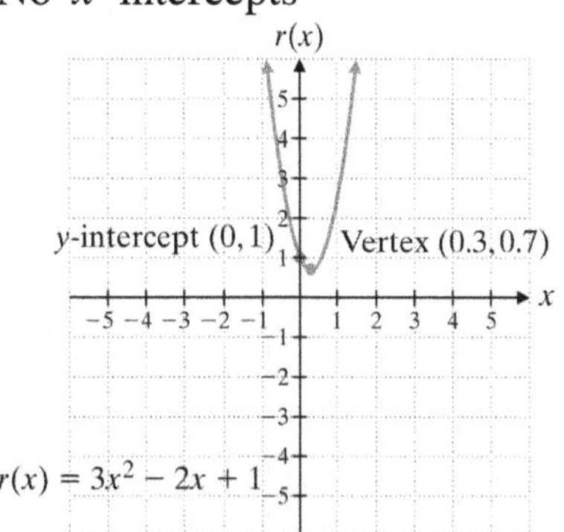

4. vertex $= (1, 0)$;

   y- intercept $= (0, -1)$;

   x- intercepts $= (1, 0)$

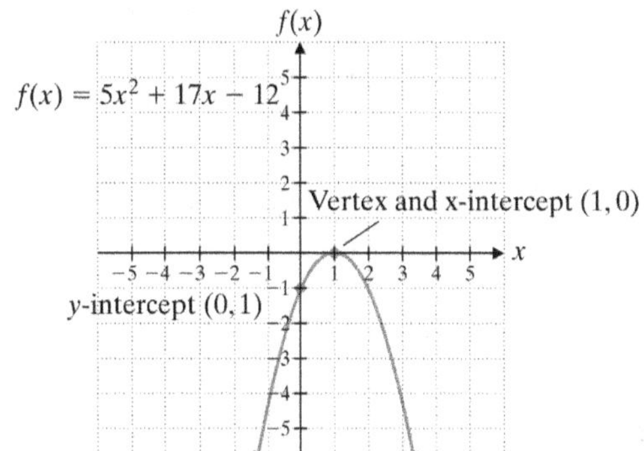

Concept Check

Answers may vary. Possible solution: The function is in standard form, $f(x) = ax^2 + bx + c$ with $a = 4$, $b = -9$, and $c = -5$. The x-coordinate of the vertex is $x_{\text{vertex}} = \frac{-b}{2a} = \frac{-(-9)}{2(4)} = \frac{9}{8}$. The y-coordinate of the vertex is

$f\left(x_{\text{vertex}}\right)=f\left(\frac{9}{8}\right)$.

Topic 2

Student Practice

2\.

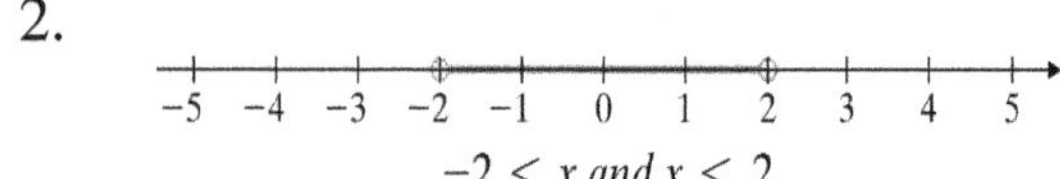

$-2 < x \text{ and } x < 2$

4\. $x < -1 \text{ or } x > 1$

6\. $x \leq 1 \text{ or } x \geq 4$

8\. $x < -3 \text{ or } x > 4$

10\. $x < -4.2 \text{ or } x > 0.2$

Extra Practice

1\. $-4 < x \leq \frac{1}{2}$

2\. $x \leq 4 \text{ or } x \geq \frac{13}{2}$

3\. $x \leq 1 \text{ or } x \geq 5$

4\. $-1 \leq x \leq 2.5$

Concept Check

Answers may vary. Possible solution: The equation $x^2+2x+8=0$ does not have any real solutions. Thus, there are no boundary points. Also, the quadratic function $f(x)=x^2+2x+8$ has no x-intercepts, so the graph lies entirely above or below the x-axis. This means that the inequality is either true for all real numbers, or has no real number solutions.

Topic 3

Student Practice

2\. $x=3,\ x=-6$

4\. $x=1,\ x=-2$

6\. $x=-2,\ x=-\frac{2}{3}$

8\. $-3 \leq x \leq 3$

10\. $-5 \leq x \leq 1$

12\. $x \leq -4 \text{ or } x \geq 4$

14\. $x \leq 1 \text{ or } x \geq 4$

Extra Practice

1\. $x=11,\ x=-20$

2\. $m=-2,\ m=\frac{2}{3}$

3\. $0 \leq x \leq 6$

4\. $1.28 \leq t \leq 1.38$

Concept Check

Answers may vary. Possible solution: Since the absolute value is always nonnegative, there is no solution. $|7x+3|$ cannot be < -4.